How glorious are the summer woods,
  Where the bright broom fork-moss grows,
With their gush of love-born melody,
  And their world of verdant boughs!

That heart is hard as the flinty rock
  That feels not the woodland's power,
With all its magic influences
  Of green leaf, bird, and flower.

The flower and the leaf with their honied breath,
  And the bird with its warbling voice,
Are holy gifts of heaven to men,
  To make their hearts rejoice.

—William Gardiner, *Twenty Lessons on British Mosses*,
Dundee, Scotland: J. Duff, 1846, p. 32

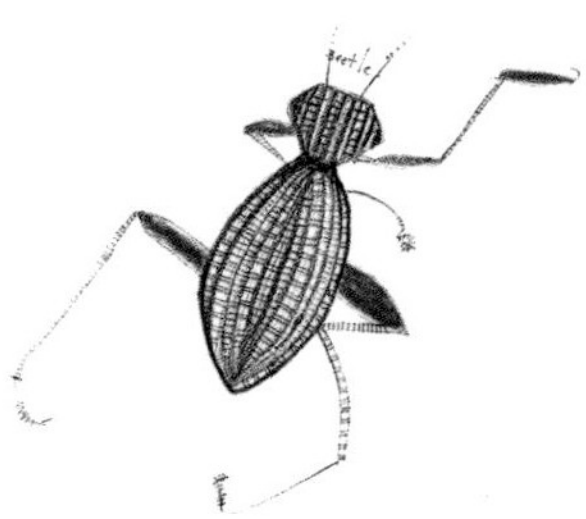

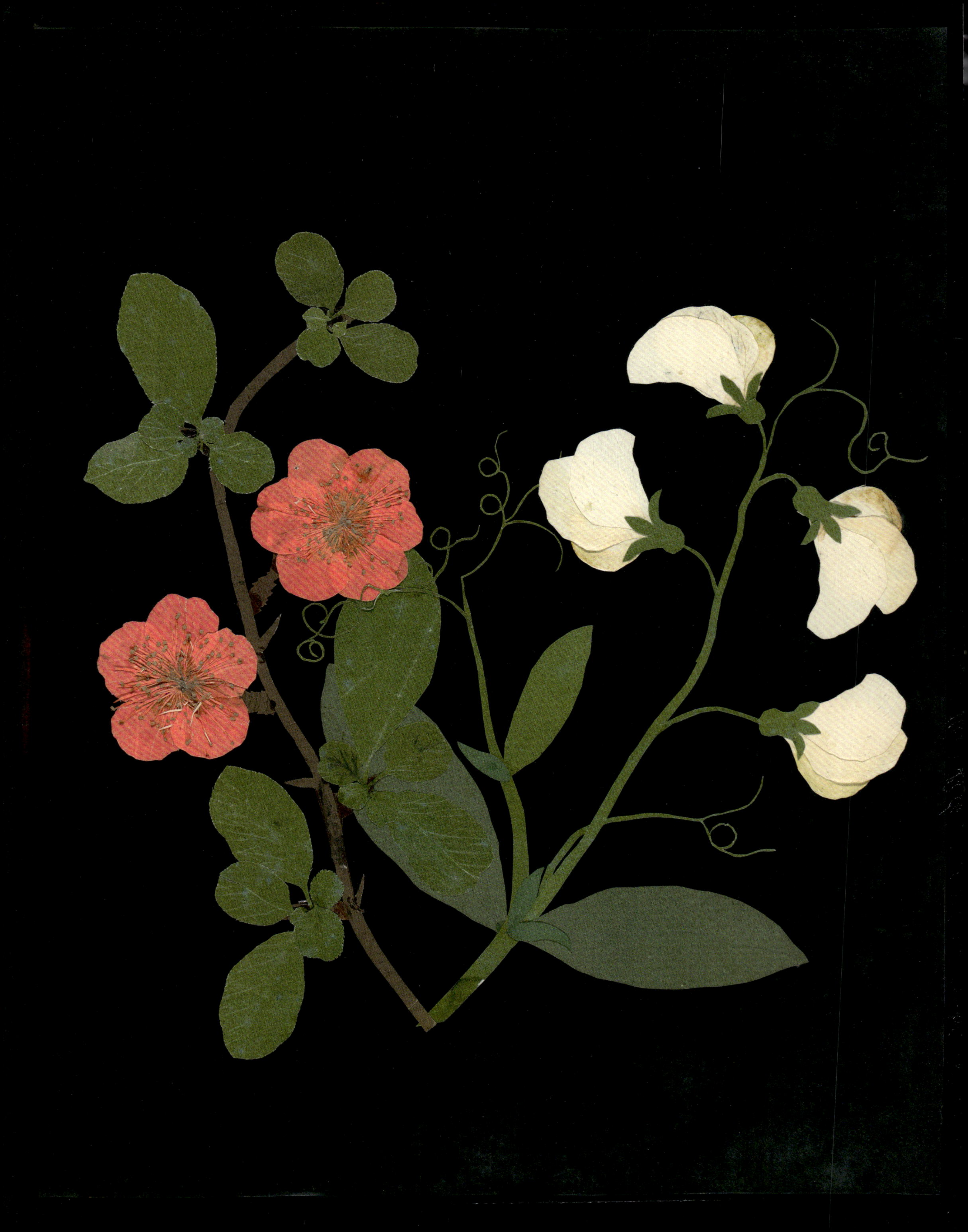

Of Green Leaf, Bird, and Flower

Artists' Books and the Natural World

Edited by Elisabeth R. Fairman

ESSAYS BY
Elisabeth R. Fairman
Robert McCracken Peck
Molly Duggins
David Burnett

CONTRIBUTIONS BY
Laurie Clark
Mandy Bonnell
Tracey Bush
Jane Hyslop
Patrick Sweeney
Dizzy Pragnell
Tim Barringer
Ron King
Clare Bryan
Lisa Ford
John Dilnot
Colin Sackett
Martin Postle
Eileen Hogan
Rebecca Salter
Clive Phillpot
Sarah Welcome

Yale Center for British Art, New Haven
Yale University Press, New Haven and London

This publication accompanies the exhibition *"Of Green Leaf, Bird, and Flower": Artists' Books and the Natural World*, organized by the Yale Center for British Art, New Haven, and on view from May 15 to August 10, 2014. Exhibition curated by Elisabeth R. Fairman

Copyedited and proofread by Christopher Lotis and A. Robin Hoffman

LIBRARY OF CONGRESS CATALOGING-IN-PUBLICATION DATA
Of green leaf, bird, and flower: artists' books and the natural world / edited by Elisabeth R. Fairman.
pages cm
Issued in connection with an exhibition held May 15, 2014–August 10, 2014, Yale Center for British Art, New Haven, Connecticut.
Includes bibliographical references and index.
ISBN 978-0-300-20424-7 (hardback)
1. Nature in art. 2. Natural history illustration—Great Britain. 3. Artists' books—Great Britain—Themes, motives. 4. Art and society—Great Britain. 5. Science—Social aspects—Great Britain.
I. Fairman, Elisabeth R., editor of compilation.
II. Peck, Robert McCracken, 1952– author. Natural obsessions.
III. Yale Center for British Art.
N7650.O4 2014
704.9'43—dc23
2014007108
A catalogue record for this book is available from the British Library.

FRONT COVER (DETAIL) AND PAGE 6: James Bolton, Caesalpinoid legume, possibly a species of *Caesalpinia* L.; Blackburn's Earth Boring Beetle (*Geotrupes blackburnii*), Seven-Spotted Ladybird Beetle (*Coccinella septumpunctata*), Purple Emperor *(Apatura iris)*, (Lepidoptera Hesperiidae), and (Nymphalidae cf. *Haematera*); shells from left: (*Cypraea ocellata* L. 1758), (*Conus marmoreus* L. 1758), and (*Semicassis granulate* Born, 1778), from the natural history cabinet of Anna Blackburne, ca. 1768, watercolor and gouache over graphite on parchment. Yale Center for British Art, Paul Mellon Fund, in honor of Jane and Richard C. Levin, President of Yale University (1993–2013)

BACK COVER: Sarah Morpeth, *Crow Landscape*, Elsdon, Northumberland, 2008, hand-cut paper, wire, and acrylic ink. Yale Center for British Art, Friends of British Art Fund

ENDPAPERS: Mandy Bonnell, Endpapers (detail) from *Antmothbeetlemillipedespider*, with poems by Gabriel Gbadamosi, London: EMH Arts, 2006, wood engraving. Yale Center for British Art, Friends of British Art Fund

HALF-TITLE PAGE: Mandy Bonnell, *Beetle*, 2006, graphite, Yale Center for British Art, Gift of the artist

FRONTISPIECE: Ellen W., *Album of Cut-Paper Flowers*, ca. 1835, hand-cut paper and collage. Yale Center for British Art, Paul Mellon Fund

OPPOSITE: Detail from *Album of Drawings of English Moths, Butterflies, Flowers, and Mollusks*, 1805–1822, pen and ink and watercolor. Yale Center for British Art, Paul Mellon Fund

PAGES 10–11: Miss Rowe, Grass specimens from *Collection of Botanical Specimens*, Liverpool, 1861, Yale Center for British Art, Paul Mellon Fund

In Memoriam

Rachel Lambert Mellon

1910–2014

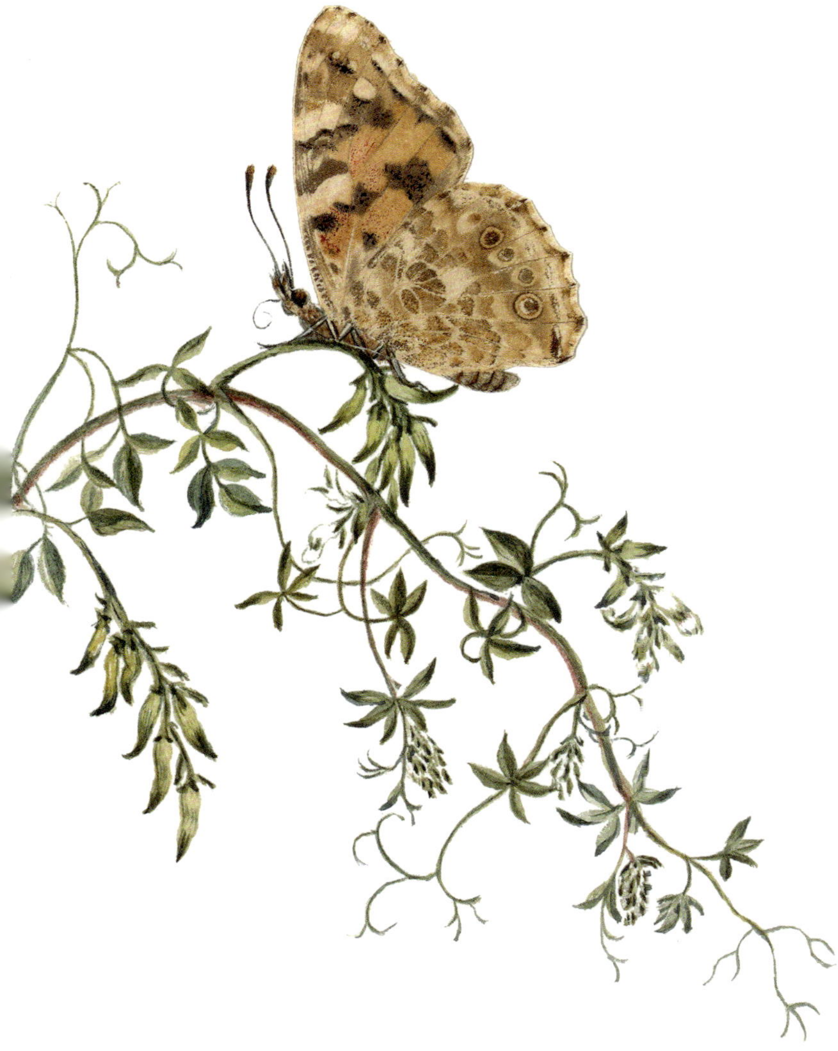

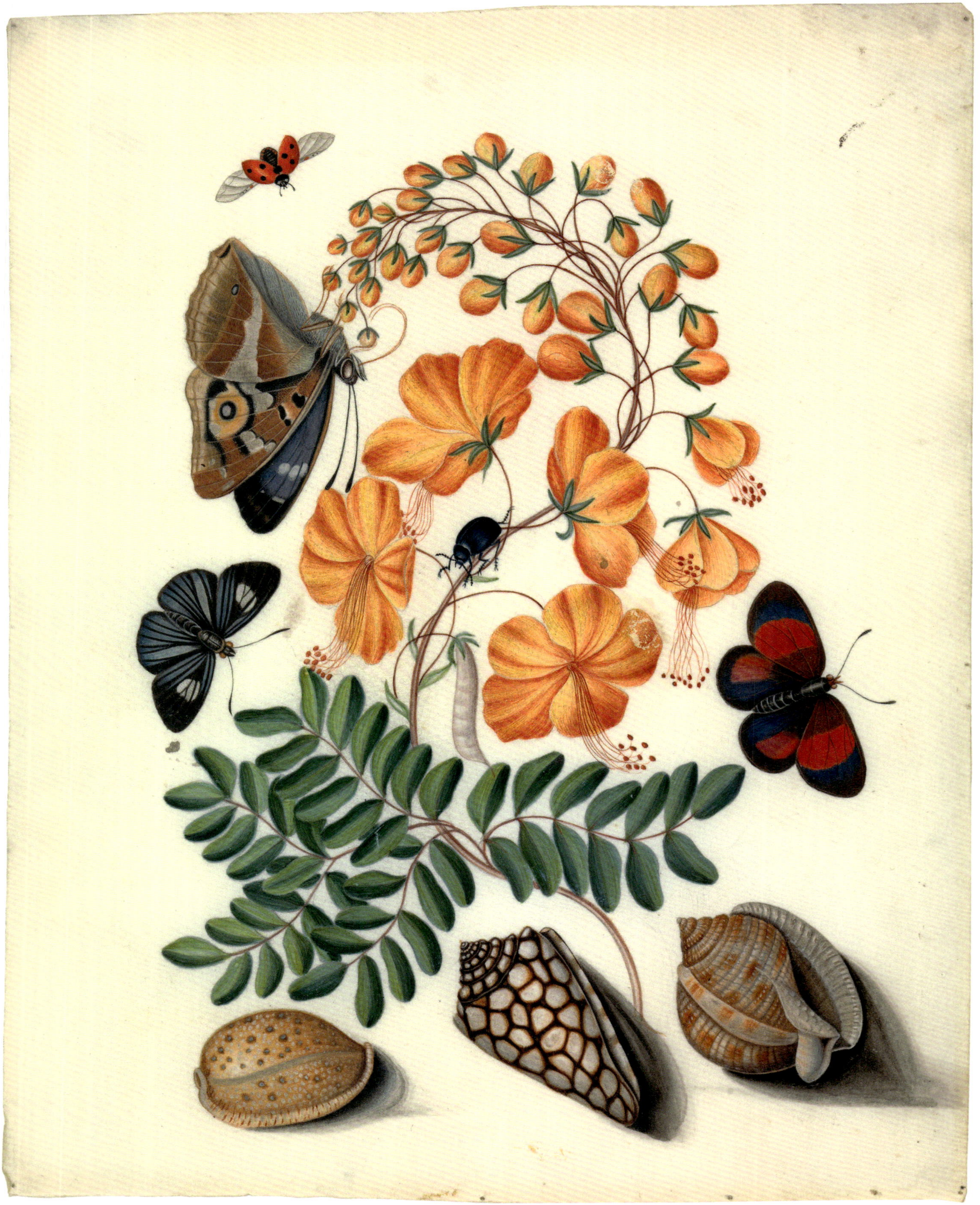

Contents

Director's Foreword

We come to know nature through a series of interdependent processes, from empirical observation in the field or laboratory, to the making of verbal notes and descriptions, and the crafting of images. The final step is the creation of a product meant to communicate knowledge of the natural world to others. As the contributors to this volume indicate so eloquently, this knowledge may be seen reflected in objects from the collections of the Yale Center for British Art, ranging in date from the fifteenth century to the present day, and made by professional and self-taught naturalists, book artists, painters, engravers, and illustrators. This book and the exhibition it accompanies are further distillations of the processes involved in coming to know nature, crafted and curated as another means of comprehending the natural world.

"Of Green Leaf, Bird, and Flower": Artists' Books and the Natural World results from the vision, passion, and knowledge of Elisabeth Fairman, Senior Curator of Rare Books and Manuscripts at the Yale Center for British Art. Building on the gift of rare books and manuscripts made to the Center by its founder, Paul Mellon (Yale College, Class of 1929), Elisabeth has collected materials relating to natural history imaginatively and assiduously, with a particular enthusiasm for works by contemporary book artists. Neither book nor exhibition are intended as a survey of artists' books, but rather as a demonstration of the ways in which historical and contemporary materials created in response to the natural world enhance and inform one another, by tracing the thread of interest in natural history that runs through the collections.

As such, this project takes its place in a series of exhibitions mounted by the Center that have explored the intersections of art and science in the global history of British art and culture, from *A New World: England's First View of America*; *"Endless forms": Charles Darwin, Natural Science, and the Visual Arts*; and *Compass and Rule: Architecture as Mathematical Practice in England, 1500–1750* to *Mrs. Delany and her Circle* and its accompanying display by the artist Jane Wildgoose, *Promiscuous Assemblage, Friendship, & The Order of Things,* as well as future exhibitions on Queen Caroline, Princess Augusta, and Queen Charlotte; the American naturalist William Bartram; and the physician and collector William Hunter.

Elisabeth has engaged an unprecedented array of contributors to this volume, including the authors of the three essays that follow and the artists, scholars, and amateur naturalists who have lent their personal reflections in the form of the statements that punctuate and illuminate the field guide. We owe them all a debt of gratitude for their expertise and insights. In bringing the book to fruition, Elisabeth has worked closely with the Center's Associate Director of Exhibitions and Publications, Eleanor Hughes, and her team; Miko McGinty, designer; and Sally Salvesen, the Center's publisher with Yale University Press. We thank exhibition designer Stephen Saitas; the Center's graphic designer, Lyn Bell Rose; and our installation team, led by Rick Johnson, for their collaborative efforts in presenting the exhibition so beautifully. We also acknowledge Sarah Welcome, Senior Curatorial Assistant in the Department of Rare Books and Manuscripts, who provided invaluable assistance to Elisabeth and to the project in general.

The exhibition has been enhanced by a number of generous institutional loans, and we are deeply grateful to E. C. Shroeder, Director of the Beinecke Rare Book and Manuscript Library; Derek Briggs, Director of the Yale Peabody Museum of Natural History; and Robert Peck, Senior Fellow, Academy of Natural Sciences of Philadelphia, Drexel University, for their beneficent support of this project. Finally, we wish to acknowledge the generosity of numerous artists and collectors who have been particularly supportive and helped to make this project truly special. A number have loaned works from their own collections, and for this we would like to thank Neale and Margaret Albert, Timothy Barringer, Mandy Bonnell, David Burnett, Tracey Bush, Cathy Courtney, Bryan Fuermann, Eileen Hogan, Yoojin Kim, Randy Lerner, Robert Peck, and Charlotte Verity.

We offer you here our own version of a field guide to Britain's countryside. We hope that you will take inspiration from the naturalists and artists discussed and illustrated in the pages that follow and will take to heart the words of John Knapp in his *Journal of a Naturalist* (1829):

> The meditations of separate naturalists in fields, in wilds, in woods, may yield a similarity of ideas; yet the different aspects under which the same things are viewed, and characters considered, afford infinite variety of description and narrative: mine, I confess, are but brief and slight sketches; plain observations of nature . . . a mere outline of rural things; the journal of a traveler through the inexhaustible regions of Nature.

Amy Meyers
Director, Yale Center for British Art

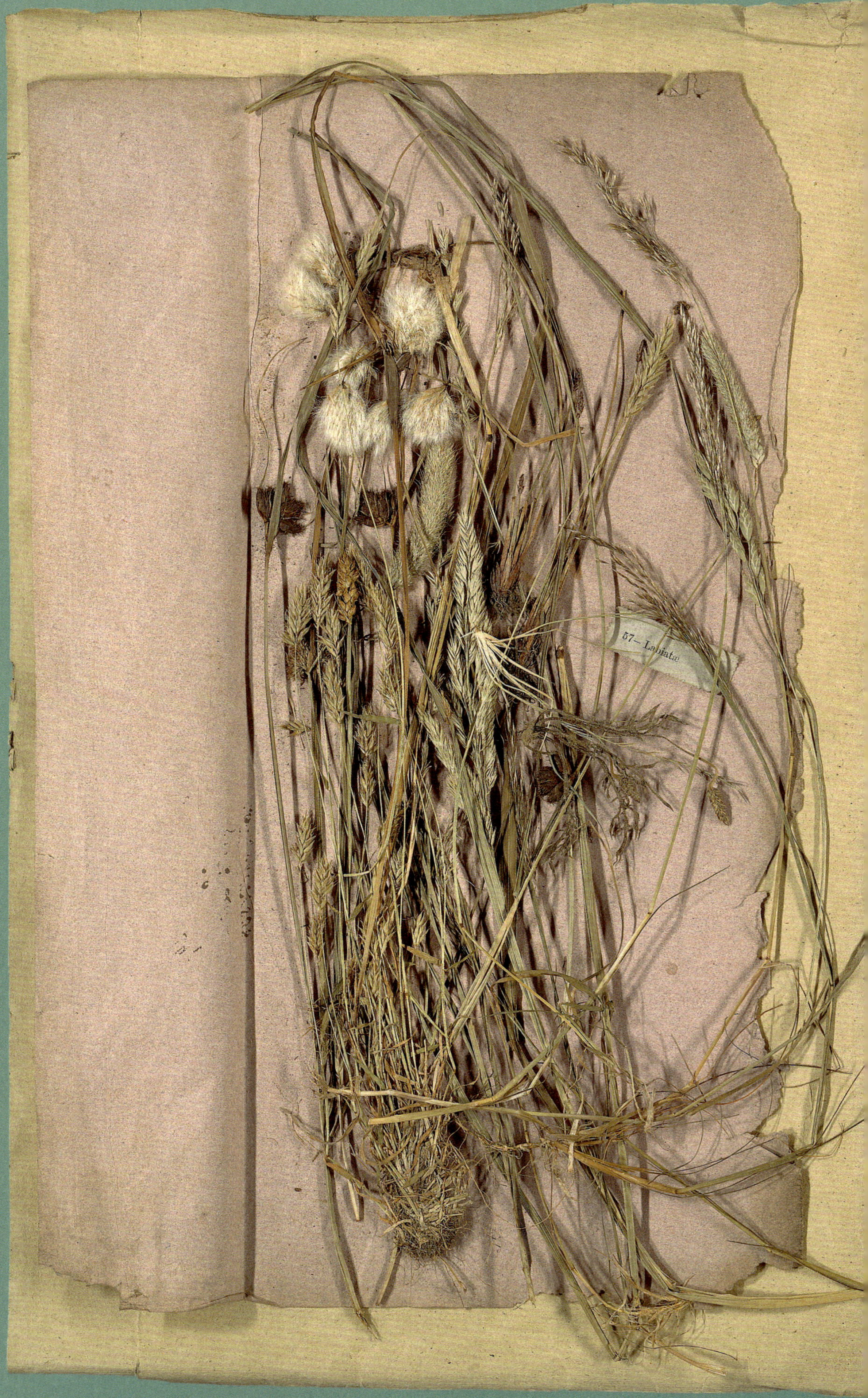
57– Labiatæ

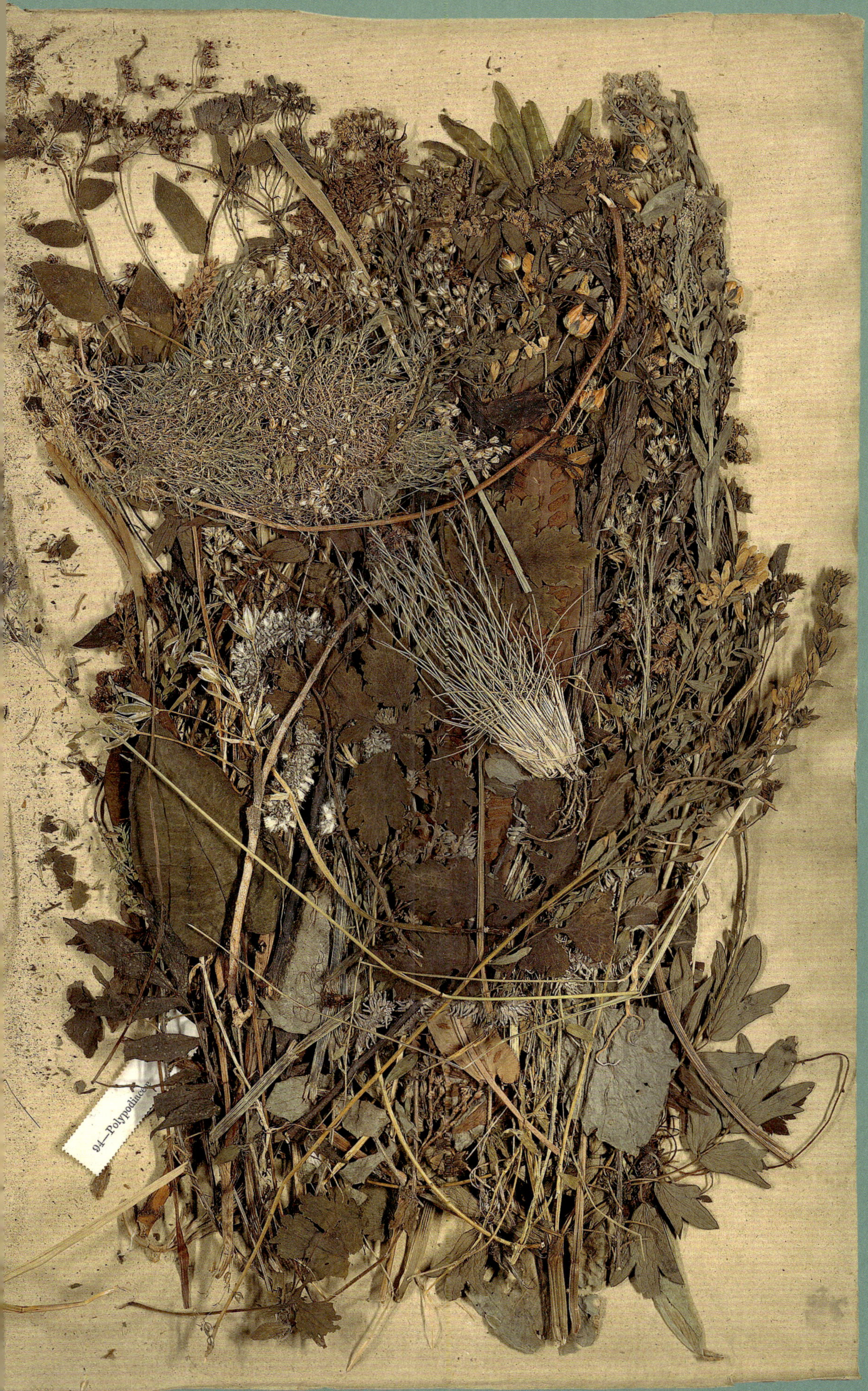
94—Polypodiace

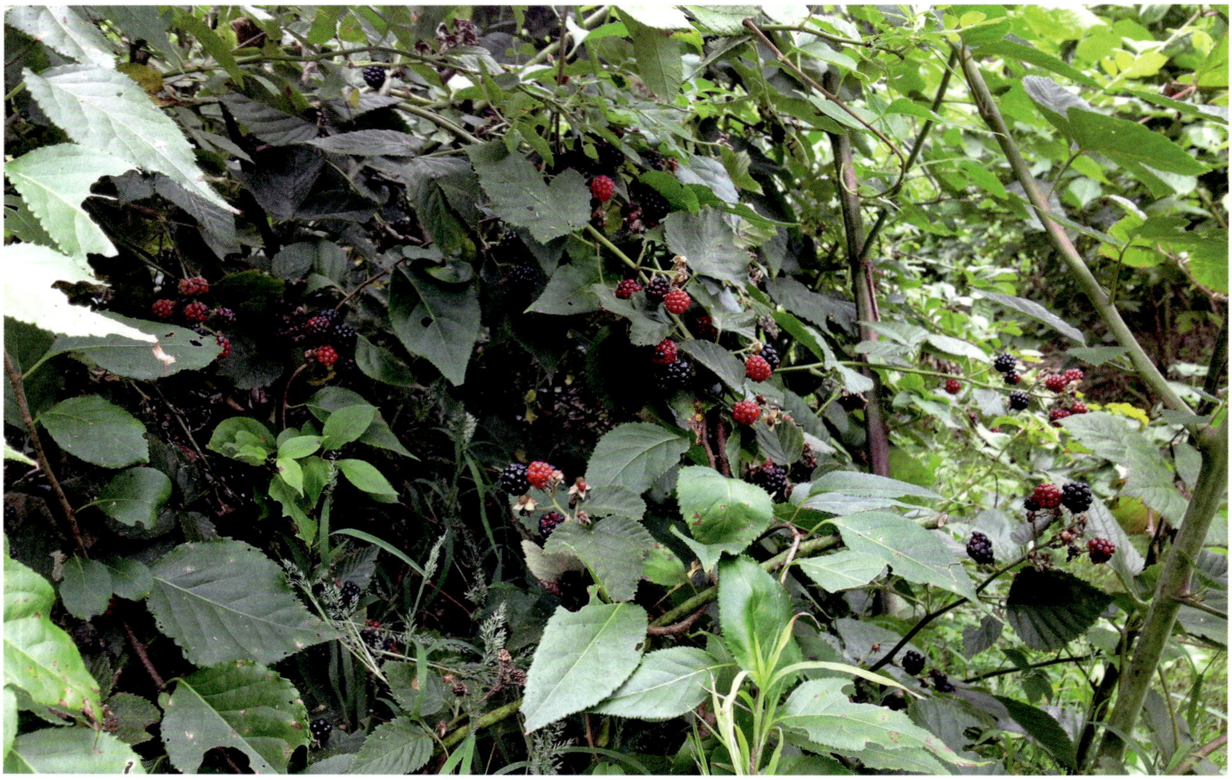

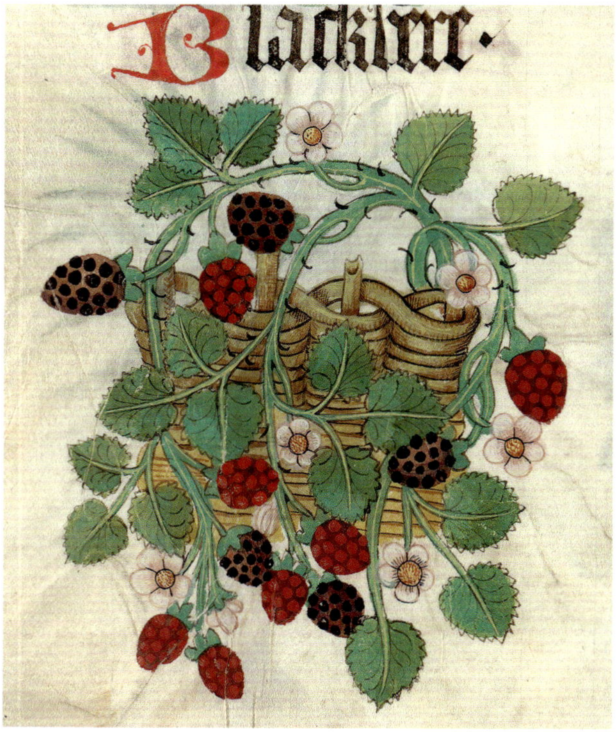

Blackbere·

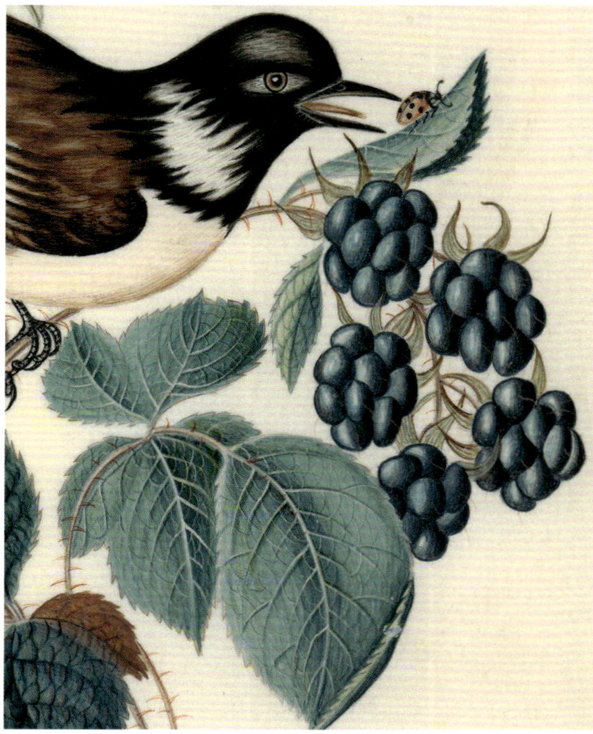

"Of Green Leaf, Bird, and Flower"
Artists' Books and the Natural World

Elisabeth R. Fairman

The exhibition *"Of Green Leaf, Bird, and Flower"* has at its heart works on natural history from the collection of Paul Mellon (Yale College, Class of 1929), founder of the Yale Center for British Art. In the Center's first exhibition catalogue, in 1977, Mr. Mellon acknowledged an intense interest that grew out of his love of the British countryside:

> From 1907 until 1914, from my first year to my seventh, my parents spent almost every summer in England, and my sister and I were invariably taken with them. I suppose it was in those summers that I first developed a taste for the English countryside, for English houses, English rivers, English parks, English skies, English clouds. . . . From those distant summers I remember huge dark trees in rolling parks, herds of small friendly deer, flotillas of white swans on the Thames, dappled tan cows in soft green fields. . . . There seemed to be a tranquility in those days that has never again been found, and a quietness as detached from life as the memory itself.[1]

Paul Mellon had been acquiring works of natural history from at least 1937, when he purchased the double elephant edition of John Audubon's *Birds of America* (London: 1828–39); in 1939 and 1947 he acquired two of Pierre Joseph Redouté's colorplate books. Many of those early acquisitions remained with his wife, Rachel Lambert Mellon, as part of her remarkable Oak Spring Garden Library at Upperville, Virginia. However, a number of important works of natural history have made their way to the Yale Center for British Art, which holds the largest and most comprehensive collection of British art outside the United Kingdom, largely the gift of Mr. Mellon. Presenting the development of British art and culture from the Elizabethan period to the present day, the Center's collections of paintings, sculpture, drawings and watercolors, prints, and rare books and manuscripts provide an exceptional resource for understanding the story of British art, life, society, and culture in its richness and depth.

The core of the collection of rare books is the material amassed by Major J. R. Abbey, one of the first collectors of British colorplate books. Acquired as a whole by Paul Mellon in the 1950s, it comprises more than two thousand volumes describing British life, customs, scenery, and travel during the period of 1770 to 1860. The often lavish illustrations in these books are the work of Britain's finest landscape artists. They form a coherent picture of the natural history, local topography, architecture, and sights encountered by British travelers on the Grand Tour in Europe and on more exotic travels to the South Seas, Africa, and India. In addition, hundreds of books were acquired by Paul Mellon in the years before the Center opened in 1977, primarily illustrated books from the fifteenth through the nineteenth centuries, including art instruction and drawing manuals, children's games, ephemera related to different subjects, and manuscript material relating to British artists of all periods. Most of the books that were in Paul Mellon's possession at the time of his death in February 1999 were bequeathed to Yale University, with the majority specifically given to the Yale Center for British Art. The bequest to the Center included just over five thousand titles, or nearly seventy-five hundred volumes, and

OPPOSITE: Late summer garden, Chester, Connecticut, 2013; "Blackbere" from *Helmingham Herbal and Bestiary* (detail), ca. 1500, watercolor and gouache on parchment. Yale Center for British Art, Paul Mellon Collection; James Bolton, ?White-cheeked Starling with Bramble and Seven-Spotted Ladybird Beetle (detail), ca. 1768, watercolor and gouache over graphite on parchment. Yale Center for British Art, Paul Mellon Fund, in honor of Jane and Richard C. Levin, President of Yale University (1993–2013)

Elisabeth R. Fairman is Senior Curator of Rare Books and Manuscripts at the Yale Center for British Art.

encompassed a wide range of subjects which reflected his many interests. It included sporting books and manuscripts; a complete set of Kelmscott Press and other British fine printing; and maps and atlases related to early discovery and exploration; and natural history.[2]

Among these, one of the most significant works of natural history is the archival collection related to James Bruce (1730–94), a Scottish diplomat and explorer who, between 1767 and 1773, attempted to discover the source of the Nile in Abyssinia (now Ethiopia and Eritrea). He was accompanied by a talented Italian draftsman, Luigi Balugani, who recorded the flora and fauna of the region in beautifully rendered watercolors (fig. 1).[3]

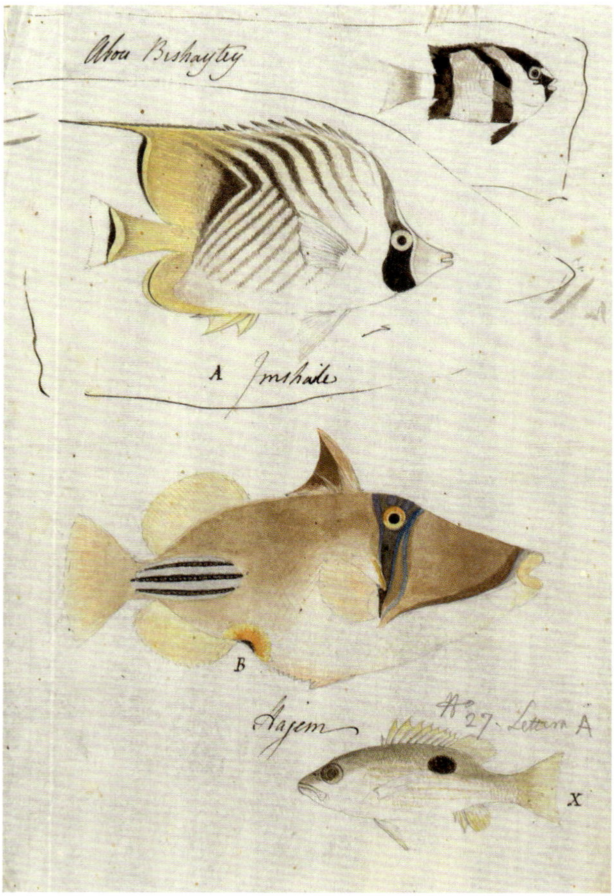

Figure 1. Luigi Balugani, *Four Fish*, ca. 1767–1773, watercolor and graphite with pen and ink. Yale Center for British Art, Paul Mellon Collection

The exquisitely drawn pattern book of plants and animals known as the *Helmingham Herbal and Bestiary*, completed around 1500, is one of the great treasures in the Center's collection and also provides the cornerstone of the exhibition. Acquired by Mr. Mellon in 1961, the manuscript, with its beautiful drawings of animals, birds, flowers, and trees, gives a remarkable picture of the depth of English knowledge of natural history in the Tudor period (fig. 2). Mr. Mellon wrote in the preface to the published facsimile:

> Although very much an amateur bookman and a far cry from even an elementary botanist, my interest in this manuscript was originally aroused more by the beasts of the Bestiary than the flowers, herbs or trees of the Herbal. It has been my wife, a professional in the fields of horticulture and silviculture, who has shared with me her enthusiasm for the floral and arborial drawings. . . . [T]his picture-book is . . . an aesthetically delightful rendering of what in fifteenth-century England was known about the inhabitants of her fields and woodlands, or what was imagined about the denizens of faraway lands. Whether it might often have been in the hands of children for study or entertainment, whether it guided the artisan in the limning of fabric or painted surfaces, and whether or not it was derived from earlier continental models, I see it simply as a charmingly natural and a thoroughly English work. I have also been impressed by the stark simplicity and directness of the drawing and colouring of the objects, as though there were a mysterious aesthetic kinship between these fifteenth-century artists or designers and our own twentieth-century artists.[4]

The present publication (and the exhibition it accompanies) explores that idea of aesthetic kinship further, by looking at the ways in which self-taught naturalists and artists recorded and observed the natural world around them from the sixteenth century to the present, examining the intersections of artistic and scientific interest. It highlights the scientific pursuits of the eighteenth and nineteenth centuries that resulted in the collecting and cataloguing of the natural world, and that informed the aesthetically oriented activities of the self-taught naturalist of the Victorian era, particularly those of women who collected and drew specimens of butterflies, ferns, grasses, feathers, seaweed, and shells, and then assembled them into albums and commonplace books. Work by contemporary artists in the exhibition, such as Mandy Bonnell, Tracey Bush, John Dilnot, Helen Douglas, Eileen Hogan, Jane Hyslop, Sarah Morpeth, Dizzy Pragnell, and Chisato Tamabayashi, reveal a shared impetus to document, interpret, and celebrate nature as in the work of their predecessors—while at the same time broadening their visions of the natural world to incorporate its interaction with consumer culture and with modern technologies.

The exhibition suggests a tantalizing continuity of past and present aesthetics that can be explored by juxtaposing contemporary works with historic ones, by putting works from the Center's historic collections side by side with relevant contemporary artists' books and prints acquired in the past two decades. Indeed, the inspiration for the exhibition comes from two specific works, made some five hundred years apart. We recently acquired *A Printmaker's Flora: An Anthology of the Names of British Wild Flowers*, published in 1996. Paging through the work—an extraordinary artists' book made up of a series of prints by different artists in a variety of techniques (etchings, lithographs, aquatints, wood engravings, relief etchings, linocuts, and collotypes) depicting British wild flowers and their common names, I was immediately struck by Rosaleen Wain's beautiful print of

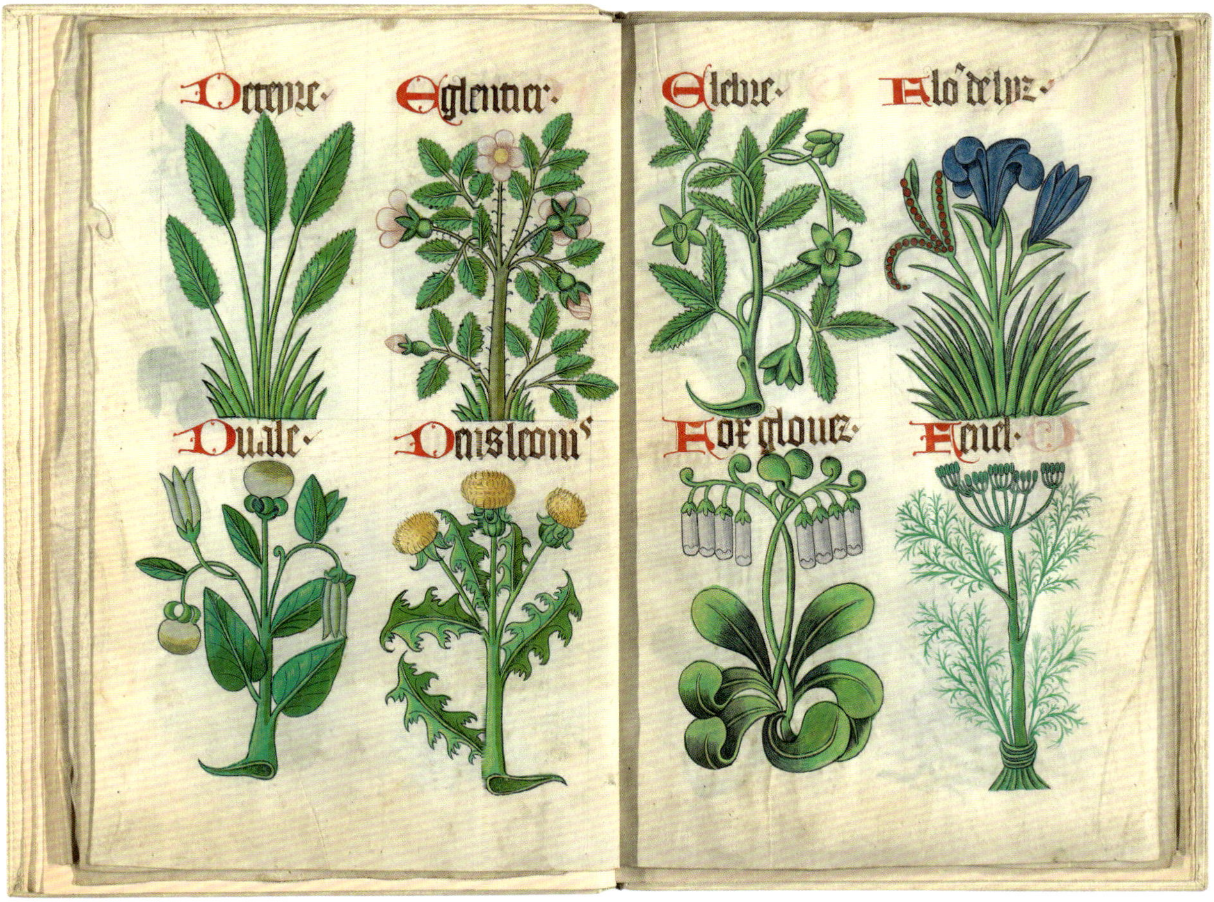

Figure 2. *Helmingham Herbal and Bestiary* (leaves 2v–3r), England, ca. 1500, gouache and watercolor with pen and ink on parchment. Yale Center for British Art, Paul Mellon Collection

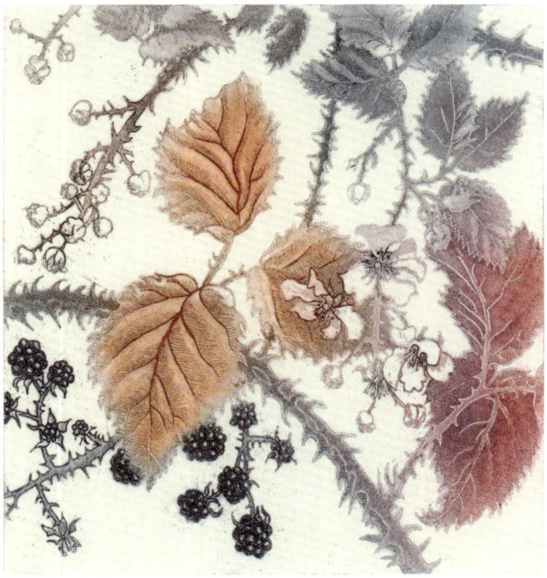

Figure 3. Rosaleen Wain, "Blackberry" (detail), etching with colored aquatint, from *A Printmaker's Flora: An Anthology of the Names of British Wild Flowers*, Dartington Hall, Devon: Dartington Printmakers, 1996. Yale Center for British Art, Friends of British Art Fund

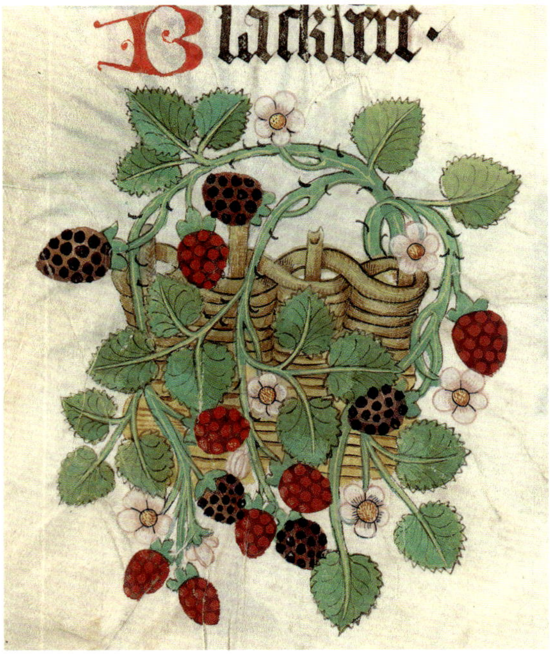

Figure 4. "Blackbere" from *Helmingham Herbal and Bestiary* (detail), ca. 1500, watercolor and gouache on parchment. Yale Center for British Art, Paul Mellon Collection

blackberries (fig. 3), recognizing that it echoed the lovely blackberries drawn in the *Helmingham Herbal* (fig. 4). These two remarkable images, separated by some five hundred years, seem to me to illustrate precisely that "mysterious aesthetic kinship" described by Paul Mellon.

We can trace this notion of aesthetic kinship further. Writing in his *Journal of a Naturalist* (1829), the naturalist John Leonard Knapp fiercely defends the blackberry bramble for its tenacity, noting that "perhaps no other shrubby plant . . . will retain its verdure through the year, preserving, by a peculiar construction of its vessels, a portion of foliage unseared by frosts, and contending with gales that destroy and strip away all the honours of its neighbours."[5] He continues in this vein for several pages, with observations about the construction of the leaves and stems and the odd insects that inhabit the hedges, remarking on the badgers who feed on the ripe berries and ending, most importantly, with an admonition to his readers:

> Notices of such incidents may perhaps be considered as too trifling to record; but the naturalist, from the habit of observing, sees many things not obvious to all persons: his province is to investigate all the operations of nature, and if he record them truly, he has done his duty; prolix and dull as his remarks will be to some, yet to another they may afford information, or tend to elucidate a conjecture.[6]

Knapp's words inform Samuel Palmer's delicate pencil sketch of blackberry brambles (1856; fig. 5), likely—as described by his son—"for want of a better term," one of the artist's "shorthand notes from nature" consisting of "innumerable little blots and scratches and hastily scribbled hints or impressions, which, slight as they were, bore a far more important part in the artist's professional career and in the growth of his best works than might be supposed."[7]

The aesthetic kinship that links the drawing of the Helmingham blackberries to the Palmer sketch also connects it to Laurie Clark's twentieth-century line block engraving made in 1979 (fig. 6), illustrating Thomas Clark's poem, and to a photograph of a late summer garden (2013; fig. 7).

A similar affinity exists between the specimen sheet of *Papaver rhoeas,* the common poppy, collected by the

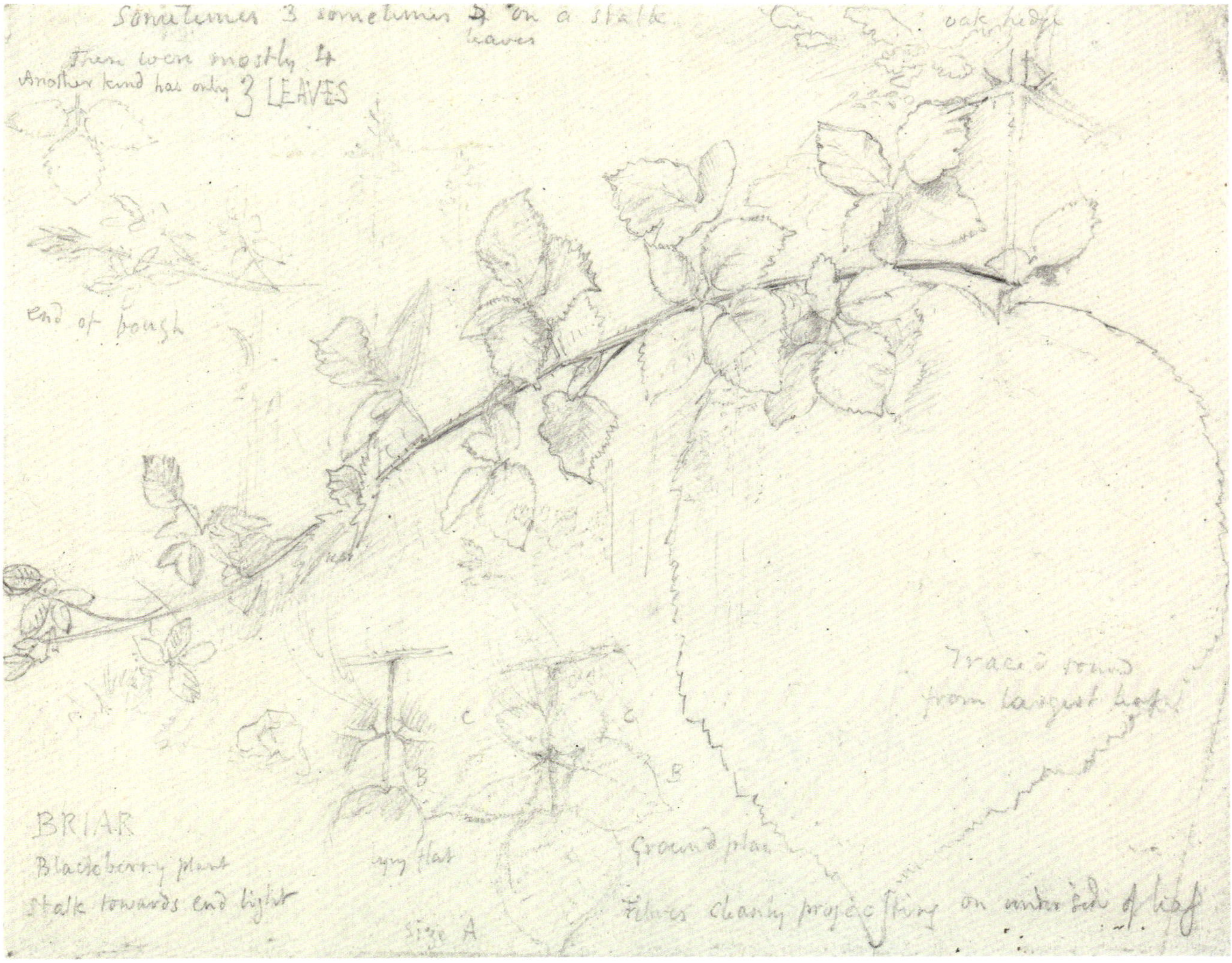

Figure 5. Samuel Palmer, *Brambles*, 1856, graphite. Yale Center for British Art, Paul Mellon Collection

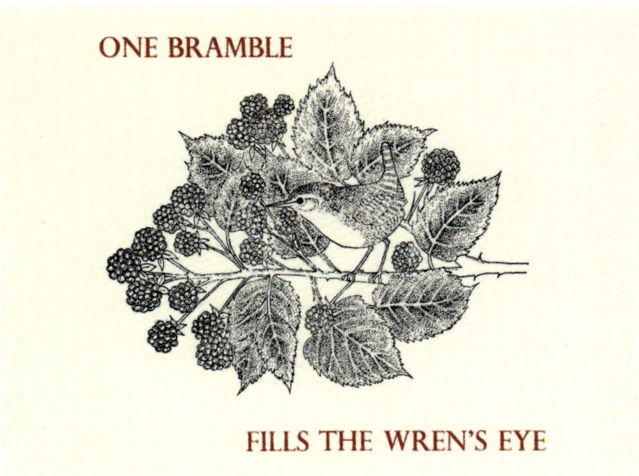

Figure 6. Thomas A. Clark and Laurie Clark, "One bramble fills the wren's eye" (detail) from *Proverbs of the Meadow*, Nailsworth, Gloucestershire: Moschatel Press, 1979, line block engraving. Yale Center for British Art, Friends of British Art Fund

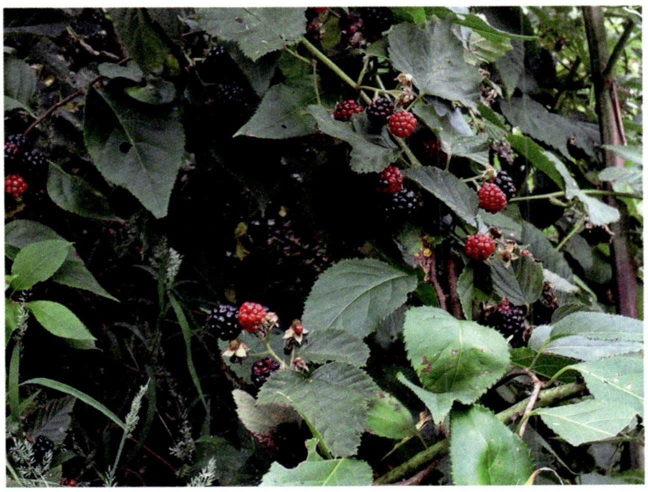

Figure 7. Late summer garden, Chester, Connecticut, 2013

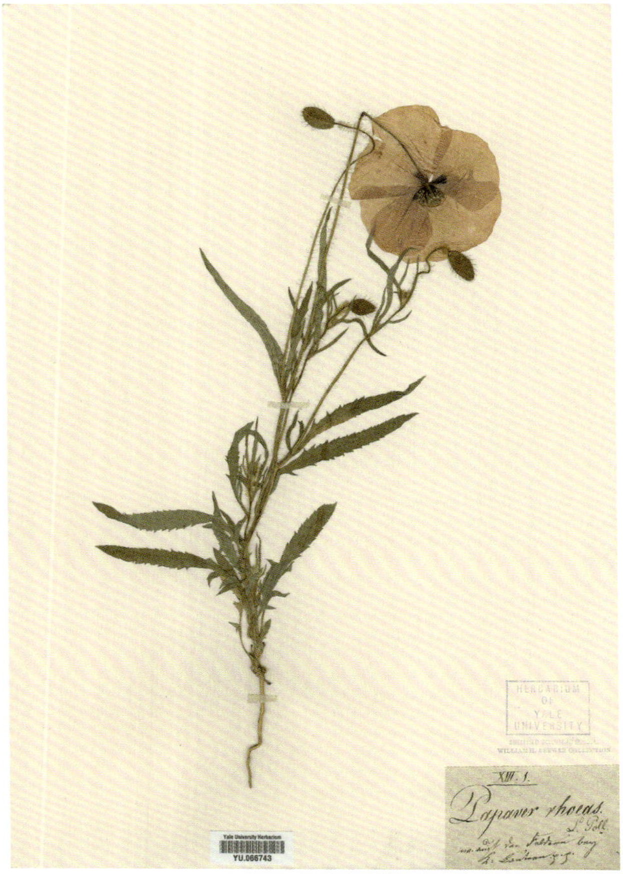

Figure 8. *Papaver rhoeas (Corn Poppy)*, collected by Gottlieb Wilhelm Bischoff or colleague, somewhere in Europe before 1850, natural specimen mounted on paper. Division of Botany, Peabody Museum of Natural History, Yale University

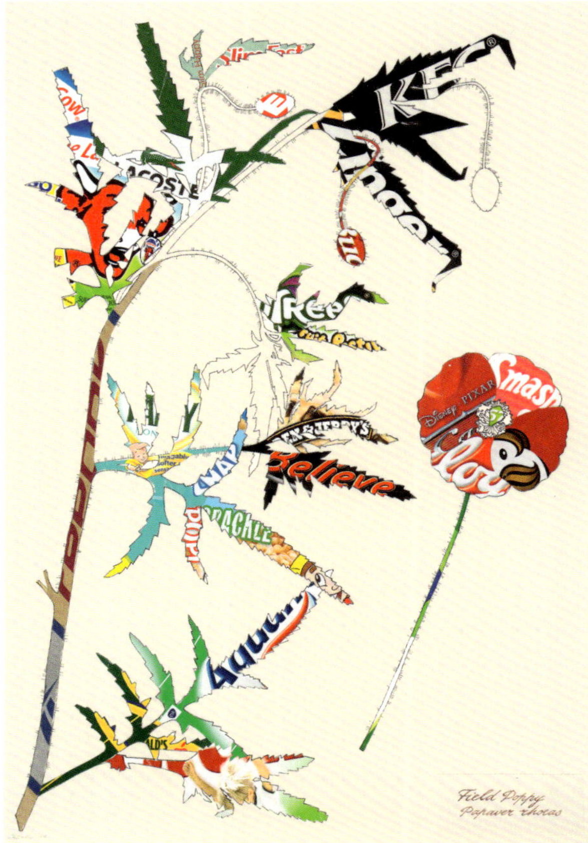

Figure 9. Tracey Bush, Herbarium sheet for *Nine Wild Plants: Common Poppy*, 2006, hand-cut paper collage with pen and ink. Yale Center for British Art, Friends of British Art Fund

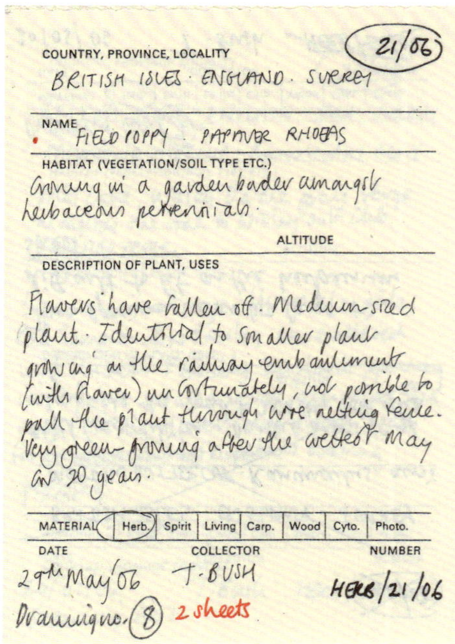

Figure 10. Tracey Bush, *Herbarium Notebook* (detail), 2005, pen and ink with natural specimens and cut paper collages. Collection of the artist

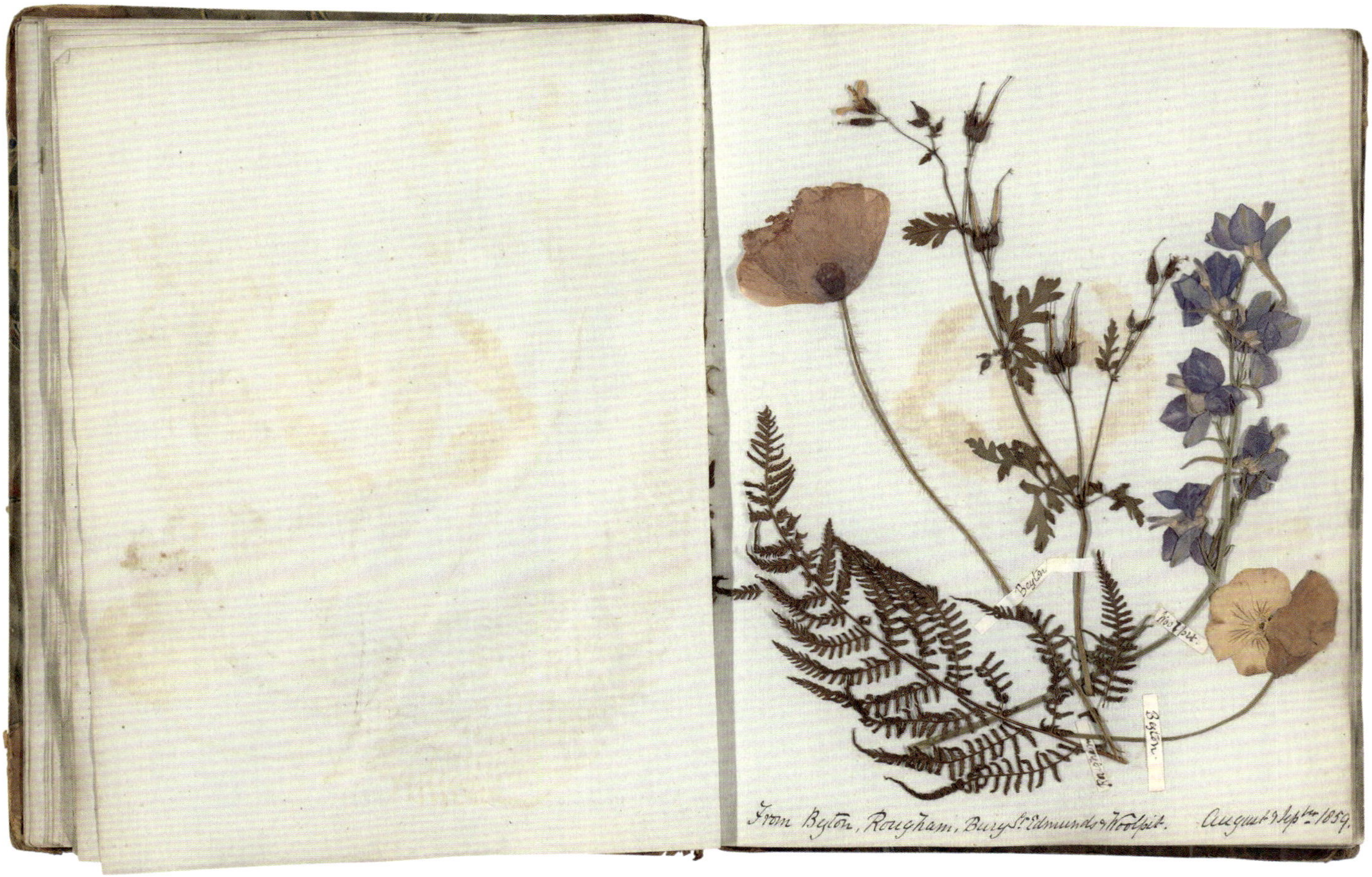

Figure 11. *Album of Dried Flower and Seaweed Specimens*, 1856–1863, natural specimens with pen and ink. Yale Center for British Art, Paul Mellon Fund

German botanist Gottlieb Wilhelm Bischoff or one of his colleagues around 1850 (fig. 8), and the collaged specimen sheet of the same plant made by the artist Tracey Bush in 2006 (fig. 9), working with notes she recorded in her own field guide (fig. 10).

The impulse of the unidentified woman who collected her poppy while walking through the fields near Bury St. Edmunds in Suffolk in 1859, carefully pressing and labeling it (fig. 11), was no different from that of the young Miss Rowe of Liverpool (likely Elizabeth Rowe, aged nineteen), who joined the newly formed Liverpool Naturalists' Field Club in 1861. In this first year, the club attracted over five hundred members; a report in the progressive *Popular Science Review* of 1862 noted that "one of the chief causes of their rapid rise and great popularity is the admission of lady members to all their meetings, whether in the open air or in the lecture-hall."[8] The first order of business at the first meeting was to distribute "amongst the ladies a printed list, headed 'L. N. F. C. Names of Natural Orders from Dr. Dickinson's "Flora of Liverpool,"'" published in 1851. The list included:

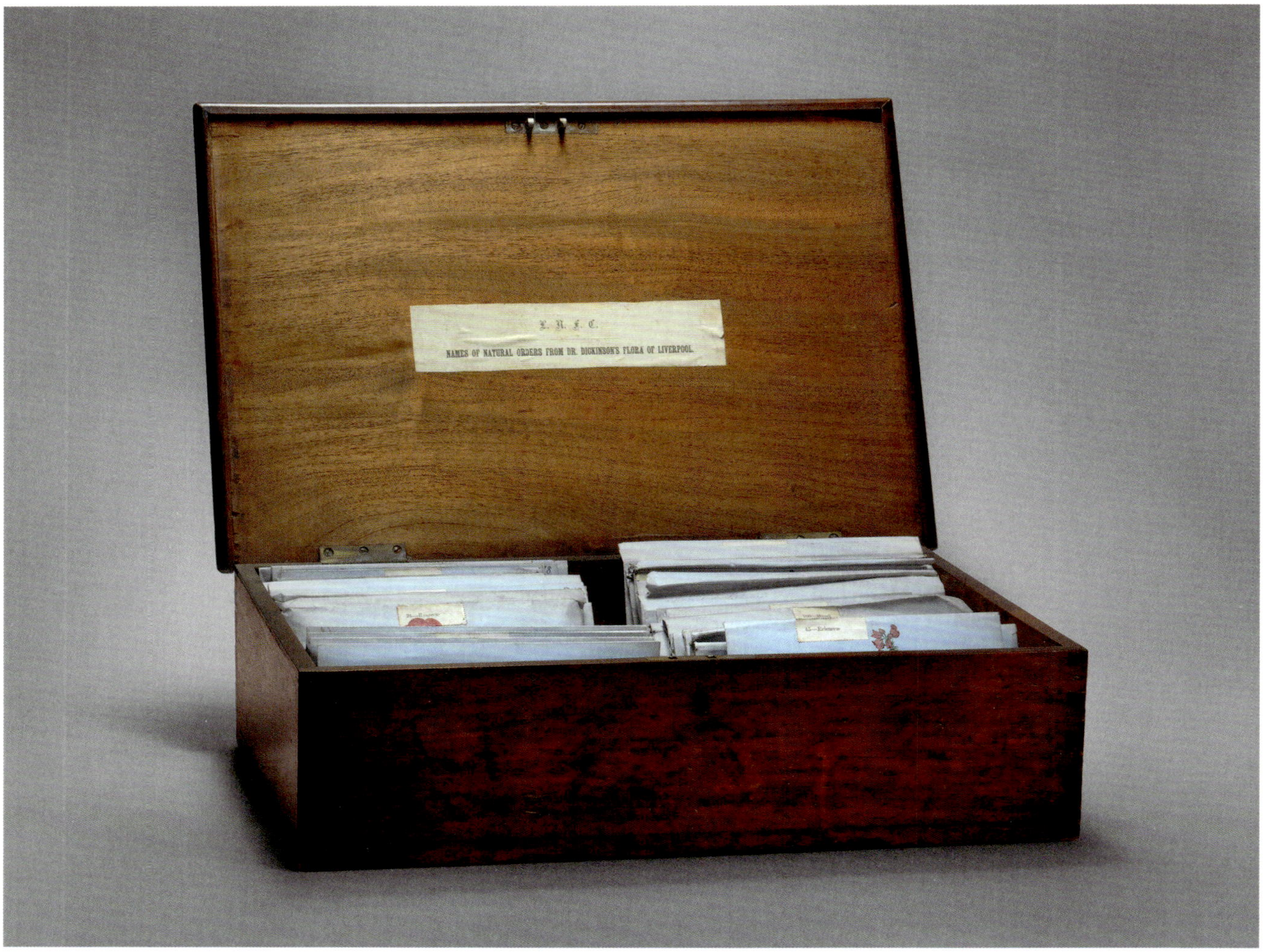

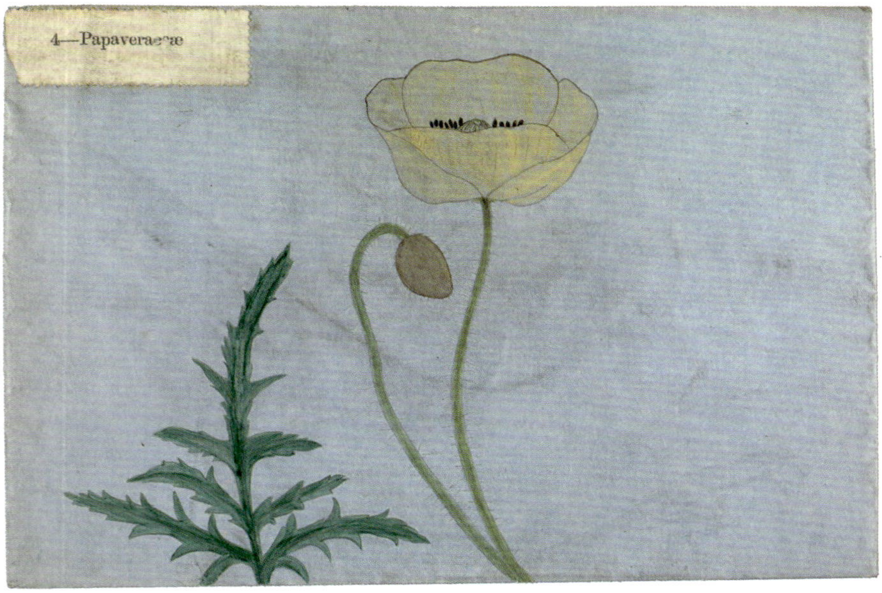

Figure 12. Miss Rowe, *Collection of Botanical Specimens*, Liverpool, 1861, mahogany box with ninety envelopes of five hundred natural specimens. Yale Center for British Art, Paul Mellon Fund

Figure 13. Miss Rowe, "4. *Papaveracea*" from *Collection of Botanical Specimens*, Liverpool, 1861, envelope with watercolor and printed label. Yale Center for British Art, Paul Mellon Fund

101 orders, beginning with "Ranunculaceae," . . . each order being separated from the adjoining ones by perforated lines, so that the greatest facility was afforded for tearing off the names of the orders. The object of this proceeding was to enable such of the ladies as desired it to compete for the "Botanical Prize" (a book value 10s. 6d.) . . . awarded . . . "to the lady who collects and arranges, according to the natural orders, the largest number of species in flower."[9]

Sadly, we do not know if the estimable Miss Rowe won the prize in the end, because at the meeting where the decision was to be announced, the judges ran out of time (owing to the length of the "excellent address" by the invited speaker, Mr. Grindon, Honorary Secretary of the Manchester Naturalists' Field Club) and so "were unable to award the prize to the successful competitress. This part of the proceedings was, therefore, postponed."[10] It must have been incredibly frustrating to the women in the room. But one hopes that Miss Rowe won. There can be no doubt that she deserved it. Housed in an elegant mahogany box (fig. 12), her collection of five hundred specimens were carefully mounted on thin writing paper, labeled in pen and ink, and sorted into genus and species. She placed them into ordinary blue stationery envelopes and drew a watercolor of one of the plant specimens on the outside, adding the correct perforated label from Dr. Dickinson's *Flora* (fig. 13). Although there is in fact no official notice of her receiving the prize, we do know she was singled out a year later in a report on the Field Club's activities as "a young lady member of this club [who] is remarkably successful in these competitions, and possesses very extensive knowledge in systematic botany."[11]

Miss Rowe and her amazing collection inspired the contemporary artist Mandy Bonnell, who, one summer at the Center, spent weeks examining the work of Miss Rowe and other Victorian self-taught naturalists. Bonnell produced two unique boxes of drawings. The first, *Wild Flowers Worth Notice* (the title taken from Phebe Lankester's field guide of the same name, published in 1861), includes detailed drawings in graphite, many of them close copies of Miss Rowe's specimens (figs. 14 and 15). It is clear she is paying homage to the young Victorian naturalist. Bonnell's *Lichen*, an assemblage of graphite drawings and cut paper

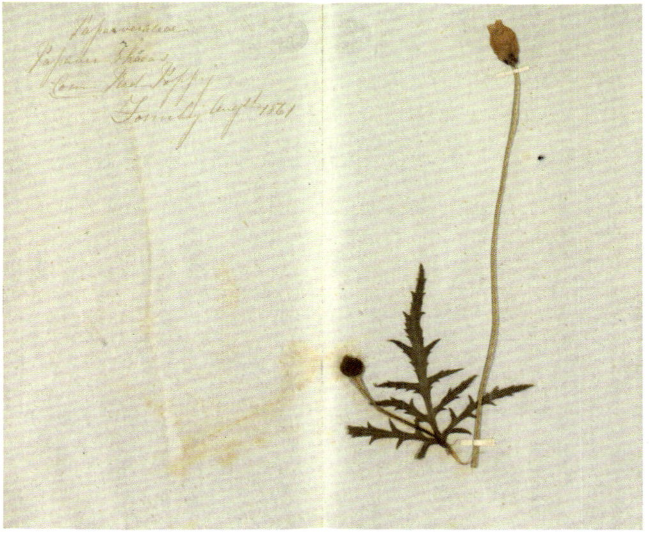

Figure 14. Miss Rowe, "*Papaver rhoeas*" from *Collection of Botanical Specimens*, Liverpool, 1861, natural specimen with pen and ink annotation. Yale Center for British Art, Paul Mellon Fund

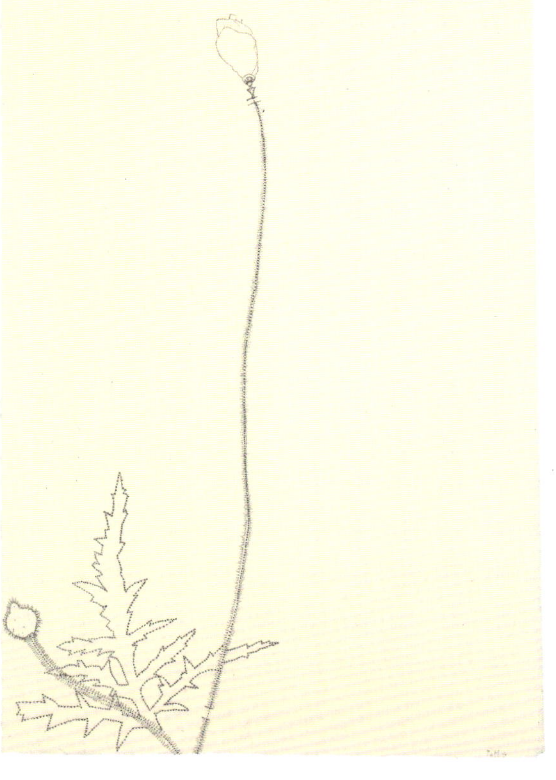

Figure 15. Mandy Bonnell, "Poppy" from *Wild Flowers Worth Notice*, London, 2012, graphite. Yale Center for British Art, Gift of the artist

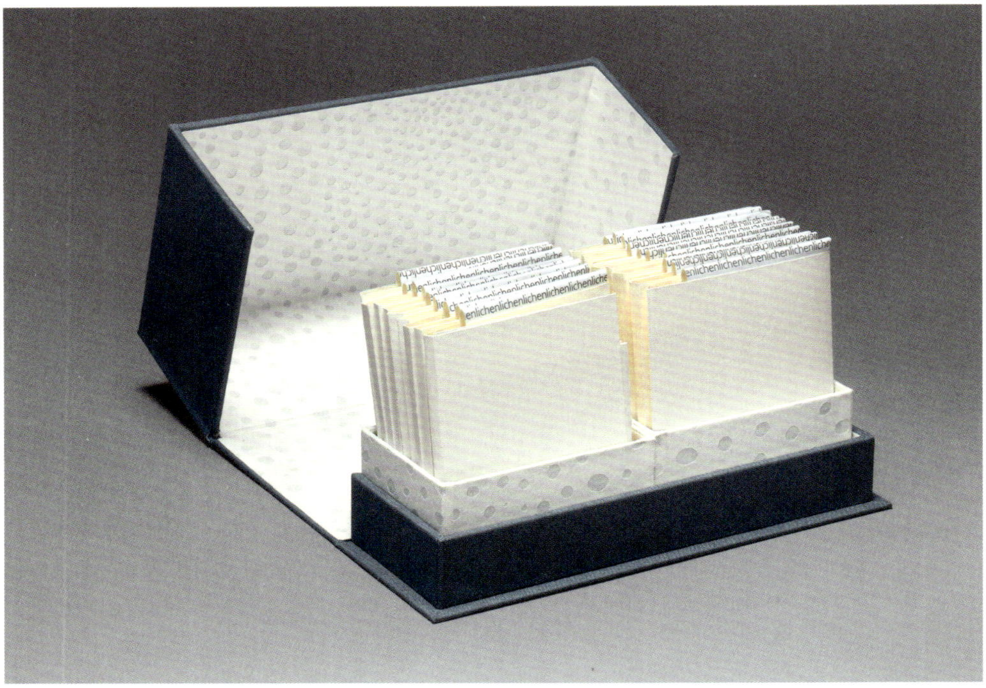

Figure 16. Mandy Bonnell, *Lichen*, with poem by Gabriel Gbadamosi, London, 2012, graphite and hand-cut paper in twenty folded packets, in cloth-covered box with hand-cut paper interior. Yale Center for British Art, Friends of British Art

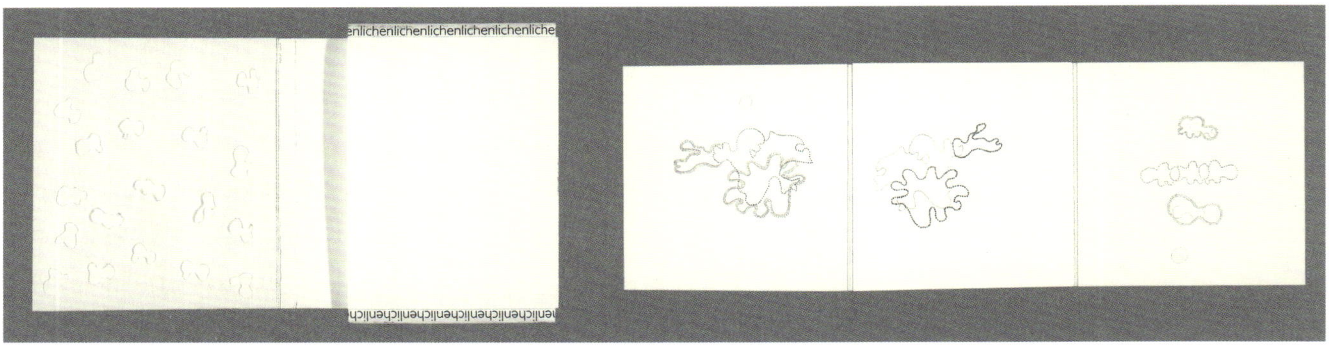

in folded packets (fig. 16), evokes Miss Rowe's envelopes of specimens.

The interest in the common flower appears to remain constant in the work of many of the contemporary artists discussed here. The artist Liz Machin explains her relationship with the past beautifully and straightforwardly in her artist's book *Working Class Botanists* (2010), illustrated with a digital photograph of a stalk of cow parsley (also known as Queen Anne's lace). "In the 19th century," she writes,

> Botany was simple—the Linnaean system of classifying plants made identifying them much easier. Botany was affordable—requiring only a few simple tools: a magnifying glass and handbook to identify plants. A box or vasculum to store samples. Clean paper to dry and press specimens. Botany was virtuous—being close to nature was considered as being close to godliness; an acknowledgement of the Lord's good works and evidence of his very existence. Botany was refined—a pastime befitting of women and children, similar to painting and flower arranging. . . . Botany was healthy—an outdoor activity beneficial to physical and mental health and an opportunity for appropriate socialising with members of the opposite sex. Botany was relaxation—an escape for the artisan lacemaker, silk and cotton weaver working from home,

> tending their plants and learning the Latin names to the rhythm of the shuttle.
>
> And I too walk the fields near my home, connecting to the past, observing, collecting, preserving and creating.[12]

These connections are demonstrated throughout the publication and the exhibition, "devoted," as poet David Burnett recently wrote, "to the green world which we ignore and ravage."[13]

Burnett's own essay introduces the work of a little-known wood engraver, Sister Margaret Tournour. This extraordinary woman returned to wood engraving in her late 70s, and, while confined to her room in a wheelchair, with her pet hedgehog as company, made exquisite engravings of the natural world. As Burnett writes, echoing the words of naturalist John Leonard Knapp, Sister Margaret's "confinement did not constrain but on the contrary focused and sharpened her observation of the natural world, which was also informed by lifelong study and teaching. She noticed how much we can observe about us every day, often phenomena which might at first sight appear slight or insignificant" (p. 68).

Robert Peck's essay focuses on efforts of the self-taught naturalist in the eighteenth and nineteenth centuries to interpret and understand nature, looking at published field guides, specimen collecting, and the drawing and recording of the natural world. He offers insights into the "interweaving of science and art, and the many ways in which humans have recorded, interpreted, and celebrated nature for the past three hundred years" (p. 44).

Molly Duggins explores the growth of interest in the nineteenth century, particularly among women, to transform nature through craft into an "artificial kingdom." Cutting and folding paper into birds and flowers, collecting and mounting specimens of seaweed and leaves into albums reveal the "particularly intimate relationship [that] existed among handicraft production, natural history practices, and the Victorian book and album" (p. 47).

These three essays provide the reader with a solid footing for the journey through the flora and fauna of the British countryside that follows. The field guide gathers together drawn, painted, and cut paper versions of flowers; blackberries and brambles; fruit and vegetables; grasses, seeds, and seedpods; trees and lichen; birds; butterflies, moths and caterpillars; beetles, bees, and spiders; a hedgehog or two, and the odd sheep—the sort of thing one might encounter while walking in the gardens and tramping over the moorlands and hills of Britain. Inhabitants of ponds and streams (kingfishers, swans, frogs, and dragonflies) are considered, along with those creatures that live along the shore (fish and crabs).

In the true spirit of a field guide, readers are invited to heed the advice of Mrs. Lankester to "observe, collect, and preserve," and record their own observations (and to slip their own poppy or daisy or feather into the collecting pocket at the back).[14] Throughout the field guide, via images and text, we are able to glimpse those self-taught naturalists, both past and present, who reveal to us a shared vision of the natural world, and it ends with a reminder of the need for close observation, the root of all their activities. The guide meanders, diverges, and overlaps in places, as only a nineteenth-century naturalist might truly understand—but as a twenty-first century naturalist might well appreciate.

1. Paul Mellon, "A Collector Recollects," foreword to *Selected Paintings, Drawings & Books* (New Haven: Yale Center for British Art, 1977), v–vi.
2. For a full discussion of Paul Mellon's book collecting, see William Reese, "Paul Mellon as a Book Collector,"in *Paul Mellon's Legacy: A Passion for British Art: Masterpieces from the Yale Center for British Art* (New Haven and London: Yale Univ. Press, 2007), 57–71.
3. For a discussion of the Center's natural history drawings by Luigi Balugani, see Paul Hulton, *Luigi Balugani's Drawings of African Plants: from the Collection made by James Bruce of Kinnaird on his Travels to Discover the Source of the Nile, 1767–1773* (New Haven: Yale Center for British Art, 1991).
4. Nicolas Barker, ed., *Two East Anglian Picture Books: A Facsimile of the Helmingham Herbal and Bestiary and Bodleian Ms. Ashmole 1504* (London: Printed for presentation to the members of the Roxburghe Club, 1988), xiii.
5. John Leonard Knapp, *The Journal of a Naturalist* (London: John Murray, 1829), 105–106.
6. Knapp, *The Journal of a Naturalist*, 107.
7. Alfred Herbert Palmer, *The Life and Letters of Samuel Palmer, Painter and Etcher* (London: Seeley, 1892), xii.
8. "The Liverpool and Manchester Field Naturalists' Societies," *The Popular Science Review: A Quarterly Miscellany of Entertaining and Instructive Articles on Scientific Subjects* 1 (1862): 123.
9. "The Liverpool and Manchester Field Naturalists' Societies," 123.
10. "The Liverpool and Manchester Field Naturalists' Societies," 123.
11. "Naturalists' Field Clubs," *The Popular Science Review: A Quarterly Miscellany of Entertaining and Instructive Articles on Scientific Subjects* 2 (1863): 94.
12. Liz Machin, *Working Class Botanists* (Manchester: L. Machin, 2010), 2–7.
13. Personal correspondence with the author, November 2013.
14. Phebe Lankester, *Wild Flowers Worth Notice* (London: R. Hardwicke, 1861), ix.

Legitima, Alis integris.

S. Atropos. Bee-tiger H.

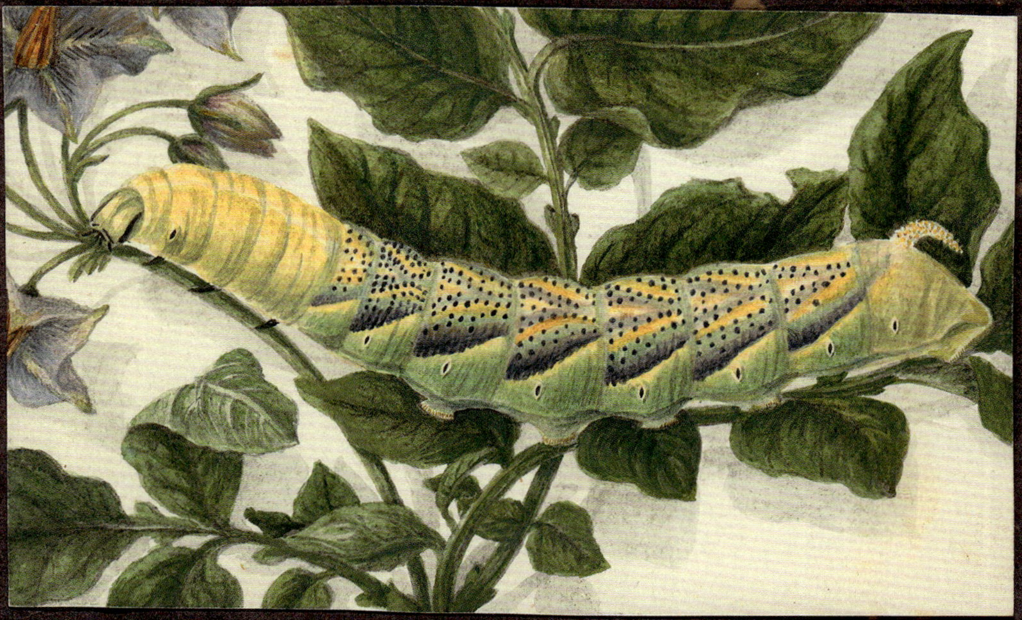

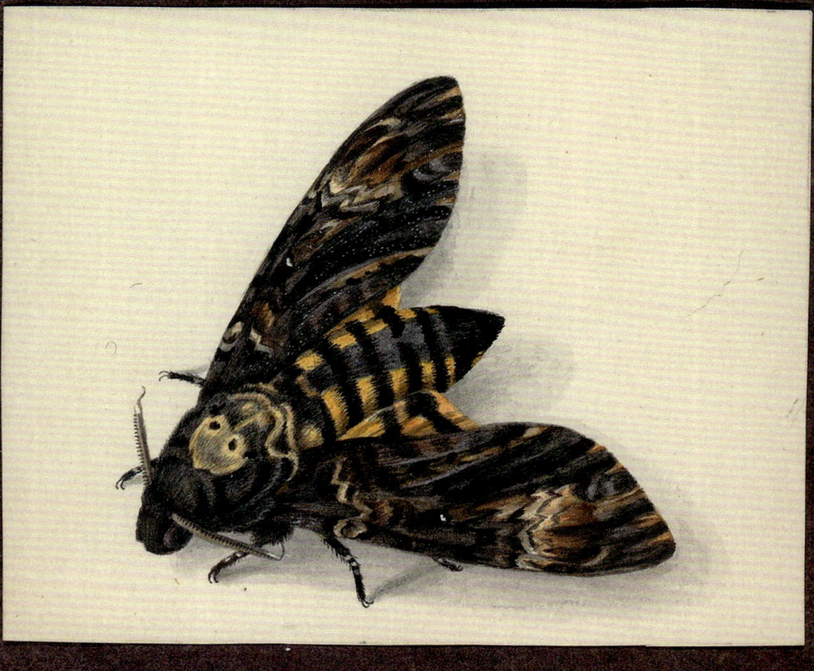

Natural Obsessions
From Specimens to Books

Robert McCracken Peck

In January 1819, George Cruikshank published a narrative print entitled *Landing the Treasure, or the Results of the Polar Expedition!!!* (fig. 17). The caricature features a group of sailors, recently returned from an exploratory trip to the Arctic regions, carrying the findings of their expedition through the streets of London to the British Museum (Montagu House), while a crowd of irreverent spectators offer derisive comments from the sidelines. The motley assemblage of Arctic veterans, some with false noses to emphasize their recent exposure to frostbite, are led by the expedition's rotund commander, Captain John Ross. They carry an eclectic assortment of natural artifacts collected on their travels, including an odiferous (decaying) polar bear, a driftwood stump labeled "Esquimeux wood," a cask with "worms from the Intestines of a seal," a lance made of the spiraling horn of a "sea unicorn" (a narwhale tooth, long a valued addition to every *wunderkammer*), and a container of "red snow," containing a rare species of cold-tolerant algae that Ross found growing on a glacier in Greenland.

To add some ethnic humor to the scene, Cruikshank placed a gangly man toward the end of the parade. Despite having features that appear more African than Asian, the non-British figure was intended to represent John Sackheuse, a native Greenlander whom Ross had employed as a guide and translator during the expedition. He is shown carrying a portfolio of his own sketches, one of which was subsequently engraved and published in Ross's report of the expedition, along with many of the captain's own equally charming and equally naïve drawings of the landscape, wildlife, and people encountered during the trip.[1] Sackheuse's portfolio might be considered one of the first artists' books created in the Arctic to be shared with the English public. Because of the quality of his drawings, the affable Inughuit was seen by those familiar with his work as an artistic prodigy. Few were aware that he had received training in drawing in England during an earlier visit to the UK (in 1817), after stowing away on an English whaler bound from Greenland to Leith.[2] The remarkable contents of Sackheuse's Arctic portfolio—which

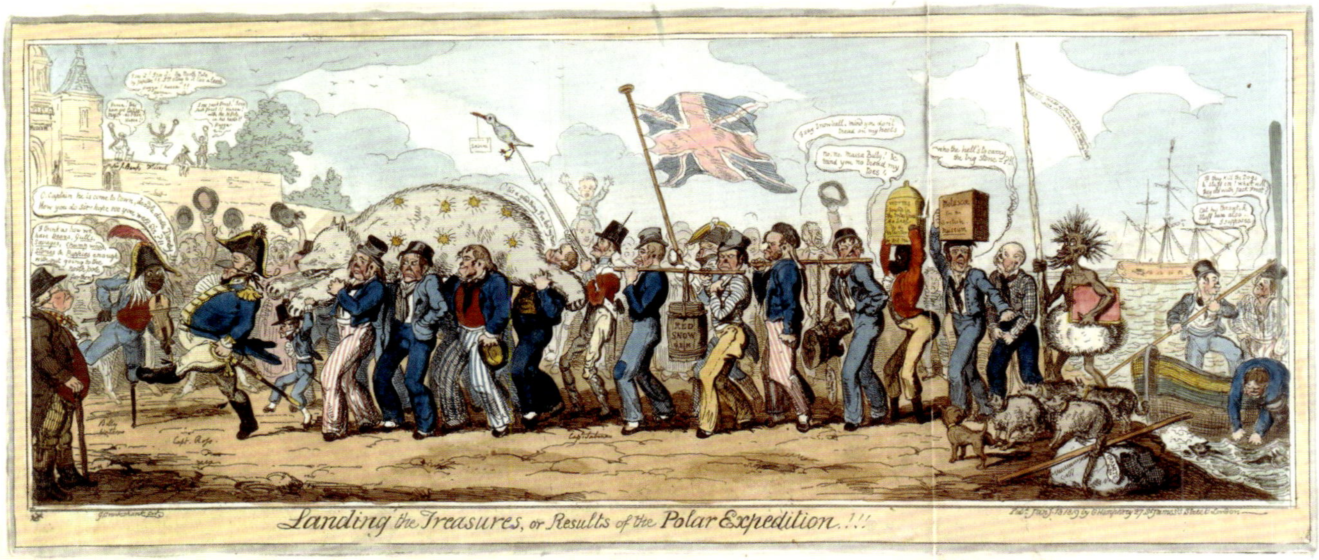

Figure 17. George Cruikshank, *Landing the Treasure, or The Results of the Polar Expedition!!!*, 1819, hand-colored etching. British Museum, London

OPPOSITE: Mary Anne Burges, "Legitima Alis integris. S. Atropos—Bee Tiger," watercolor and gouache with pen and ink, in *Lepidoptera of the British Isles*, ca. 1810. Yale Center for British Art, Paul Mellon Fund

Robert McCracken Peck is Senior Fellow of the Academy of Natural Sciences of Philadelphia, Drexel University.

included a widely reproduced view of the expedition leader, his second in command, William Parry, and a sailor from one of their ships exchanging trinkets for natural artifacts with a group of Inuit on the Arctic shore—were soon dispersed. But the spirit of inquiry and artistic creativity that inspired Sackheuse to create them survives in countless other less famous sketchbooks and volumes containing impressions of nature at home and abroad.

Cruikshank's ridicule notwithstanding, the timeless and apparently universal instinct to seek, find, collect, organize, name, and record natural artifacts had been encouraged in England as a worthy pursuit from at least the end of the seventeenth century, when Enlightenment ideals called for a systematic approach to science as a way of both improving the state of society and reinforcing individual devotion to the creator of life on Earth. While most of this activity was pursued privately, some of it was undertaken with government support. Exploratory expeditions, such as the one satirized in Cruikshank's caricature, served the dual purpose of reinforcing Great Britain's dominance as a global power and providing new areas of investigation among academics at home. Beginning with Captain Cook and continuing well into the nineteenth century, the Royal Navy did more than any other government agency to support and encourage this kind of scientific research, often including professional artists on its expeditions. It further helped to foster the integration of science and the visual arts by providing naval officers with training in drawing. This ensured that they would return from their exploratory voyages with not only accurate navigational charts and shoreline profiles, but also interesting visual records of what they had seen abroad. In the late eighteenth century and for the next hundred years, a steady stream of illustrated expedition reports helped to foster a public appreciation for landscape drawing and natural history illustration. This, in turn, encouraged countless individuals to create visual records of their own with a focus on subjects closer to home.

The political humor of Cruikshank's engraving was directed, in part, at the government's expenditure of funds for a seemingly worthless investigation, but it was also a gentle poke at the increasingly popular hobbies of nature study, collecting, and drawing that would become grist for cartoonists' mills for at least another century. Nor was the theme a new one. Forty-seven years before, the caricaturist "Whipcord" (Matthew Darly) had poked fun at international collecting by lampooning the influential naturalist Joseph Banks as "The Fly Catching Macaroni" with one foot in the Arctic and one in the Antarctic (fig. 18). "I rove from Pole to Pole, you ask me why, I tell you Truth, to catch a ___ Fly!" reads the sarcastic caption.[3] When the engraving was issued in 1772, Banks had recently returned from accompanying Captain Cook on his first circumnavigation of the world (1768–1771). Although they were made at Banks's own expense, his extensive natural history and ethnographic collections, like those made on many subsequent British expeditions, were eventually given to the British Museum, thus helping to institutionalize what had begun as a personal interest.[4] Later, as president of the Royal Society, where he actively encouraged the study of science at home and throughout the empire, Banks was among the most visible of scholars and collectors of natural history. But there were many others as well. The caricaturist Thomas Rowlandson lampooned another well-dressed insect collector of the era by showing him adding specimens to a home collection (fig. 19). In fact, in the late eighteenth and early nineteenth centuries, few British families, regardless of social class, educational background and personal wealth, were *not* involved in some way with the collection of scientific specimens or the study of nature. As early as 1763, *The Critical Review* observed that natural history had become "the favorite study of the times."[5] This interest would deepen and broaden in succeeding decades until it became something of a national obsession during the Victorian era.

Critical to the expansion of the study of natural history by academics and amateurs alike was the development of universally acceptable systems of nomenclature. A commonly agreed upon method of classifying and naming organisms allowed naturalists in one part of the country—or the world—to converse with those from other places and to know how the plants and animals they were discussing related to each other. Having an interest in natural history was one thing, but being able to talk about it, in person and in writing, allowed naturalists from all walks of life to share their hobby with others, thus expanding the numbers of people actively engaged in looking for, collecting, and recording different aspects of the natural world.

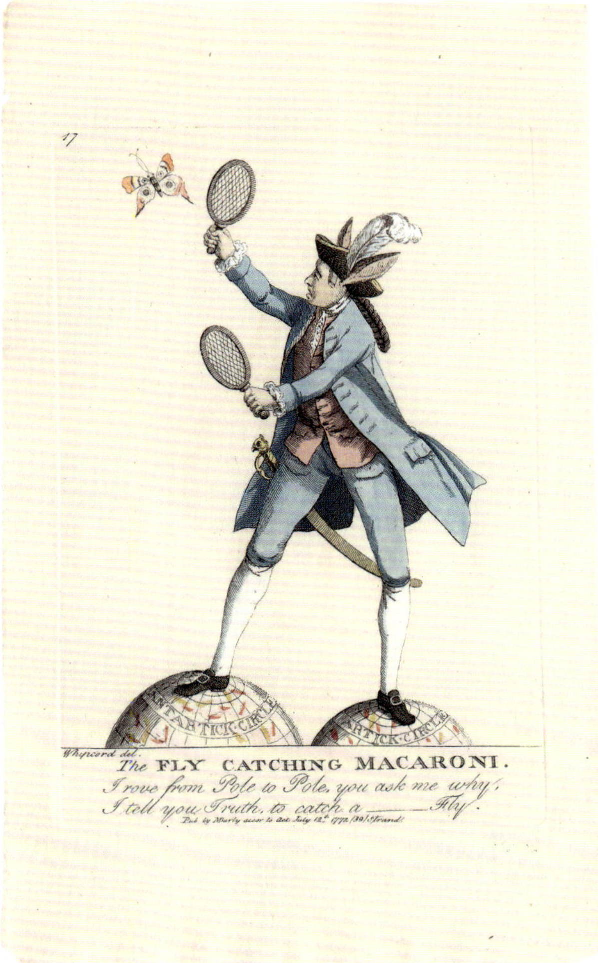

Figure 18. Whipcord, *The Fly Catching Macaroni*, 1772, hand-colored etching. The Lewis Walpole Library, Yale University

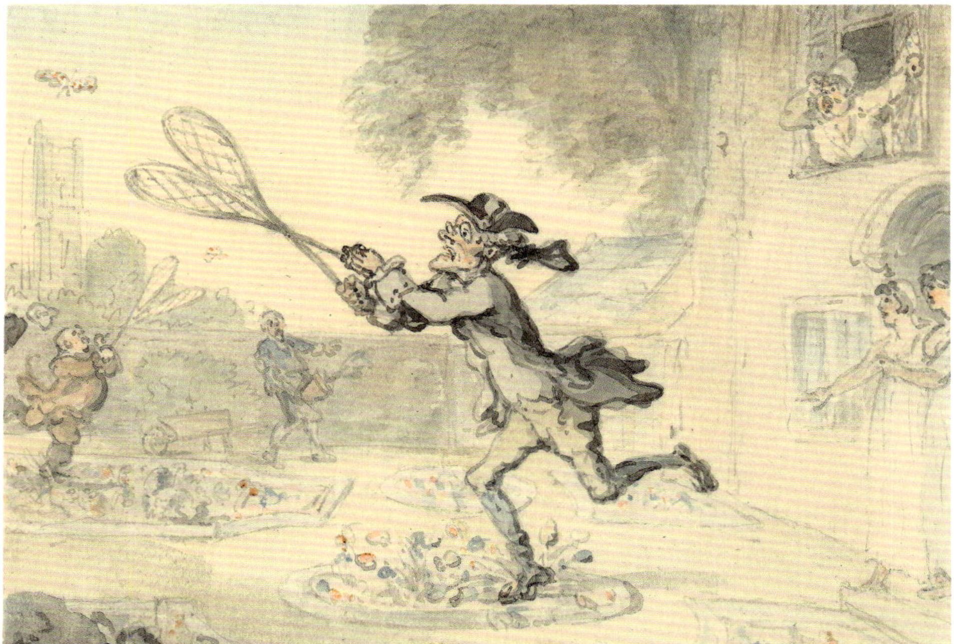

Figure 19. Thomas Rowlandson, *Butterfly Hunting*, ca. 1806, watercolor with pen and gray ink over graphite. Yale Center for British Art, Paul Mellon Collection

The earliest efforts to organize plants and animals into categories of classification were based on the utility of the organisms to humans.[6] In medieval guidebooks, subjects were often arranged alphabetically. Animals were identified as either edible or inedible; wild or tame; useful or useless; and then subdivided into how best they could serve man. These systems were usually hierarchical, with the lion, the eagle, and the whale serving as the "kings" of various orders of creation, but always placing man at the pinnacle of a divinely ordered world.[7] Some authors ranked animals by how physically attractive or handsome they were. "Noble" creatures, like horses (domestic) or deer (wild), were placed at the top of the list, and "base" ones, such as toads, snakes, and spiders, were placed at the bottom. Bestiaries often used animals as vehicles for conveying religious instruction. Many were loaded with messages of moral guidance.[8] Often, in this pre-scientific time, illustrations of real animals shared space with imaginary or apocryphal creatures whose reality had not yet been disproved, thus horses, dogs, and hares shared space with dragons and griffons (fig. 20).

In pre-Enlightenment times, plants too were organized in relation to their usefulness to man. They usually were categorized by their nutritional or medicinal qualities and given local, common names that were tied either to their use or their appearance: adder tongue, flea bane, toad flax, etc. (fig. 21). Such vernacular identifications could be confusing (and even dangerous), for there was no consistency in how names were assigned to any given organism. A health-

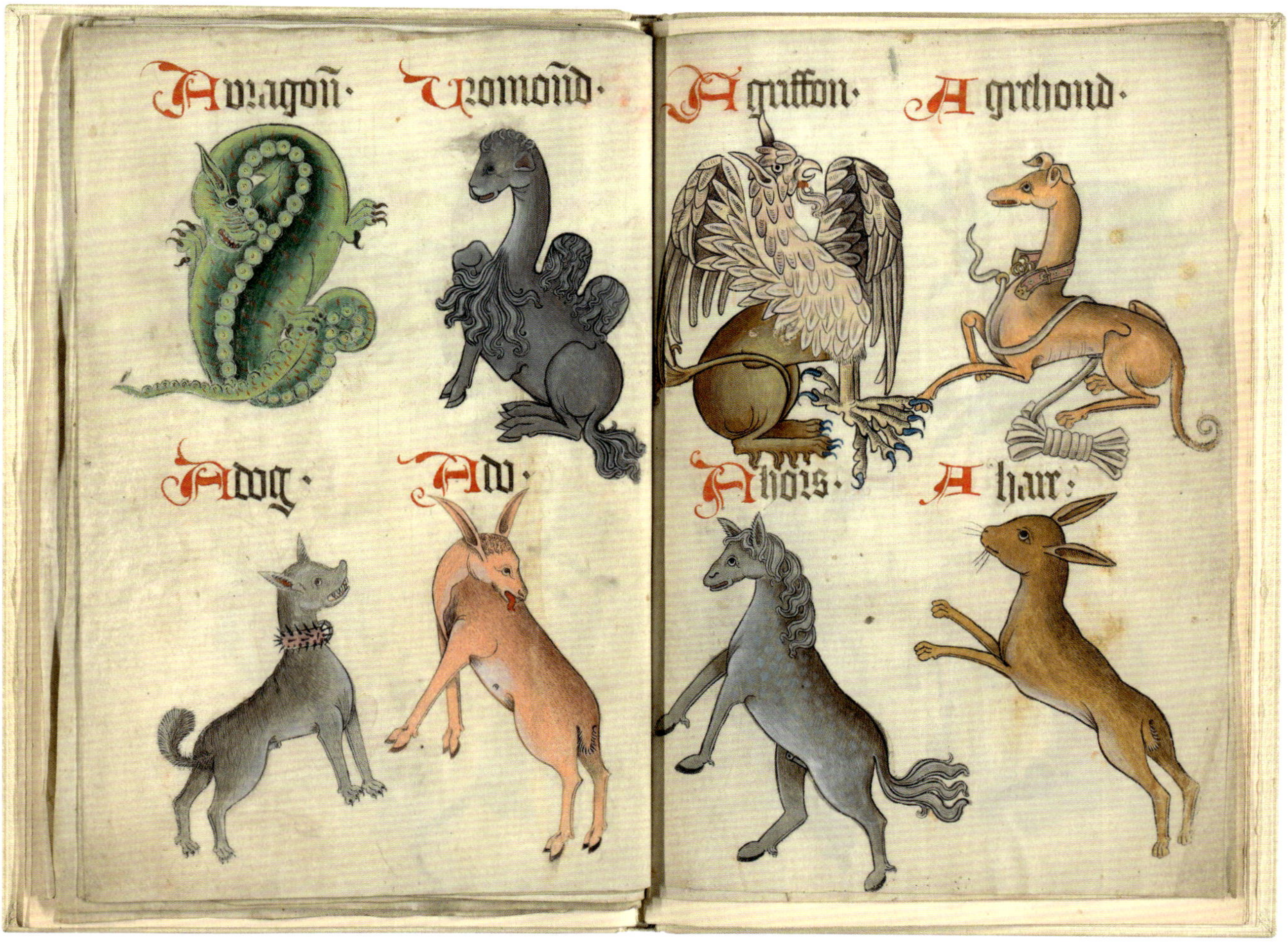

Figure 20. *Helmingham Herbal and Bestiary* (leaves 17v–18r), ca. 1500, gouache and watercolor with pen and ink on parchment. Yale Center for British Art, Paul Mellon Collection

ful plant from one region might share a name with a poisonous one from another. In the sixteenth century, European and British naturalists began to experiment with more objective, less homocentric methods of classification in an attempt to impose a rational order on the natural world. They employed both "artificial" systems, using selected physical features to classify plants and other organisms, and "natural" systems that incorporated a more holistic approach. The seventeenth-century English naturalist John Ray developed one system of botanical classification that gained relatively widespread support in his own country during the late seventeenth and early eighteenth centuries. It was based primarily on the seeds or fruiting bodies of plants (an artificial system), but also took into account a number of other features associated with a plant's overall structure and habits of growth (a natural system).[9] In the 1730s the Swedish botanist Carl Linnaeus incorporated some of Ray's ideas with those of others to develop a slightly simpler system that eventually gained precedence among botanists. In this system the sexual parts of flowers, not the fruits, were used to distinguish differences between plants. Linnaeus made a compelling case for his method in *Systema Naturae* (1735), a slim volume of fourteen pages that was soon expanded and went through many subsequent editions. While some considered its explicit references to sex offensive ("obscenity is the very basis of the Linnean system," complained the *Encyclopaedia Britannica*), the easy applicability of his method, and his simplified system of

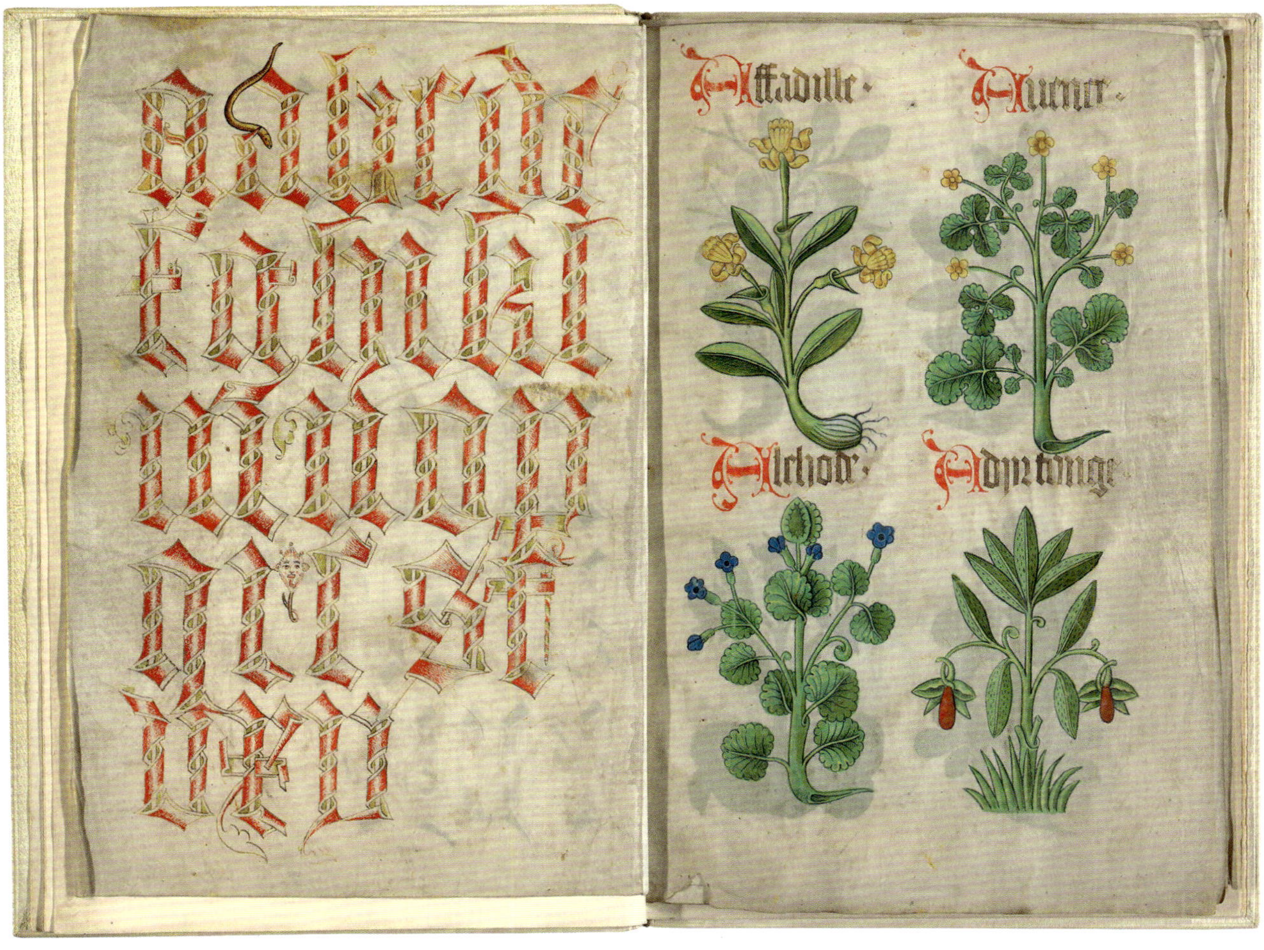

Figure 21. *Helmingham Herbal and Bestiary* (leaves 1v–2r), ca. 1500, gouache and watercolor with pen and ink on parchment. Yale Center for British Art, Paul Mellon Collection

nomenclature in which a genus and a species name were applied to each plant, proved very appealing to the many plant lovers and would-be botanists who had struggled with other, more complicated systems.[10] Linnaeus's suggestion that binomials also could be applied to other organisms was equally well received. In 1799 there were more than fifty classification systems in use, but it was Linnaeus's system that ultimately prevailed.[11]

In addition to being an innovative theoretician and charismatic teacher, Linnaeus was a practicing botanist. He built a large personal collection of plants and animals through his own international travels and with contributions from a wide circle of students and correspondents from around the world (fig. 22). In a series of influential publications, of which *Species Plantarum* (1753) and *Genera Plantarum* (tenth edition, 1758) were the most significant, he applied his classification system to what was then known of the vegetative world by boldly renaming all of the plants that had been described up to that time. In these and other publications, he renamed and described some 7,700 species of plants and 4,400 species of animals. While the classification of vertebrates and invertebrates proved somewhat more complicated and ultimately incorporated ideas generated by a smattering of other naturalists ranging from Aristotle to Ray, Linnaeus's system of botanical classification was eventually accepted by botanists the world over as the starting point for all botanical nomenclature. Thus, botanical names published before 1753 have no standing in modern nomenclature unless they were adopted by Linneaus in 1753 or reused at a later date by other authors.[12]

Shortly after Linnaeus's death in 1778 his widow contacted Sir Joseph Banks, who earlier had expressed interest in acquiring her late husband's library and scientific collections. Banks declined the invitation to buy the collections for himself, but, not wishing to miss the opportunity to secure them for Great Britain, he convinced a wealthy young friend and protégé named James Edward Smith to make the purchase instead. Thanks to Banks's encouragement, for a mere 1,000 guineas Smith became the owner of a collection that stands at the very foundation of modern taxonomy.[13]

In 1788, Smith founded a society to honor the great classifier and to encourage "the cultivation of the Science of Natural History in all its branches, and more especially of the Natural History of Great Britain and Ireland."[14] The Linnean Society of London has been committed to this mission ever since.[15] The society became an important gathering place for naturalists in England, a clearing house for botanical research and natural history discoveries of all kinds, and a model for more local groups, such as the Liverpool Naturalists' Field Club and many similar organizations throughout Great Britain and around the world. With the help of the Linnean Society and its many sister institutions, the once solitary activity of nature study became a very social enterprise.

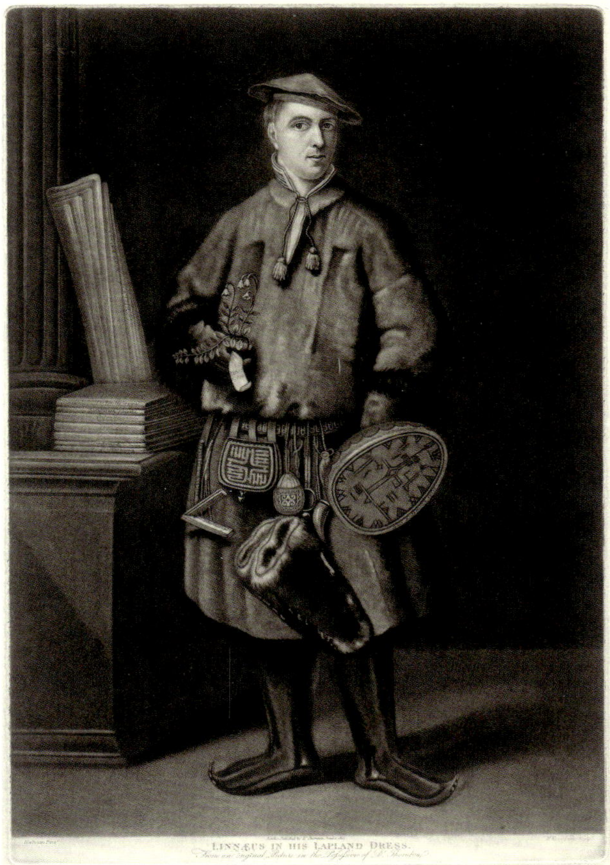

Figure 22. Henry Kingsbury after Martin Hoffman, *Linnaeus in Laplander Costume*, 1805, mezzotint. Archives of the Academy of Natural Sciences of Philadelphia, Drexel University (ANSP Coll. 2009-005)

What made the subject so popular was its physical, intellectual, and social accessibility. As William Swainson wrote in 1834, "Natural history has this peculiar advantage—that it can be prosecuted, in one shape or other, by almost every body, and under every ordinary circumstance.

Of all sciences, it is that which requires, in most of its departments, the fewest materials. It is as much within reach of the cottager as of the professor."[16] Another advantage, as George Cruikshank had pointed out in *Landing the Treasure*, was that interesting discoveries did not require expensive overseas expeditions. There were plenty of unstudied or misunderstood subjects to investigate in England, as a stout, John Bull-like figure standing near the front of the parade in Cruikshank's returning polar expedition caricature observes: "I think as how we have bears, gulls, savages, chump wood, stones, and puppies enough [to study here] without going to the North Pole for them."

Gilbert White, a country minister from Hampshire, demonstrated that, by careful observation, anyone could make a meaningful record of the local flora and fauna in any neighborhood, and that doing so could be a respectable, even reverential way of spending one's time. White's simply written account of his daily forays into the fields and woodlots of his parish, *The Natural History and Antiquities of Selborne* (first published in 1789 under this title), became a classic in the study of natural history, inspiring generations of would-be naturalists up to the present day to take to the countryside (fig. 23). Similarly, Thomas Bewick, a gifted wood engraver from Northumberland, created simple and appealing books describing British birds, mammals, and fish that anyone could understand and enjoy. His *General History of Quadrupeds* (1790) (fig. 24) and *History of British Birds* (*Land Birds*, 1797 and *Water Birds*, 1804) provided not only readable descriptions of local animals, but charming vignettes of rural life that served as a welcome counterpoint to the growing industrialization and urbanization of the nation.[17] Many of his elegant line engravings were copied by amateur artists seeking to hone their skills of observation and rendering. Some also found their way onto popular ceramic tableware being manufactured in Staffordshire for a growing middle class.

Freed from the confusion of local names and the cumbersome combination of adjectives that had characterized the scientific names favored by earlier pre-Linnaean systems, naturalists began to engage in the shared pleasures of identifying the world around them. At the most basic level they could simply increase their enjoyment of a day by the seaside or a ramble in the country by learning the names of

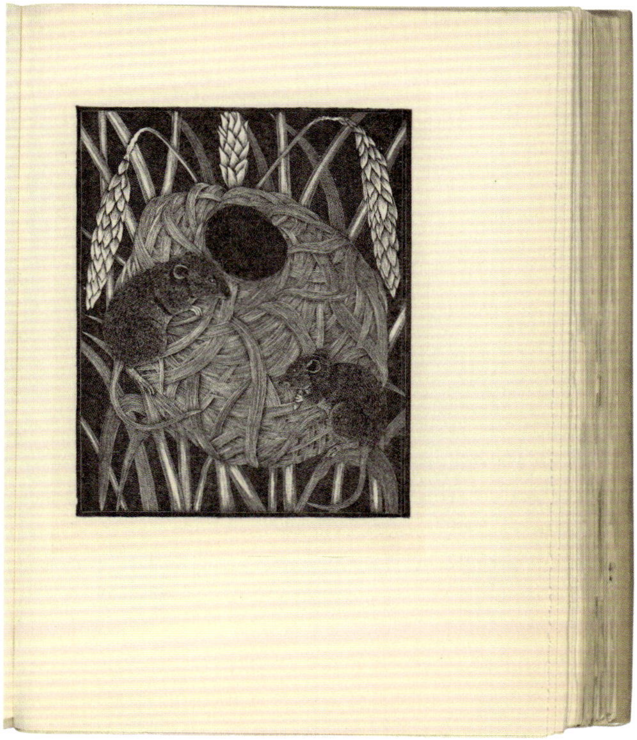

Figure 23. Eric Fitch Daglish, "Harvest Mice," wood engraving, in Gilbert White, *The Natural History of Selborne*, New York: E. P. Dutton & Co., 1930. Yale Center for British Art, Paul Mellon Collection

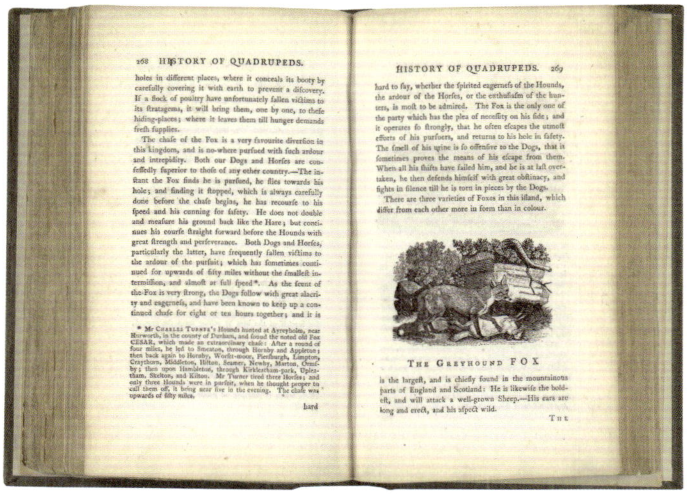

268 HISTORY OF QUADRUPEDS.

holes in different places, where it conceals its booty by carefully covering it with earth to prevent a difcovery. If a flock of poultry have unfortunately fallen victims to its ftratagems, it will bring them, one by one, to thefe hiding-places; where it leaves them till hunger demands frefh fupplies.

The chafe of the Fox is a very favourite diverfion in this kingdom, and is no-where purfued with fuch ardour and intrepidity. Both our Dogs and Horfes are confeffedly fuperior to thofe of any other country.—The inftant the Fox finds he is purfued, he flies towards his hole; and finding it ftopped, which is always carefully done before the chafe begins, he has recourfe to his fpeed and his cunning for fafety. He does not double and meafure his ground back like the Hare; but continues his courfe ftraight forward before the Hounds with great ftrength and perfeverance. Both Dogs and Horfes, particularly the latter, have frequently fallen victims to the ardour of the purfuit; which has fometimes continued for upwards of fifty miles without the fmalleft intermiffion, and almoft at full fpeed*. As the fcent of the Fox is very ftrong, the Dogs follow with great alacrity and eagernefs, and have been known to keep up a continued chafe for eight or ten hours together; and it is

* Mr Charles Turner's Hounds hunted at Ayreyholm, near Hurworth, in the county of Durham, and found the noted old Fox CESAR, which made an extraordinary chafe: After a round of four miles, he led to Smeaton, through Hornby and Appleton; then back again to Hornby, Worfet-moor, Pierfburgh, Limpton, Craythorn, Middleton, Hilton, Seamer, Newby, Marton, Ormfby; then upon Hambleton, through Kirkleatham-park, Upleatham, Skelton, and Kilton. Mr Turner tired three Horfes; and only three Hounds were in purfuit, when he thought proper to call them off, it being near five in the evening. The chafe was upwards of fifty miles.

hard

HISTORY OF QUADRUPEDS. 269

hard to fay, whether the fpirited eagernefs of the Hounds, the ardour of the Horfes, or the enthufiafm of the hunters, is moft to be admired. The Fox is the only one of the party which has the plea of neceffity on his fide; and it operates fo ftrongly, that he often efcapes the utmoft efforts of his purfuers, and returns to his hole in fafety. The fmell of his urine is fo offenfive to the Dogs, that it fometimes proves the means of his efcape from them. When all his fhifts have failed him, and he is at laft overtaken, he then defends himfelf with great obftinacy, and fights in filence till he is torn in pieces by the Dogs.

There are three varieties of Foxes in this ifland, which differ from each other more in form than in colour.

The Greyhound FOX

is the largeft, and is chiefly found in the mountainous parts of England and Scotland: He is likewife the boldeft, and will attack a well-grown Sheep.—His ears are long and erect, and his afpect wild.

The

Figure 24. Thomas Bewick, "The Greyhound Fox," wood engraving, in *A General History of Quadrupeds*, Newcastle upon Tyne: S. Hodgson, R. Beilby & T. Bewick, 1790. Yale Center for British Art, Paul Mellon Collection

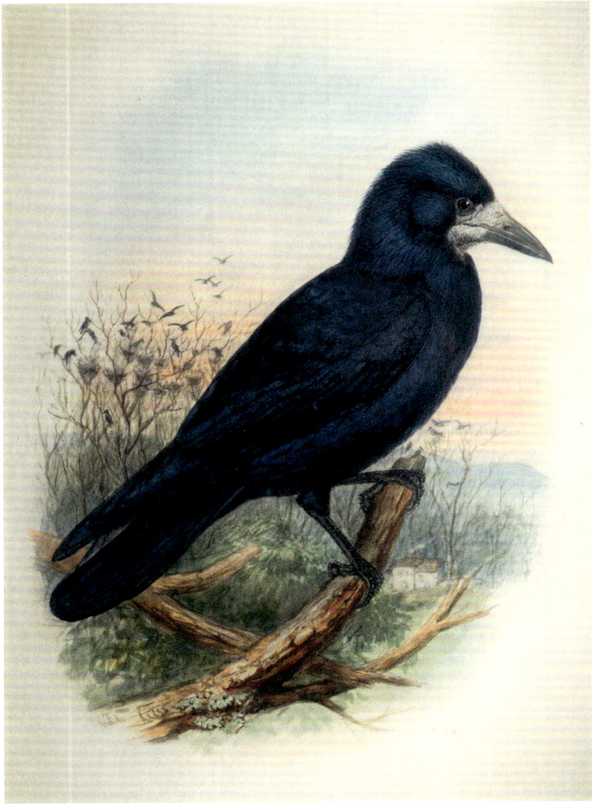

Figure 25. John Gould and Henry Constantine Richter, *Rook* (Corvus frugilegus), ca. 1862, watercolor and gouache, for John Gould's *Birds of Great Britain*, London: Taylor, 1873. Private collection

common local shells, flowers, butterflies, and birds. If their aspirations were more ambitiously academic, they could contribute meaningful information to a full range of subjects and help to build encyclopedic inventories of wildlife in Great Britain and elsewhere. Many chose to record their findings by annotating published works from their libraries or by creating written and visual records of their own.

As public interest in nature grew, there was an explosion of local nature clubs and larger regional and national institutions devoted to the study of plants and animals. Their activities were encouraged in the nineteenth century by an increasingly affordable, accessible, and prolific popular press.[18] At the high end of the market were the lavishly illustrated color plate books like John Gould's *Birds of Great Britain* (1862–1873) with its charming tableaux of family groups or individual birds set in recognizably British landscapes (fig. 25). But there were also much more affordable books such as those included in the *Naturalist's Library* (1835–1852), edited by Sir William Jardine. This multivolume work provided a more global perspective (and larger pool for subjects) by combining local coverage with illustrations and descriptions of exotic discoveries from throughout the British Empire. Even the popular press included regular articles about natural history, and readers took great pleasure in exchanging news of their findings through letters to their local newspapers. *Curtis's Botanical Magazine*, begun in 1787, was an accessibly written, beautifully illustrated periodical that had wide public appeal.[19] Its publisher was William Curtis, an enterprising apothecary who took particular interest in the plants and insects that he saw all around him in London (fig. 26). A demonstrator of plants and *Praefectus Horti* at the Chelsea Physic Garden from 1772 to 1777, Curtis established his own London Botanic Garden at Lambeth (in 1779), and later at Brompton (in 1789). His book *Flora Londinensis* (six volumes, 1777–1798) was among the first to focus on urban nature.

In all areas of study, from botany to entomology, and especially in new disciplines like vertebrate and invertebrate paleontology, women were just as likely to make important discoveries as their male counterparts. Though she lacked a

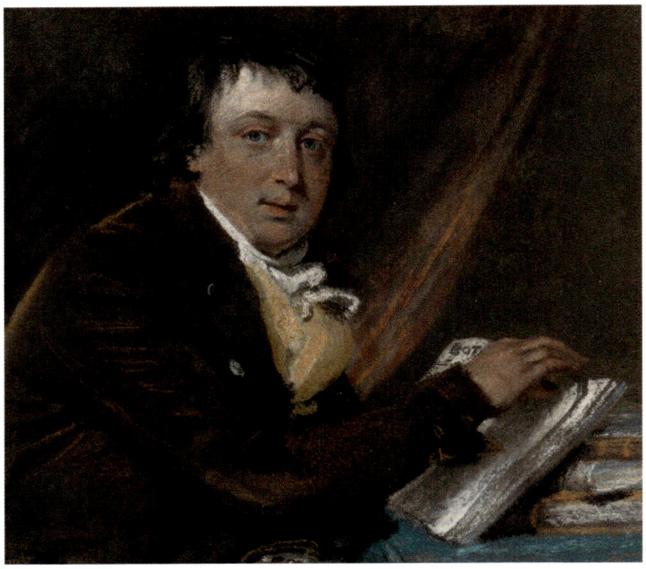

Figure 26. John Raphael Smith, *William Curtis (1746–1799), Entomologist and Botanist*, ca. 1787, pastel and gouache. Yale Center for British Art, Given by Lowell Libson in honor of Mrs. Rachel Lambert Mellon and the Oak Spring Garden Library

formal education or any important social connections, and never published any of her findings, Mary Anning famously spawned a public awareness of prehistoric life through her early nineteenth-century excavations of large and impressive skeletons of ichthyosaurs ("fish lizards"), plesiosaurs (marine reptiles), and pterosaurs (flying reptiles) in Lyme Regis, Dorset. As her finds made their way into publicly accessible (institutional) collections and gained even wider notoriety through a lithograph titled *Duria Antiquior* ("an earlier Dorset") issued by George Scharf in 1830 after a design by Henry De la Beche, they helped to transform how the public thought about deep time and man's place in the universe.[20]

Anning's contemporary, Etheldred Benett from Wiltshire, focused on smaller fossils and shells. Her meticulously arranged collections are as aesthetically pleasing as they are scientifically important (fig. 27). Benett is recognized today as the founder of invertebrate paleontology in England.[21] She might also be seen as an artist who used fossils as a base for her didactic displays.

Many women delighted in combining drawing, painting, scrapbooking, collage, and other artistic pursuits traditionally associated with their gender (a topic discussed in more detail in Molly Duggins's essay) with the newly sanctioned study of science to create books that were both beautiful and substantive. Some were strictly personal; others were intended for wider dissemination. The work of Margaret Gatty, a clergyman's wife from Yorkshire who wrote an authoritative book on British seaweeds, is a good example of the latter. Another is Sarah Bowdich, an adventurous traveler (to France, West Africa, and elsewhere) and a prolific writer with a deep interest in the natural world. In 1828 she began work on a book on the fresh water fishes of Great Britain, which contained a letterpress (printed) text and forty-four original watercolor illustrations (fig. 28).[22] Mrs. Bowdich was scrupulous in her efforts to achieve accuracy in her illustrations. "My object," she explained, "has been to give, rather a correct representation of the individual fish than to form a [beautiful, but less scientifically accurate] picture; and by so doing, I trust I have satisfied the Naturalist, without offending the Amateur."[23] In many of her plates she used gold and silver foil, covered by translucent washes of color, to achieve the reflective

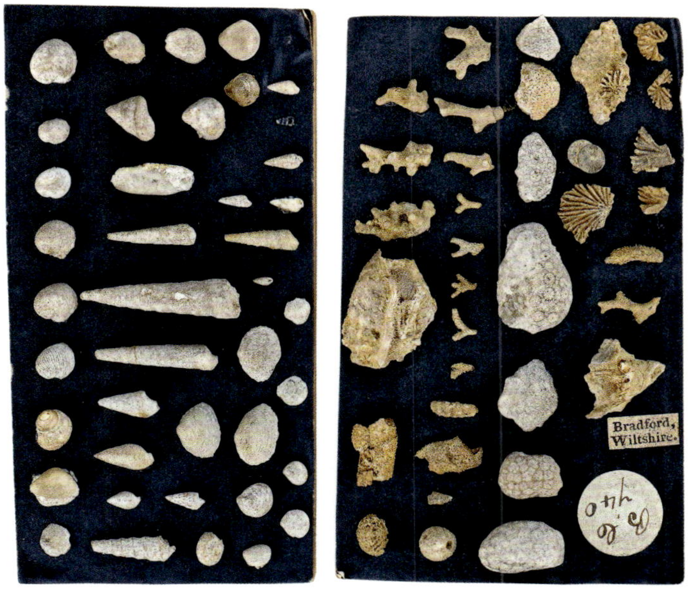

Figure 27. *Fossils and shells*, collected by Etheldred Benett at Bradford, Wiltshire, before 1831, natural specimens mounted on boards. Academy of Natural Sciences of Philadelphia, Drexel University.

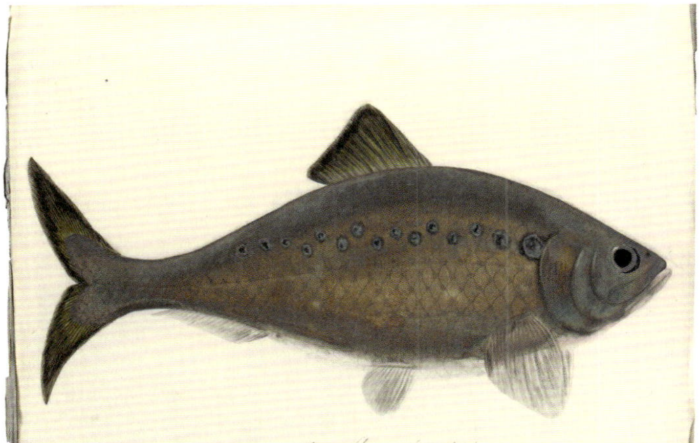

Figure 28. Sarah Bowdich, "Shad," watercolor, gouache, pen and ink, with gold and silver leaf, in *The Fresh-water Fishes of Great Britain*, London: R. Ackermann, 1828. Beinecke Rare Book and Manuscript Library, Yale University

appearance of scales. Most importantly, she insisted on working from living or freshly caught models, as she explained in the preface of her book:

> Every Drawing has been taken from the living Fish immediately [after] it came from the water it inhabited; so that no tint has been lost or deadened, either by changing the quality of that element, or by exposure to the atmosphere.[24]

While the edition size of Mrs. Bowdich's publication was understandably small, it was clearly intended for public distribution. Other naturalists' books were far more personal in nature, growing from a long tradition of commonplace books in which original observations were often combined with quotations transcribed from published sources. Such volumes, created by both men and women, were highly variable in content, ranging from nearly encyclopedic treatments of scientific subjects to albums in which pressed ferns and flowers were interspersed with literary quotations and watercolor landscapes, or with illustrations of birds, moths, butterflies, shells, and other subjects selected for their aesthetic appeal (fig. 29).

A commonplace book kept by Joshua Spencer in the late seventeenth and early eighteenth centuries (fig. 30) is typical of the kind of album kept by a serious naturalist of the early Enlightenment.[25] It is organized by subject and filled with quotations from important scientific figures, including Sir Isaac Newton, John Ray, Humphrey Prideaux, and Robert Knox. Like the text, some of its illustrations are copied from published works, but others are clearly original and based on the author-artist's first-hand observations. The contents include sections on mammals, birds, insects,

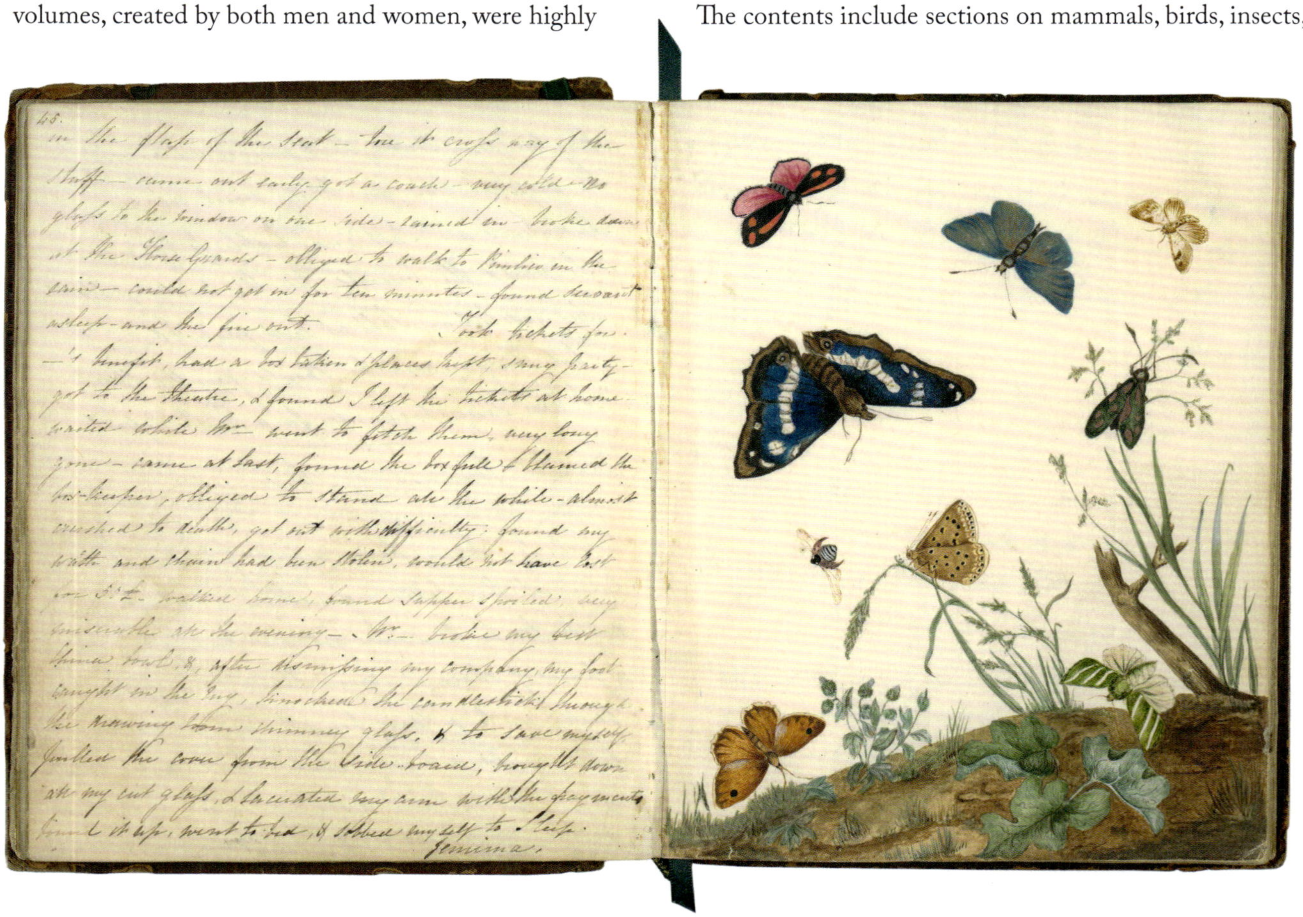

Figure 29. *Commonplace Book, with Drawings of Insects* (pages 73–74), ca. 1830, pen and ink with watercolor and gouache over graphite. Yale Center for British Art, Paul Mellon Fund

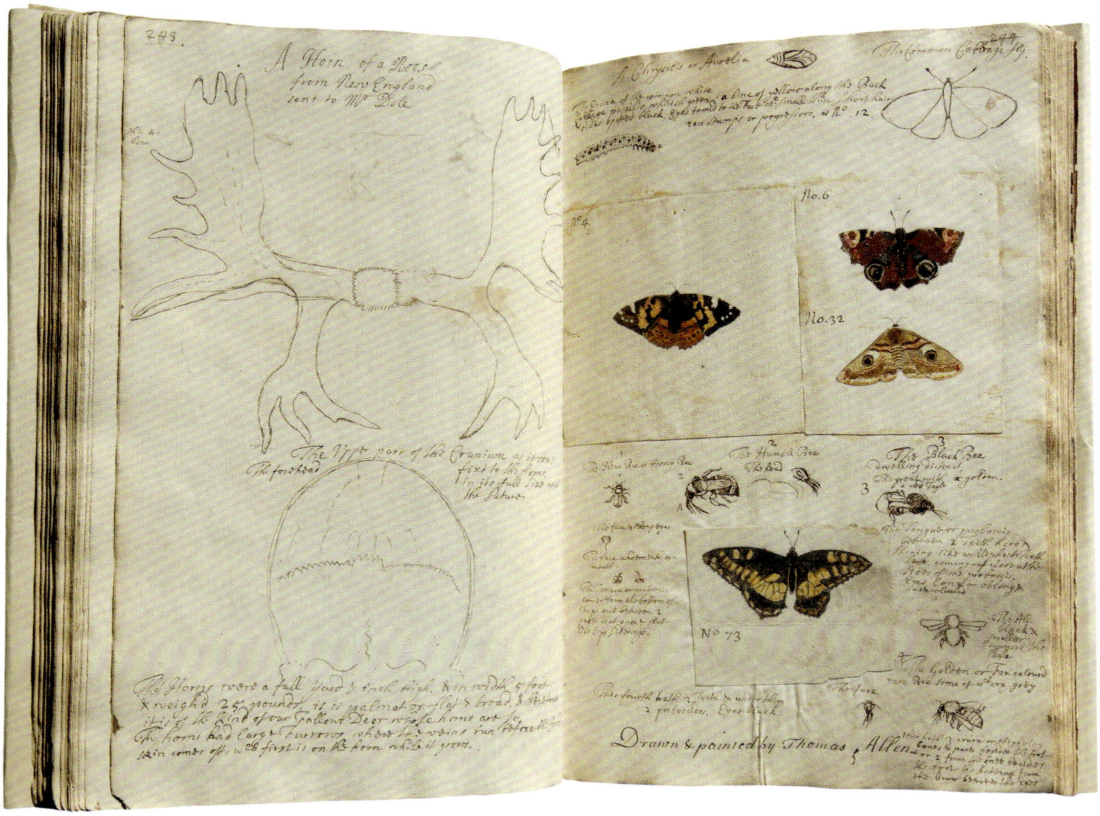

Figure 30. Joshua Spencer, "Butterflies," pen and ink with watercolor, in his *Commonplace Book*, 1692–1729. Archives of the Academy of Natural Sciences of Philadelphia, Drexel University (ANSP Coll. 426)

shells, and astronomy (including an illustration of an eclipse of the sun). It was created as a personal reference book by a person whose wide-ranging curiosity and depth of knowledge is clearly evident and impressive by any measure.

Notebooks kept by Mary Anne Burges, a knowledgeable amateur entomologist of the late eighteenth century, and James Francis Stephens, a teacher of Charles Darwin, are more narrowly focused. Each contains original observations and beautifully rendered drawings and watercolors of insects in various stages of metamorphosis (fig. 31). These were journals kept for serious purposes. Other albums were inspired by patriotism or personal sentiment.

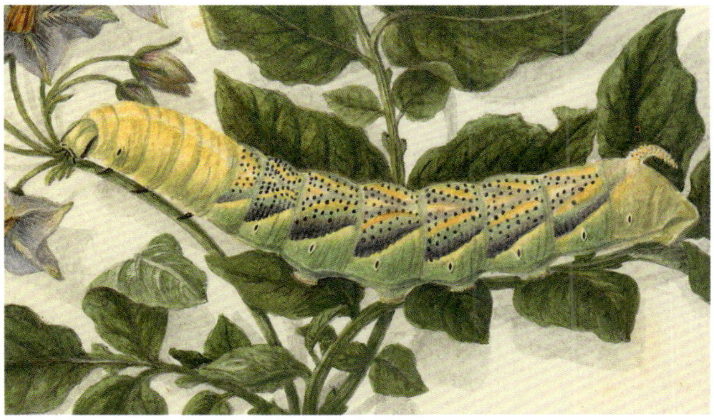

Figure 31. Mary Anne Burges, "Legitima Alis integris. S. Atropos—Bee Tiger" (detail), watercolor and gouache with pen and ink, in *Lepidoptera of the British Isles*, ca. 1810. Yale Center for British Art, Paul Mellon Fund

Plants of every kind, whether collected as a means of symbolic communication, for scientific study, as *memento mori*, souvenirs, talismans, or simply for their visual and sensual appeal, are among the most easily preserved bits of nature. Pressed botanical specimens, small and flat, are ideally suited for retention in books, sometimes as book marks, sometimes as the primary focus of artistic or scientific albums. Collectors with serious botanical interests and large numbers of plants to organize, however, generally used more flexible (and expandable) ways of preserving their specimens. For them, loose sheets, rather than bound albums, provided the best means for both reference and preservation. While institutional collections could contain hundreds of thousands of sheets, even the collections of amateurs could grow to an impressive size.

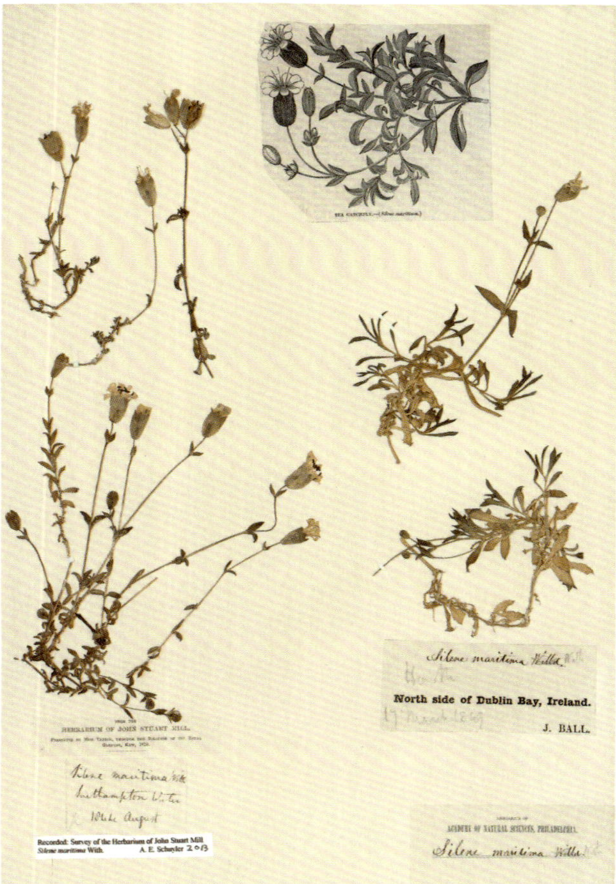

Figure 32. *Silene maritima* (Sea campion), collected by John Stuart Mill at Southampton Water, England, August 1844, natural specimens mounted on paper. Academy of Natural Sciences of Philadelphia, Drexel University

John Stuart Mill is best remembered as a philosopher and political economist, but he was also an avid botanist who spent his free time collecting plants in England and continenal Europe. His herbarium eventually contained more than twelve thousand plants, each artfully pressed on a single sheet of paper and identified with its scientific name and place of collection (fig. 32). Thus the great champion of utilitarianism attempted to advance society by advancing the field of botany and satisfied his own physical and aesthetic appetites by collecting and assembling plants for study and display.

Mill, like most collectors, developed a network of like-minded individuals through which he could expand his collection by exchange and purchase. Thus his herbarium ultimately contained plants from many parts of the world that he himself had never visited. Nor were plants the only parts of nature available for purchase, for, with the expansion of its political and economic influence in the eighteenth and nineteenth centuries, England became an important hub for the international trade in natural history artifacts. When the American artist Rubens Peale visited London in 1802, he noted both the large number of natural history specimens that were on public display at such places as the Leverian Museum and the ease with which similar items could be purchased from dealers elsewhere in the city. "A very good collection of [natural] Curiosities might be bought in London," he reported to his father, the museum impresario Charles Willson Peale. "Preserved Birds and Beasts are in windows all over town and Pictures in abundance, valuable stones, coins, Medals &c."[26]

Some thirty years later, William Swainson observed:

> It is now a well-known fact, that collections of South American insects can be purchased cheaper in London than in the country they actually inhabit. Birds are still more salable here. We have purchased, in fact, hundreds below the average price that their mere stuffing would cost in England. Shells [from around the world are so commonly available that they] are now a complete "drug on the market."[27]

To aid those wishing to collect scientific specimens for themselves, book sellers of the late eighteenth and nineteenth centuries offered a variety of "how to" manuals, advising on everything from the equipment needed to collect the organisms in the wild to the chemicals needed to preserve them.[28] For most amateur naturalists, the aesthetic appeal of natural objects and the desire to understand their ways of life were sufficient justifications for spending time in the country studying, collecting, and drawing wild subjects. But in case more serious excuses were needed for such activity, the authors of these popular guides were happy to provide them. In his handbook *Instructions for Collecting, Rearing, and Preserving British and Foreign Insects* (1839), for example, the English entomologist Abel Ingpen laid out a positive case for the study of insects. Defensively rejecting any suggestion that entomology was "a frivolous amusement (as some have ignorantly asserted)" or a dilettante's hobby "suffering in importance by comparison with other studies," Ingpen believed that the study of insects was worthy of standing "in the foremost rank of Natural Science!"[29] He went on to point out that, in addition to their "gratification of our mental and ocular faculties," insects provided humans with invaluable products for the enhancement of life, ranging from clothing to ink. This, he insisted, more than justified time spent on their collection, illustration, and careful study (fig. 33). To make his point, Ingpen reminded his readers:

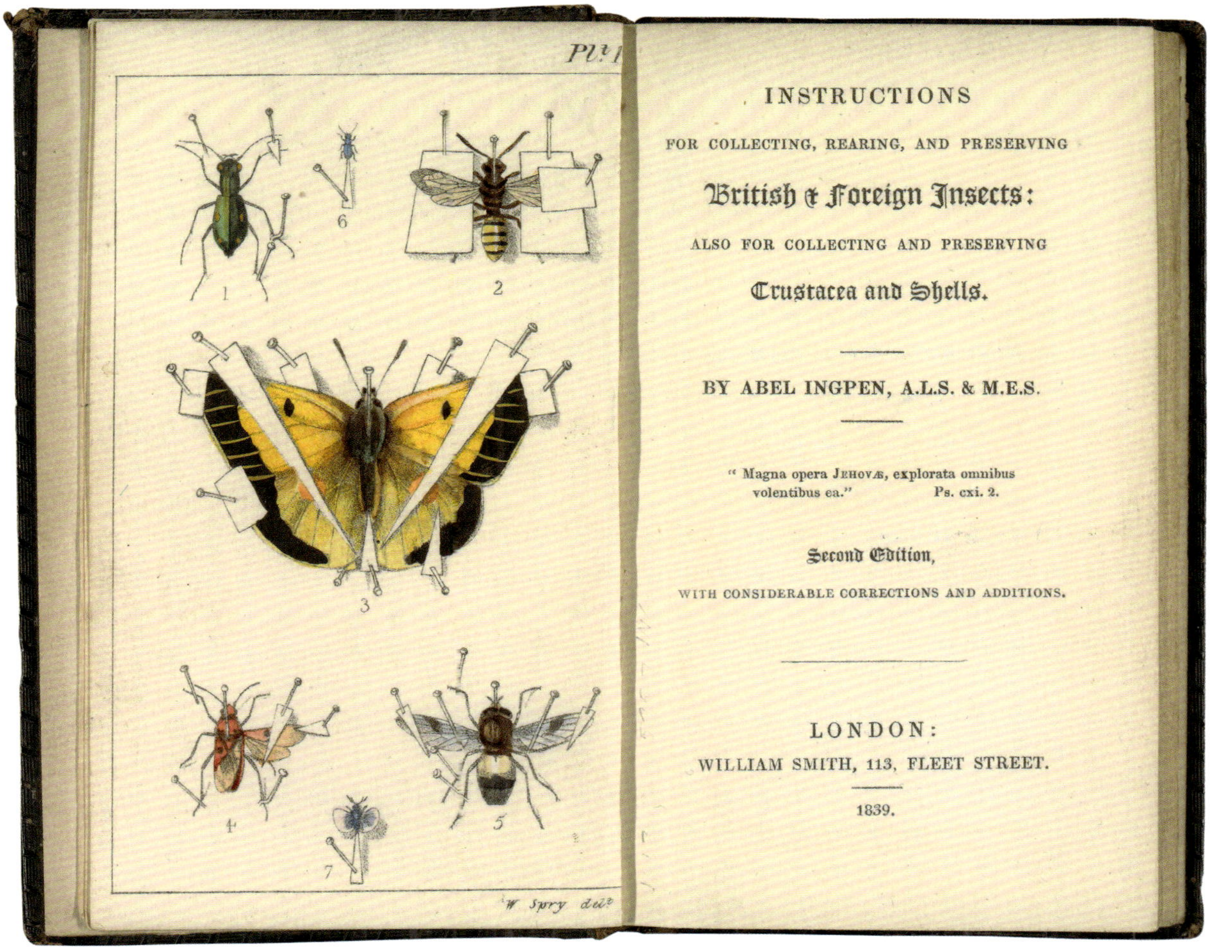

Figure 33. Abel Ingpen, *Instructions for Collecting, Rearing, and Preserving British and Foreign Insects*, London: William Smith, 1839, hand-colored lithograph. Yale Center for British Art, Paul Mellon Fund

> The delicious luxury furnished by the *bee*, and the beautiful dye of the *coccus*: the materials for an exquisite fabric, prepared by the *silk worm*, which gives employment to millions! and [ink] the ingredient produced by the *gall-fly*, to which mankind is deeply indebted for the promulgation of knowledge! are all real benefits. And from others, lessons of industry and economy; virtue and morality; perfection in various arts, and even civil government, may be learned.[30]

As with so many naturalists in the late eighteenth and early nineteenth century, Ingpen's trump card in justifying nature study was the uplifting religious insights such investigations would provide. By inventorying the world's diversity, he explained, one was glorifying God: "the contemplation of the works of the CREATOR is the highest delight of the rational mind."[31] Ingpen then went on to offer detailed instruction for collecting insects and other organisms. Sarah Bowdich produced a similar book, focusing her instruction on "collecting, preparing, and mounting objects of natural history."[32] In 1840, William Swainson published another "how to" guide for natural history collectors, in which he assured his readers that they didn't need to be "scientific naturalists" as long as they had a "discriminating eye" and a "courageous, persevering, and inquiring spirit."[33] In other words, it was acceptable for an amateur to collect and display products from nature for pure pleasure. To Swainson and his readers, aesthetic considerations were just as valuable as scientific ones.

Before commercially published books on the subject were available—such as those by Ingpen, Bowdich, and Swainson—information about collecting and preserving specimens was passed along in more limited, personal ways. In 1776, the London shell dealer George Humphrey created a lengthy collector's guide covering every field of natural history (fig. 34).[34] The handwritten and illustrated volume was probably intended for use by the captain of a whaling ship headed on an around-the-world voyage, for there are specific instructions for its user to be on the lookout for desirable specimens in locations as distant as the Cape of Good Hope, the Falkland Islands, Brazil, Surinam, "the Spanish Main," and Sierra Leon. A similar manual was created by the surgeon and scientist John Hunter in 1772 for Sir Joseph Banks to use on his voyage with Captain Cook.[35]

Home collections ranged in seriousness and scale from modest and decorative to encyclopedic. One of the largest and most comprehensive natural history collections of the eighteenth century was that assembled by the London physician Sir Hans Sloane. He bequeathed over 1,000 bird skins and eggs, over 1,500 fish, 5,500 insects, and nearly

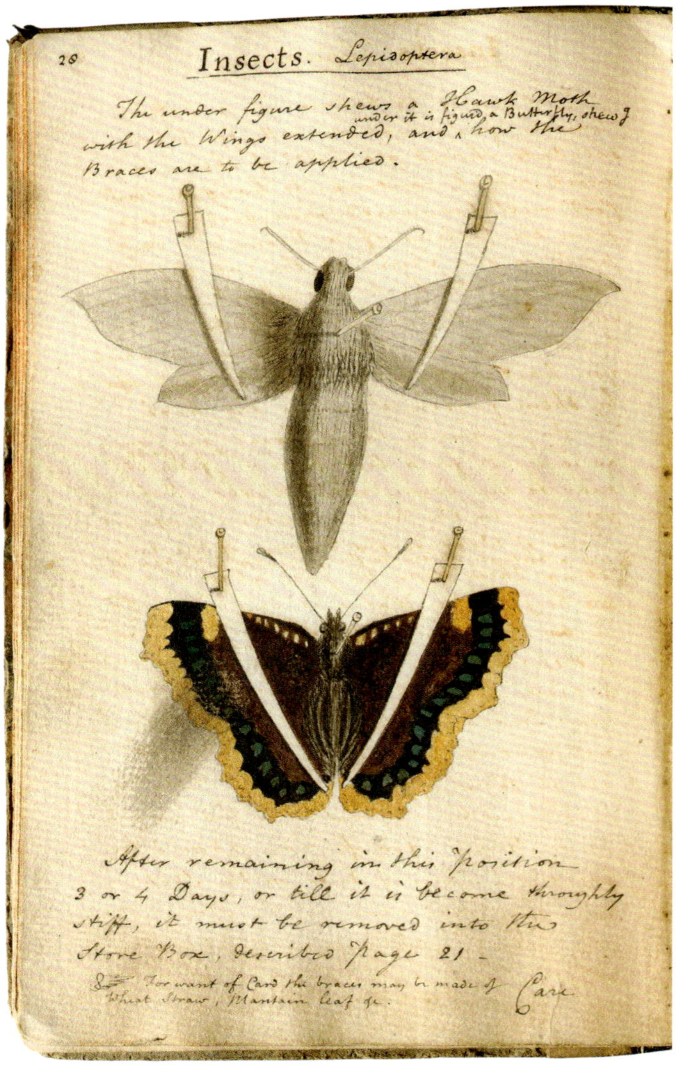

Figure 34. George Humphrey, "Insects Lepidoptera (p. 20)," pen and ink with watercolor, in *Collecting and Preserving all Kinds of Natural Curiosities*, 1776. Archives of the Academy of Natural Sciences of Philadelphia, Drexel University (ANSP Coll. 371)

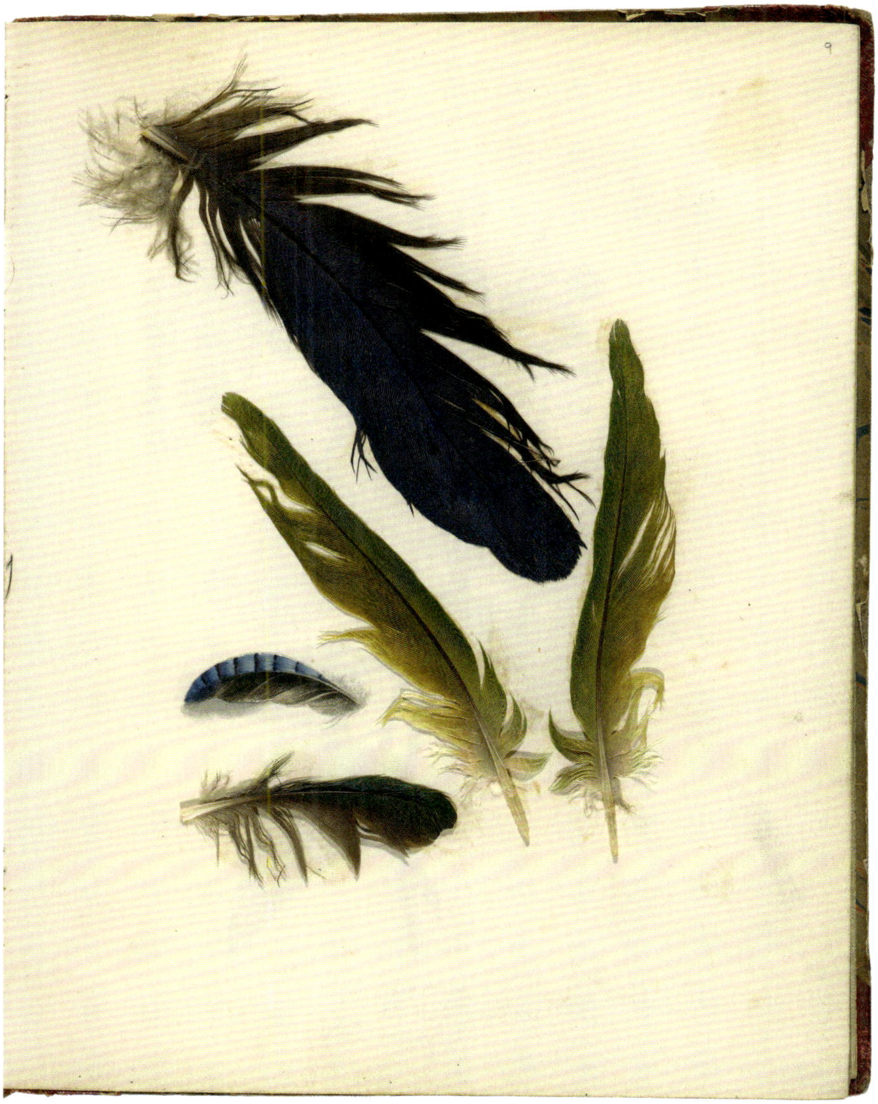

Figure 35. Edward Lear, Feathers from *Family Album* (fol. 8), 1828–1836, natural specimens and watercolor. MS Typ 55.4, Houghton Library, Harvard University

2,000 preserved quadrupeds to the nation, forming the basis for the natural history collections in the British Museum.[36]

While collecting actual specimens—plants, insects, shells, minerals, etc.—was the first instinct of any naturalist, creating drawings of them was often the next. Some especially creative artists combined the two. Early in his career, the English nonsense writer and landscape painter Edward Lear made beautifully detailed watercolors of birds and other animals, and even a number of feather studies.[37] In one family album assembled by Lear and his sister, there is a page in which several real feathers are pasted in and one painted beside them. The painted feather is so accurately rendered, at first glance one assumes all of them are real—or all are painted. Although the real feathers are now in disarray, when the album was first created it would have been hard to know which was which (fig. 35). Thus art and science merge in a trompe l'oeil through which Lear could demonstrate both his talent as an artist and his wit.

In a later nineteenth-century album, the English ornithologist Henry Seebohm used bird feathers in a slightly different way. On the surface, the distribution of feathers

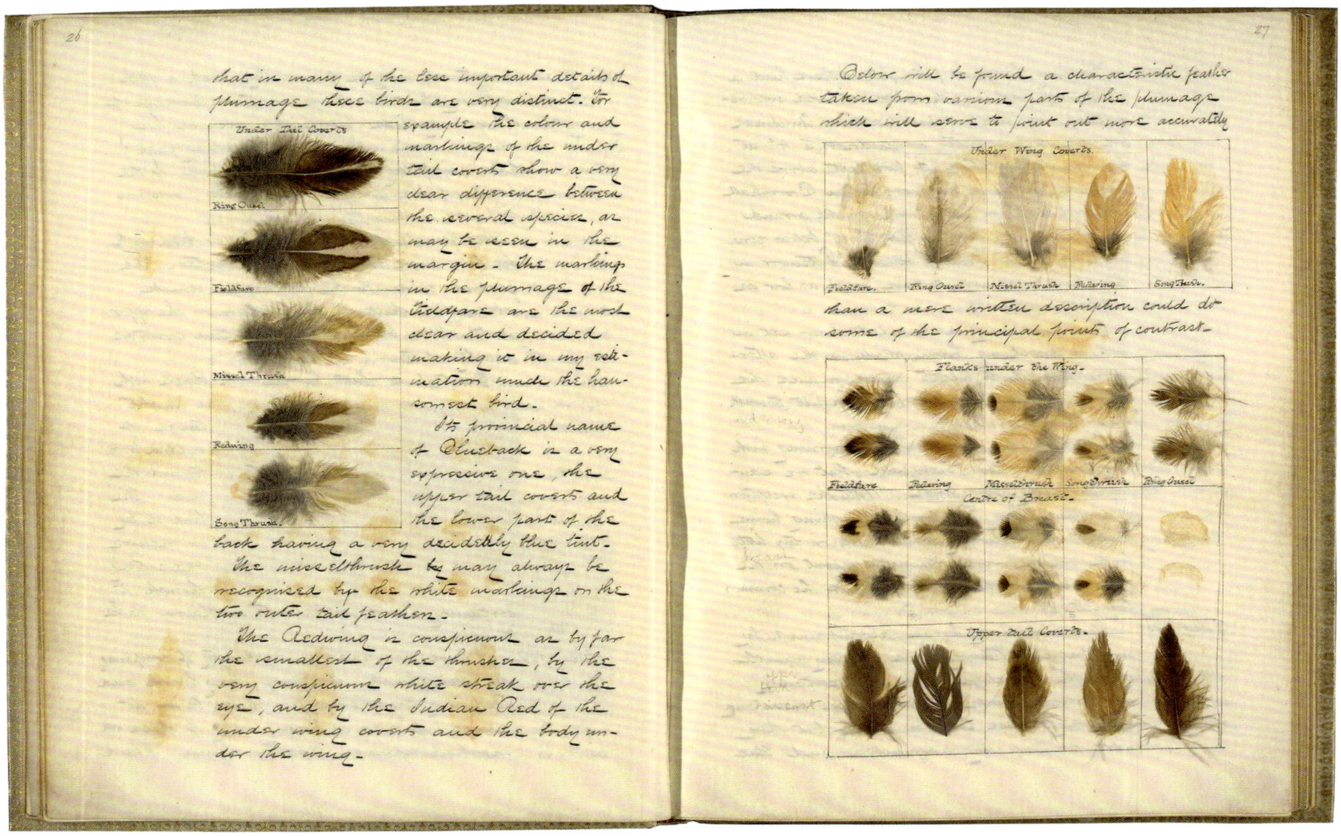

26

that in many of the less important details of plumage these birds are very distinct. For example the colour and markings of the under tail coverts show a very clear difference between the several species, as may be seen in the margin. The markings in the plumage of the Fieldfare are the most clear and decided making it in my estimation much the handsomest bird.

Its provincial name of Blueback is a very expressive one, the upper tail coverts and the lower part of the back having a very decidedly blue tint.

The Missel thrush may always be recognized by the white markings on the two outer tail feathers.

The Redwing is conspicuous as by far the smallest of the thrushes, by the very conspicuous white streak over the eye, and by the Indian Red of the under wing coverts and the body under the wing.

Under Tail Coverts

Ring Ouzel

Fieldfare

Missel Thrush

Redwing

Song Thrush

27

Below will be found a characteristic feather taken from various parts of the plumage which will serve to point out more accurately

Under Wing Coverts.

Fieldfare. | Ring Ouzel | Missel Thrush | Redwing | Song Thrush.

than a mere written description could do some of the principal points of contrast.

Flanks under the Wing.

Fieldfare | Redwing | Missel Thrush | Song Thrush | Ring Ouzel

Centre of Breast.

Upper Tail Coverts.

Figure 36. Henry Seebohm, *Ornithological Notebook and Diary* (pages 27–28), Sheffield?, 1860–1866, pen and ink with natural specimens. Yale Center for British Art, Paul Mellon Fund

throughout his album provides variety and visual continuity. The subtle colors and patterns created by their display make every page of the book a delight for the eye (fig. 36). But the author's handwritten notes reveal that these three-dimensional illustrations were more than decorative embellishments or souvenir trophies from the author's frequent shooting excursions. Seebohm included them to reinforce his observations about minute details in the plumages of the many similar but distinct bird species described in his text. At times, the matter-of-fact tone of his narrative seems at odds with the sensitivity demonstrated in the creation of his album. He can describe the beauty of a bird's appearance and song in one sentence, remark on the joy he feels in seeing and hearing the bird in the next, then, apparently without regret, record shooting it for closer examination. The feathers thus become part of his narrative.

To Seebohm, and to most other nineteenth-century naturalists, the beauty and utilitarian aspects of the natural world were never in conflict. The albums they created could include sentimental poems or essays, or be strictly scientific in content. Many combined both. In the case of Henry Seebohm, an aesthetic arrangement of feathers was also a didactic tool. The feathers, beautiful in their own right, and organized with a keen sense of design, were also irrefutable testaments to the veracity of his identifications.

For naturalists and artists of the eighteenth and nineteenth centuries, the line between objective observation and artistic interpretation was frequently blurred. This was especially true when the specimens themselves became part of the visualization process. Seebohm's feather album provides one such example. Books that incorporate nature printing and photography provide others.

Nature printing, a process of transferring the inked impressions of actual specimens to paper, was first explained and illustrated by Leonardo da Vinci in *Codex Atlanticus* (1490–1505). W. Curtis explained how to create a nature print in *The Naturalist's and Traveller's Companion* in 1772:

> The impressions of plants well taken off upon paper, look very little inferior to the best drawings, and may be done with very little trouble. For this purpose, some printer's ink and a pair of printer's bosses, such as are used for laying the ink on types, are necessary. After rubbing these bosses with a little of the ink, lay the plant betwixt them, and press it so as to give it sufficient colour, then take the plant and lay it carefully on a sheet of paper, and press it with the hand, to give the impression of the plant to the paper, which may be afterwards coloured according to nature.[38]

The technique was used for both scientific and artistic purposes throughout the eighteenth and nineteenth centuries, and remains in use to the present day. It may have achieved its most sophisticated and impressive use as an illustrative technique in Thomas Moore and John Lindley's sumptuous publication, *The Ferns of Great Britain and Ireland, Nature-Printed* (1855). This book incorporated a process, patented by Henry Bradbury, in which ferns were pressed into soft metal to create delicate molds from which harder metal printing plates were made. These plates were then used in an offset printing process to create exquisitely detailed, color impressions of the life-sized fern fronds (fig. 37). *The Ferns of Great Britain and Ireland* was a technical, but popularly marketed book on a plant group that, along with seaweeds, briefly fired collecting passions in the British Isles and elsewhere.[39] The book reviewer for *The Phytologist* (a popular mid-nineteenth-century fern magazine) was undoubtedly thinking of this and several other books by Moore that incorporated Bradbury's nature printed plates when the reviewer expressed his admiration:

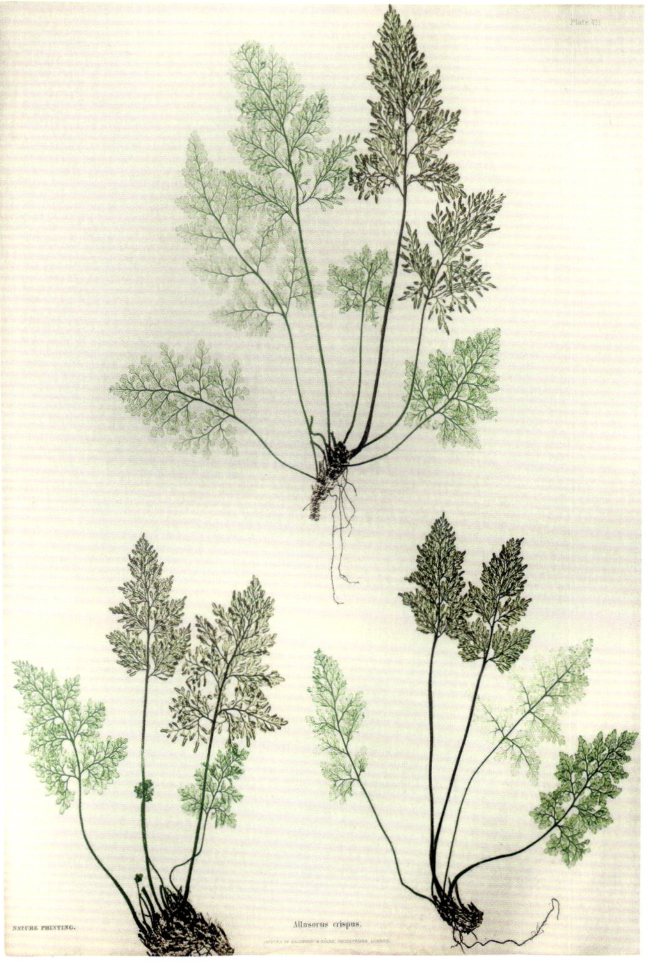

Figure 37. Henry Bradbury, *"Allosorus crispus,"* colored nature print from Thomas Moore, *The Ferns of Great Britain and Ireland*, London: Bradbury and Evans, 1855. Private collection

> The literature of ferns is in a palmy state. Judging by the distinct publications on this branch of botanical science, it surpasses that of all the other branches together. There are now . . . at least a score of works on the British Ferns, and all of them enjoying what may be called triumphant success. Among works of science their celebrity is unprecedented.[40]

Although photography initially lacked the important element of color offered in nature printing, despite sometimes also presenting its subjects in negative form it was a reproductive medium through which it seemed possible for nature to record itself. William Henry Fox Talbot, the British photographer who pioneered the use of paper negatives in the 1830s, had a special interest in plants.[41] He collected their leaves and their seeds, and made a series of what he

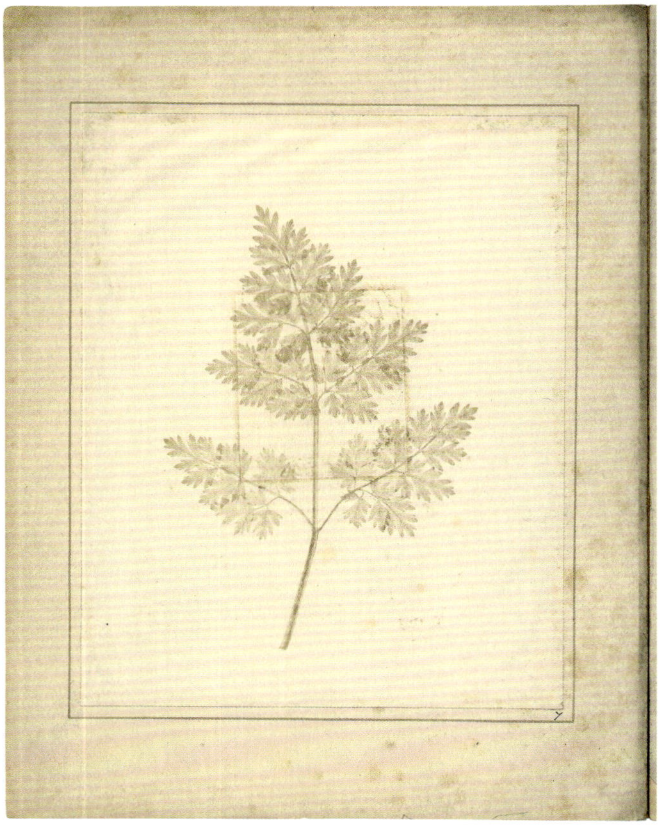

Figure 38. William Henry Fox Talbot, "Leaf of a Plant" (part 2, plate VII), salt paper print from calotype negative, in *The Pencil of Nature*, London: Longman, Brown, Green, and Longmans, 1844–46. Yale Center for British Art, Paul Mellon Fund

called "photogenic drawings," in which he allowed leaves to record themselves, in all their intricate beauty, on light-sensitive paper, without the use of a camera.[42] He called these leaf photographs "specimens," thus reinforcing the interchangeable nature of the pictures and the leaves themselves (fig. 38).[43] Talbot also photographed entire tree silhouettes using paper negatives and a view camera. Like his leaf prints, the resulting images seem to take the place of herbarium sheets for the artist-naturalist. In each he could capture the essence of a tree, conveniently reduced to a manageable and uniform size. In this way, Talbot adapted the newest "species" of art—photography—to the service of science.[44]

Anna Atkins, a contemporary of Talbot's, and the daughter of John George Children, Secretary of the Royal Society and Keeper of Natural History at the British Museum, also employed a cameraless photographic process to record botanical specimens. In the 1840s she created a series of striking blue "cyanotypes" or "photograms" in which she placed delicate specimens of algae and seaweed directly upon a light sensitive paper and then exposed them to the sun (fig. 39). After perfecting this technique, invented by Sir John Herschel, Atkins created and self-published a three-volume book titled *Photographs of British Algae: Cyanotype Impressions* (1843–1854). It is generally accepted as the first book to use photographic illustrations.[45]

Because of the time and labor involved in producing the almost four hundred individual "photograms" contained in each of her books, Atkins created only a very small number of *Photographs of British Algae* to give to friends and institutions, rather than selling them commercially. Fewer than twenty copies are known to exist today. Each contains the same handwritten text (also reproduced by photographic means) and approximately 390 unique cyanotypes.[46] Atkins went on to create two other books on natural history: *Cyanotypes of British and Foreign Ferns* (1853) and *Cyanotypes of British and Foreign Flowering Plants and Ferns* (1854). The creation of these was so time consuming and labor intensive, she made only one of each book.

Despite their explicitly scientific intent, Anna Atkins's books could be considered "artists' books" because of their subject matter and attractive design, and because the photographic technique used to produce them is now given artistic status. The images in Atkins's "published" books were related scientifically by their subjects—British algae, seaweeds, and ferns—but her personal albums were more eclectic and included images of feathers and lace as well as plants. These are united by the sensibilities of the artist and the techniques used, rather than by a strict adherence to subject matter. In this regard they are closer to commonplace books than the scientific monographs on shells and other invertebrates for which Atkins (then Children) had hand-drawn illustrations before experimenting with photography.[47] In the totality of her work, Atkins serves as a historical link between the authors of commonplace books, who compile miscellaneous information from other sources

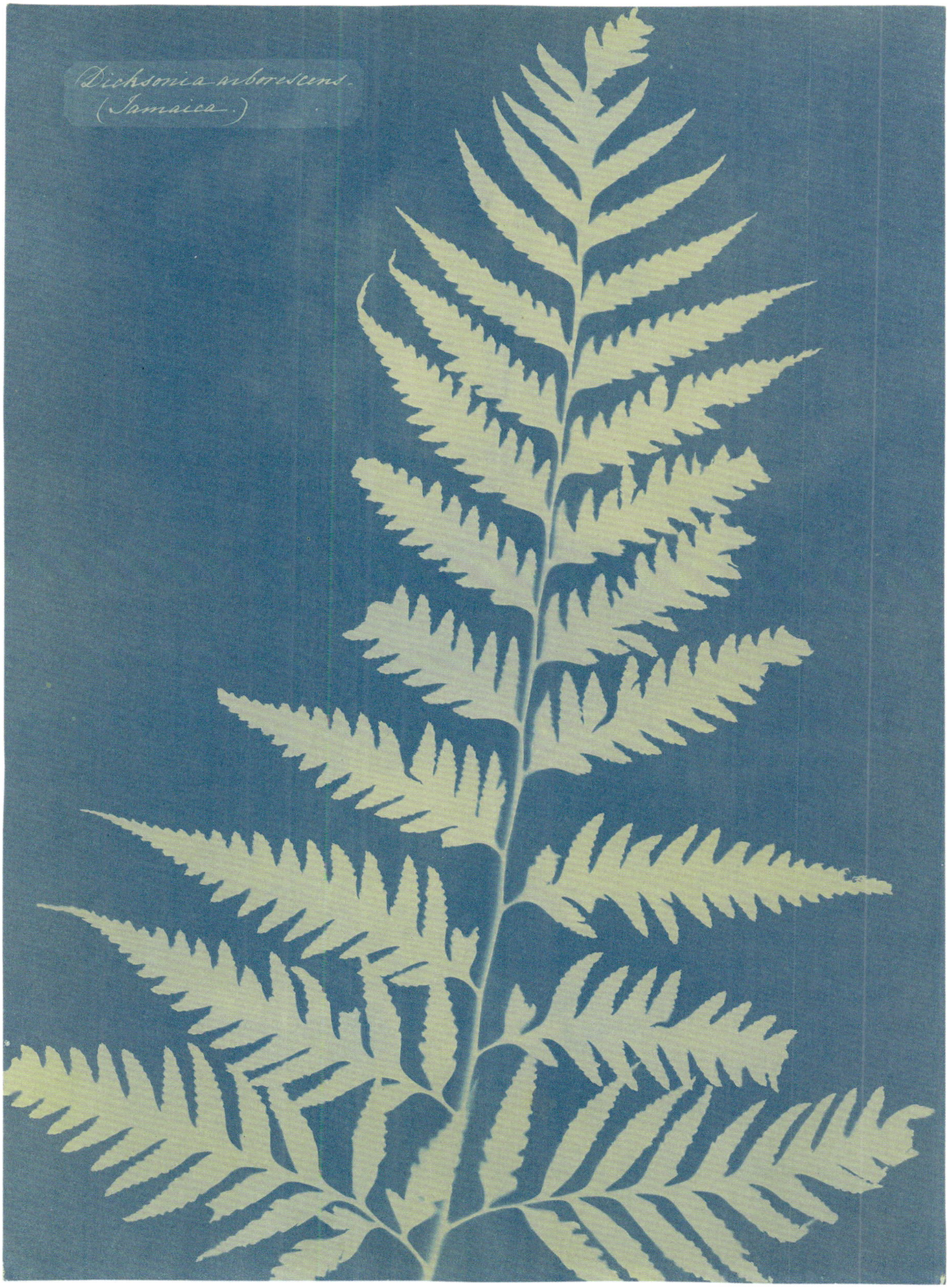

Figure 39. Anna Atkins, *Dichsonia arborescens (Jamaica)*, ca. 1850, cyanotype. Yale University Art Gallery, Gift of George Hopper Fitch, BA 1932

and assemble or transcribe it in their albums; field naturalists, who make original observations and record them in words and pictures in their notebooks; scientific illustrators, who transcribe the details of nature into a two-dimensional medium; and book artists who sometimes try to do it all (as, for example, in Tracey Bush's notebook, fig. 40). Although some of today's book artists may be unaware of the historical precedents, they are clearly inspired by the same impulses to record, interpret, and celebrate nature as their predecessors.

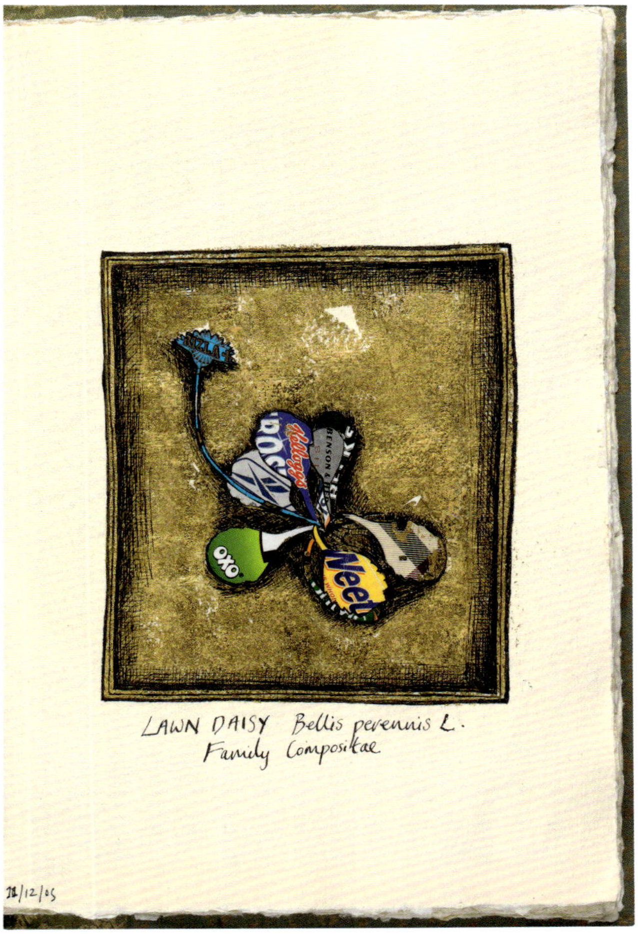

Figure 40. Tracey Bush, "Lawn Daisy" (leaf 4r), 2005, paper packaging collage with gold leaf and pen and ink, in *Florilegium*. Collection of the artist

In an address at the opening of the Birmingham Free Library in 1866, the English nonconformist preacher George Dawson observed that "a great library contains the diary of the human race."[48] While a library or art museum focusing on artists' books alone could never make so sweeping a claim, it would, at the very least, offer important insights into the interweaving of science and art, and the many ways in which humans have recorded, interpreted, and celebrated nature for the past three hundred years. Whether they are perceived as works of art, or records of nature, the artists' books described in the exhibition *"Of Green Leaf, Bird, and Flower": Artists' Books and the Natural World* and in this publication stand as elegant testaments to the beauty, joy, and mystery of life on earth.

1. For a reproduction and discussion of Sackheuse's published illustration of Ross meeting the Inuit, see Robert M. Peck, "The Art of the Arctic: British Painting in the Far North," *Journal for Maritime Research* 14, no. 2 (2012): 67–93.
2. Ross explained that Sackheuse received his artistic training from "Mr Nasmyth, the artist, who introduced him to Sir James Hall." Sir James, in turn, introduced the talented Greenlander to Captain Basil Hall, who introduced him to the Admiralty, by whom he was "engaged as an interpreter" for Ross's expedition. See John Ross, *A Voyage of Discovery . . . for the Purpose of Exploring Baffin's Bay and Inquiring into the Probability of a North-west Passage* (London: John Murray, 1819), xxii. For Sackheuse, see Ann Savours, *The Search for the North West Passage* (New York: St. Martin's Press, 1999) 50–53.
3. For a reproduction and analysis of a similar caricature by Darly portraying Banks as "The Botanic Macaroni," see Patricia Fara, *Sex, Botany & Empire: The Story of Carl Linnaeus and Joseph Banks* (New York: Columbia Univ. Press, 2003), 8–16.
4. After 1872, it became government policy to deposit all collections made at government expense in the British Museum. Lynn Barber, *The Heyday of Natural History, 1820–1870* (Garden City, NY: Doubleday & Co., 1980), 162.
5. *Critical Review* 16 (1763): 312, quoted in Keith Thomas, *Man and the Natural World: A History of the Modern Sensibility* (New York: Pantheon Books, 1983), 283.
6. For a detailed discussion of this, see Thomas, *Man and the Natural World*, 51–70.
7. Thomas, *Man and the Natural World*, 60.
8. Richard Barber, introduction to *Bestiary* (London: The Folio Society, 1992); Florence Fearrington, *Rooms of Wonder From Wunderkammer to Museum 1599–1899* (New York: The Grolier Club, 2012).
9. Thomas, *Man and the Natural World*, 65.
10. *Encyclopaedia Britannica*, quoted in Robert John Thornton, *The Temple of Flora*, ed. R. King (London: Weidenfeld and Nicolson, 1981), 9.
11. Patricia Fara, *Sex, Botany & Empire*, 20.
12. William T. Stearn, "Linnaean Classification, Nomenclature, and Method," in Wilfrid Blunt, *The Compleat Naturalist: A Life of Linnaeus* (New York: The Viking Press, 1971), 242. See also Lynn Barber, *The Heyday of Natural History*, 47–56; and Lisbet Koerner, *Linnaeus: Nature and Nation* (Cambridge, MA: Harvard Univ. Press, 1999).
13. The collection contained more than 19,000 of Linnaeus's pressed plants, 3,200 dried insects, 1,500 shells, 2,500 minerals, between 700 and 800 pieces of coral, 2,500 books, and about 3,000 letters and manuscripts. See Lynn Barber, *The Heyday of Natural History*, 55, and Blunt, *The Compleat Naturalist*, 236–38.
14. Preamble from the Royal Charter of the Linnean Society (1802), quoted in

Andrew T. Gage and William T. Stearn, *A Bicentenary History of the Linnean Society of London* (London: Academic Press, 1988), 2.
15. Except for the minerals, which had been sold, Smith's collection of Linnaeus's books and specimens was purchased by the Linnean Society from his estate following his death in 1828. The collections are still in the society's care.
16. William Swainson, *A Preliminary Discourse on the Study of Natural History* (London: Longman, Rees, Orme, Brown, Green & Longman, and J. Taylor, 1834), 130, quoted in Lynn Barber, *The Heyday of Natural History*, 35.
17. For more on Bewick, see Jenny Uglow, *Nature's Engraver: A Life of Thomas Bewick* (New York: Farrar, Strauss and Giroux, 2006); Iain Bain, *The Watercolours and Drawings of Thomas Bewick and his Workshop Apprentices* (London: Gordon Fraser, 1981); and Diana Donald, *The Art of Thomas Bewick* (London: Reaktion Books, 2013).
18. See Richard D. Altick, *The English Common Reader: A Social History of the Mass Reading Public 1800–1900* (Chicago: Univ. of Chicago Press, 1957), 260–348.
19. The magazine is now published by the Royal Botanic Gardens, Kew.
20. Martin J. S. Rudwick, *Scenes From Deep Time: Early Pictorial Representations of the Prehistoric World* (Chicago: Univ. of Chicago Press, 1992), 44–47.
21. See Hugh S. Torrens et al., "Etheldred Benett of Wiltshire, England," *Proceedings of the Academy of Natural Sciences of Philadelphia*, 150 (April 14, 2000): 59–123.
22. Mrs. T. Edward Bowdich, *The Fresh-water Fishes of Great Britain* (London: R. Ackermann, 1828).
23. Bowdich, *The Fresh-water Fishes of Great Britain*, preface, iii.
24. Bowdich, *The Fresh-water Fishes of Great Britain*, preface, iv.
25. Joshua Spencer, *Commomplace Book 1692–1729*, collection #426, Archives of the Academy of Natural Sciences of Philadelphia, Drexel University.
26. Rubens Peale to Charles W. Peale, Sept. 6, 1802. See Lillian B. Miller, ed., *The Selected Papers of Charles Willson Peale and his Family: Volume 2, The Artist as Museum Keeper, 1791–1810* (New Haven and London: Yale Univ. Press, 1988), 455.
27. William Swainson, *Taxidermy: Bibliography and Biography* (London: Longman, Orme, Brown, Green & Longmans, 1840), 96–97.
28. For a detailed discussion of historical collecting and preservation techniques, see Robert McCracken Peck, "Preserving Nature for Study and Display," and "Alcohol and Arsenic, Pepper and Pitch: Brief Histories of Preservation Techniques," in *Stuffing Birds, Pressing Plants, Shaping Knowledge: Natural History in North America, 1730–1860*, ed. Sue Ann Prince (Philadelphia: American Philosophical Society, 2003), 10–25; 26–53.
29. Abel Ingpen, *Instructions for Collecting, Rearing, and Preserving British and Foreign Insects* (London: William Smith, 1839), 15.
30. Ingpen, *Instructions for Collecting*, 14.
31. Ingpen, *Instructions for Collecting*, 13–15.
32. Sarah Bowdich (Mrs. R. Lee), *Taxidermy: Or, The Art of Collecting, Preparing, and Mounting Objects of Natural History for the Use of Museums and Travelers* (London: Longman, Hurst, Rees, Orme, and Brown, 1820).
33. William Swainson, *Taxidermy: Bibliography and Biography*, 1–2.
34. George Humphrey, "Collecting and Preserving all Kinds of Natural Curiosities," 1776, Collection #371, Archives of the Academy of Natural Sciences of Philadelphia, Drexel University.
35. John Hunter, "Directions for Collecting & Preserving Animals," manuscript, The Papers of Sir Joseph Banks, State Library of New South Wales. I am indebted to Simon Chaplin for drawing this manuscript to my attention. A printed set of instructions on collecting and preserving specimens from about this same period, also prepared by John Hunter, is owned by the Royal College of Surgeons (Hunterian Museum).
36. Kim Sloan, ed., *Enlightenment: Discovering the World in the Eighteenth Century* (Washington, DC: Smithsonian Books, 2003), 85. See also Marjorie Caygill, "Sloane's Will and the Establishment of the British Museum," in *Sir Hans Sloane: Collector, Scientist, Antiquary, Founding Father of the British Museum*, ed. Arthur MacGregor (London: British Museum Press, 1994).
37. For more on Lear's natural history paintings, see Robert M. Peck, "The Natural History of Edward Lear," *Harvard Library Bulletin*, 22, no. 2–3 (Summer–Fall 2011).
38. W. Curtis, *The Naturalist's and Traveller's Companion: Containing Instructions for Discovering and Preserving Objects of Natural History. . . .* (London: George Pearch, 1772), 24–25, quoted by Peck, "Alcohol and Arsenic, Pepper and Pitch," in Prince, *Stuffing Birds*, 45.
39. See Sarah Whittingham, *Fern Fever: The Story of Pteridomania* (London: Frances Lincoln, 2012).
40. Alexander Irvine, ed., *The Phytologist. A Botanical Journal*, vol. 2, (Dec. 1857): 296.
41. Larry Schaaf, *The Photographic Art of William Henry Fox Talbot* (Princeton: Princeton Univ. Press, 2000).
42. See Carol Armstrong, "Cameraless: From Natural Illustrations and Nature Prints to Manual and Photogenic Drawings and Other Botanographs," in Carol Armstrong and Catherine de Zegher, eds., *Ocean Flowers: Impressions From Nature* (Princeton: Princeton Univ. Press, 2004), 86–179.
43. Armstrong, "Cameraless," in Armstrong and de Zegher, *Ocean Flowers*, 93.
44. For an in-depth discussion of Talbot's interest in botanical photography, see Mirjam Brusius, Katrina Dean, and Chitra Ramalingam, eds. *William Henry Fox Talbot: Beyond Photography*, Studies in British Art 23 (New Haven and London: Yale Univ. Press, 2013). In this volume, see especially: Anne Secord, "Talbot's First Lens: Botanical Vision as an Exact Science," 41–66; and Herta Wolf, "Nature as Drawing Mistress," 119–42.
45. See Larry Schaaf, *Sun Gardens: Victorian Photograms by Anna Atkins* (New York: Hans P. Kraus & Aperture, 1985). In addition to a detailed essay about Atkins, this publication illustrates the complete range of her cyanotypes. These extend beyond British algae to include other botanical subjects, including ferns, *Papaver*, *Iris*, *Asplenium*, grasses, etc.
46. Each of the existing copies of Atkins's book is unique. The number of plates—like the plates themselves—differs slightly with each set.
47. For more on the non-photographic illustrations of Anna Atkins (then Anna Children), see Armstrong, "Cameraless," in Armstrong and de Zegher, *Ocean Flowers*, 87–165.
48. George Dawson, "Inaugural Address," *Borough of Birmingham Opening of the Free Reference Library October 26, 1866*, Birmingham: Osborne, 1866.

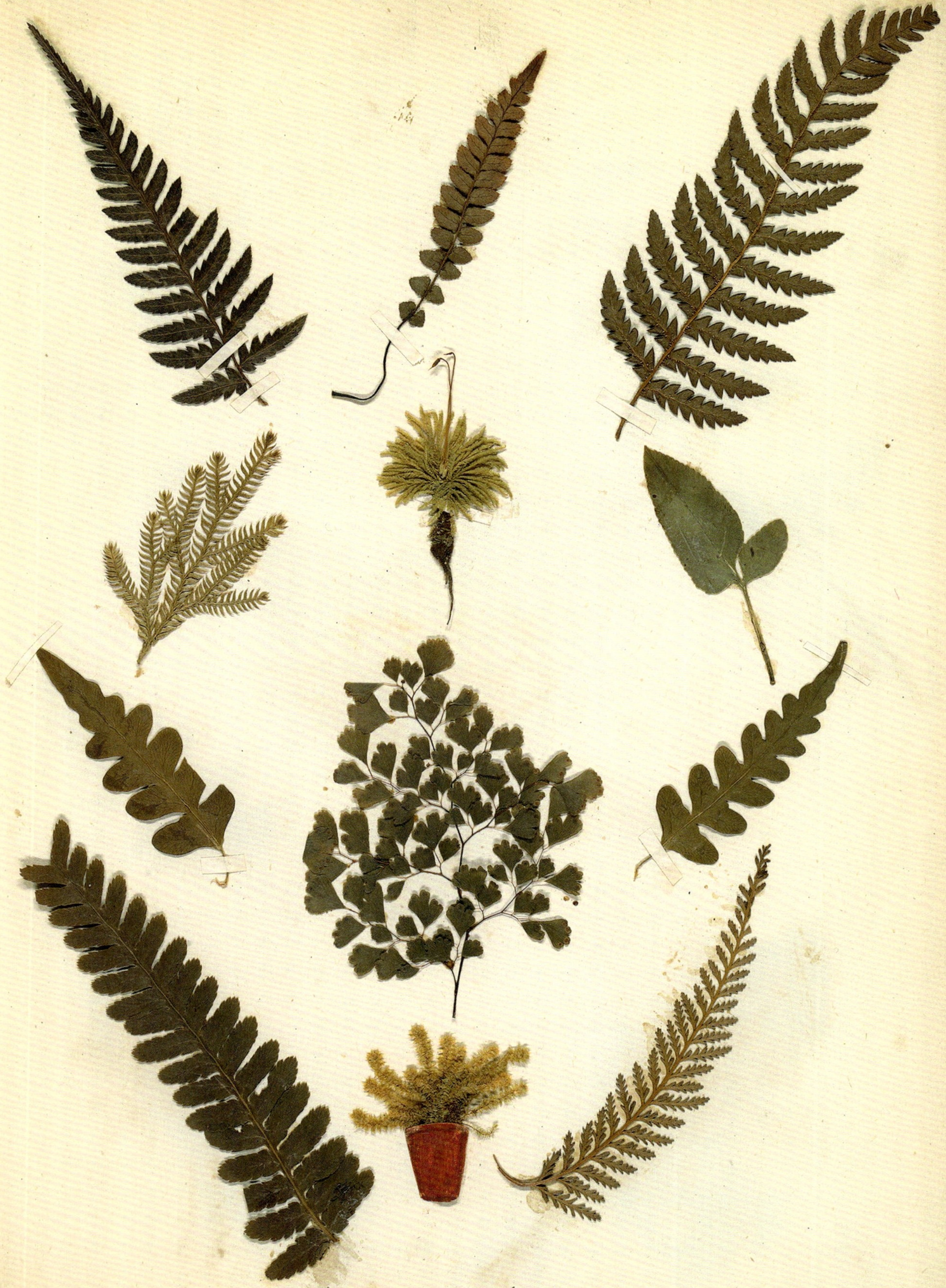

English, and New Zealand Ferns.

"Which Mimic Art Hath Made"

Crafting Nature in the Victorian Book and Album

Molly Duggins

> The flowers which grace their native beds,
> Awhile put forth their blushing heads;
> But, ere the close of parting day,
> They wither, shrink, and die away;
> But these, which mimic art hath made,
> Nor scorched by sun, nor killed by shade,
> Shall blush with less inconstant hue,
> Which art at pleasure can renew.[1]

In an age defined by industrialism and urbanism, the Victorians were consumed with replicating and processing the natural world. Art, according to *The Hand-Book of Useful and Ornamental Amusements and Accomplishments* (1845), had the power to stay the inevitable biological cycle of decay, to preserve or "renew" at will the blushing hues of botanicals, the gossamer wings of butterflies, and the bright plumage of birds. In concert with nineteenth-century natural history, an imaginative aesthetic science that combined empiricism with romanticism and rational amusement with material *voluptas*, Victorian handicraft was enlisted to extend and facilitate an intimate engagement with the natural realm.

Made by hand in the home, Victorian craft was both tactile and tactical: a gendered material practice of productive leisure through which women could demonstrate domestic skill and engage in a modern mercantile economy while cultivating an aesthetic sensibility associated with late eighteenth-century aristocratic culture.[2] Inspired by John Ruskin's spiritual homilies on the beauty of nature, which comfortably elided with middle-class notions of romantic consumption, Victorian women sought to create what Shirley Hibberd described in *Rustic Adornments for Homes of Taste* (1857) as "the union of Nature and Art in the ministration of human sympathies within doors" through the ubiquitous medium of craft.[3] Enshrined under glass, incorporated into pictures, and worked in wool and wax, beads and seeds, nature was transformed through craft into an artificial kingdom, in which it was not only dominated, domesticated, and sacralized, but also was employed to champion feminine and mechanical industry.[4]

The exhibition *"Of Green Leaf, Bird, and Flower": Artists' Books and the Natural World* reveals that a particularly intimate relationship existed between handicraft production, natural history practices, and the Victorian book and album. A rich cross-fertilization among botany and the decorative arts was fostered through the growing body of illustrated botanical and horticultural literature that provided inspiration for pattern books and drawing manuals, while handicraft guides included instructions on preserving and arranging natural specimens. Books and albums offered decorative engagement through commercially produced or handmade bindings that incorporated gilt ornamentation, lacquered boards, mother-of-pearl inlay, and embroidery, while bookbinding papers were often used in craft designs. Paper was also the most common mount for natural history illustration and was employed through artistic manipulation to replicate natural forms. Through craft, the Victorian book and album were thus actively involved in imitating, preserving, and memorializing the natural world.

OPPOSITE: Ann Paley, "English, and New Zealand Ferns" from *Fern and Flower Album*, 1873–1878, natural specimens with pen and ink. Yale Center for British Art, Paul Mellon Fund

Molly Duggins is a Lecturer in the Department of Art History and Theory at the National Art School, Sydney, Australia.

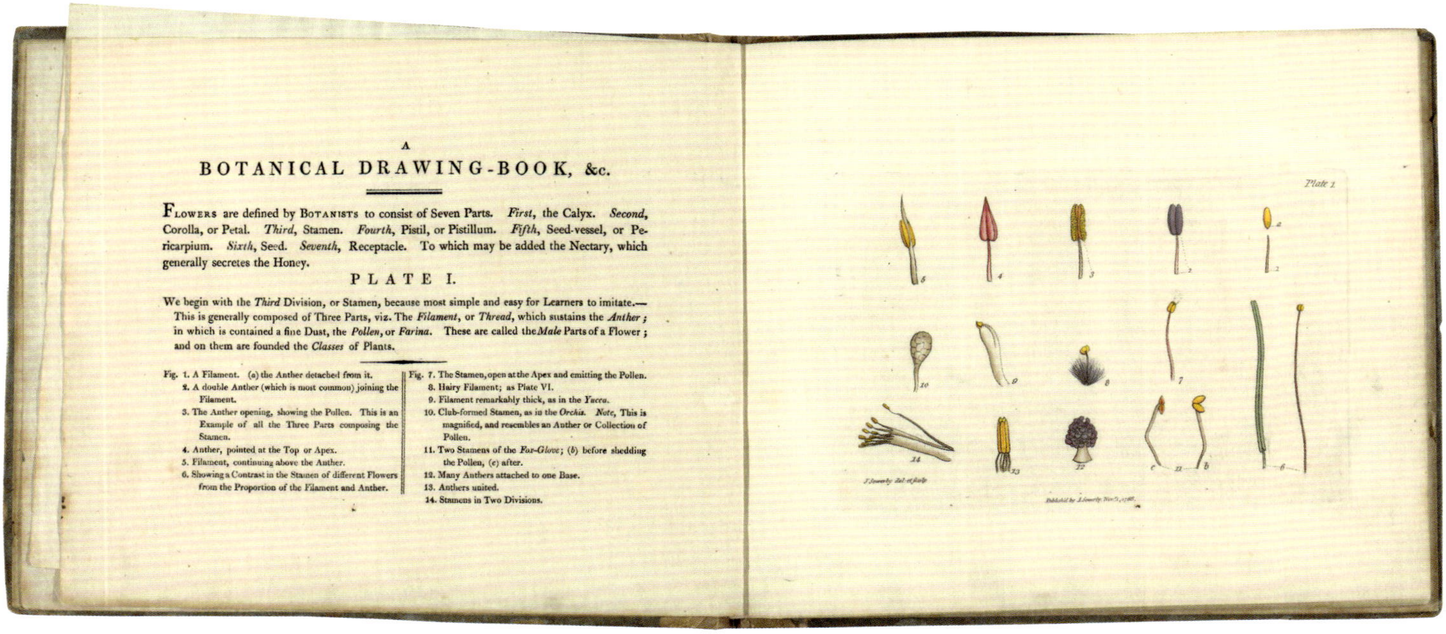

A

BOTANICAL DRAWING-BOOK, &c.

FLOWERS are defined by BOTANISTS to consist of Seven Parts. *First*, the Calyx. *Second*, Corolla, or Petal. *Third*, Stamen. *Fourth*, Pistil, or Pistillum. *Fifth*, Seed-vessel, or Pericarpium. *Sixth*, Seed. *Seventh*, Receptacle. To which may be added the Nectary, which generally secretes the Honey.

PLATE I.

We begin with the *Third* Division, or Stamen, because most simple and easy for Learners to imitate.— This is generally composed of Three Parts, viz. The *Filament*, or *Thread*, which sustains the *Anther*; in which is contained a fine Dust, the *Pollen*, or *Farina*. These are called the *Male* Parts of a Flower; and on them are founded the *Classes* of Plants.

Fig. 1. A Filament. (*a*) the Anther detached from it.
2. A double Anther (which is most common) joining the Filament.
3. The Anther opening, showing the Pollen. This is an Example of all the Three Parts composing the Stamen.
4. Anther, pointed at the Top or Apex.
5. Filament, continuing above the Anther.
6. Showing a Contrast in the Stamen of different Flowers from the Proportion of the Filament and Anther.

Fig. 7. The Stamen, open at the Apex and emitting the Pollen.
8. Hairy Filament; as Plate VI.
9. Filament remarkably thick, as in the *Yucca*.
10. Club-formed Stamen, as in the *Orchis*. *Note*, This is magnified, and resembles an Anther or Collection of Pollen.
11. Two Stamens of the *Fox-Glove*; (*b*) before shedding the Pollen, (*c*) after.
12. Many Anthers attached to one Base.
13. Anthers united.
14. Stamens in Two Divisions.

Figure 41. James Sowerby, plate 1 from *A Botanical Drawing-Book, or, An Easy Introduction to Drawing Flowers According to Nature*, London: Richard Taylor and Co., 1807, hand-colored engraving. Yale Center for British Art, Paul Mellon Fund

Nature Imitated

Described by Louisa Anne Twamley in *The Romance of Nature, or, The Flower Seasons Illustrated* (1836) as "forming one of the sweetest lines in the God-written Poetry of Nature," flowers were the most widely reproduced botanical forms in the nineteenth century.[5] Through their analysis and arrangement, the student of botany came closer to understanding the divine language of nature while exerting a sense of rational order over the natural realm. Advocating self-instruction through copying, drawing manuals such as James Sowerby's *A Botanical Drawing-Book, or, An Easy Introduction to Drawing Flowers According to Nature* (1807) presented instruction through dissection, with each lesson focusing on a particular part of the plant, from the calyx and corolla to the stamen and pistil, "in order to facilitate botanical studies, and blend amusement with improvement" (fig. 41).[6]

Such works were read and used as teaching tools in the drawing room, breakfast room, and garden. Exactingly replicated in a variety of mediums to form perfect "fac-similes of nature," the transient beauty of flowers was immobilized through handicraft judged not for its originality but for its skill of execution, fidelity as a copy, and ability to deceive the eye.[7] Embodying domestic values of discipline and industriousness, this art of arrangement was associated with a commercialized decorative affinity between women and flowers and the fashioning of feminine selfhood.[8]

The intricate detail and botanical accuracy of twenty-three cut-paper floral collages in an album compiled by Ellen W. prefigure the Victorian fascination with the imitative arts while referencing late eighteenth-century aristocratic handicraft practices that intersected with an elite form of salon science. Reminiscent in particular of Mary Delany's "paper mosaicks" of botanical specimens, constructed out of myriad pieces of hand-colored paper in the 1770s and 1780s, Ellen W.'s articulated flower parts, exemplified in the distinctive petals and stamen of her cut-paper passionflower, reflect the analytical botany promoted in Sowerby's drawing book, while her decorative groupings of specimens reveal the stylistic influence of florilegia (fig. 42).[9] Embracing the legacy of paper as a precious commodity, her collages celebrate its pliability and delicacy as an artistic medium, a theme that continues to inspire the work of contemporary book artists from Sarah Morpeth's three-dimensional hand-stitched, painted, and bound bouquet, *A Garden Book* (2011; fig. 43), to Jane Hyslop's collaged herbarium sheets *Buttercup* and *Dandelion* (both 2009), which play with the compositional conventions and artifice of cut-paper specimens through their deconstruction.[10]

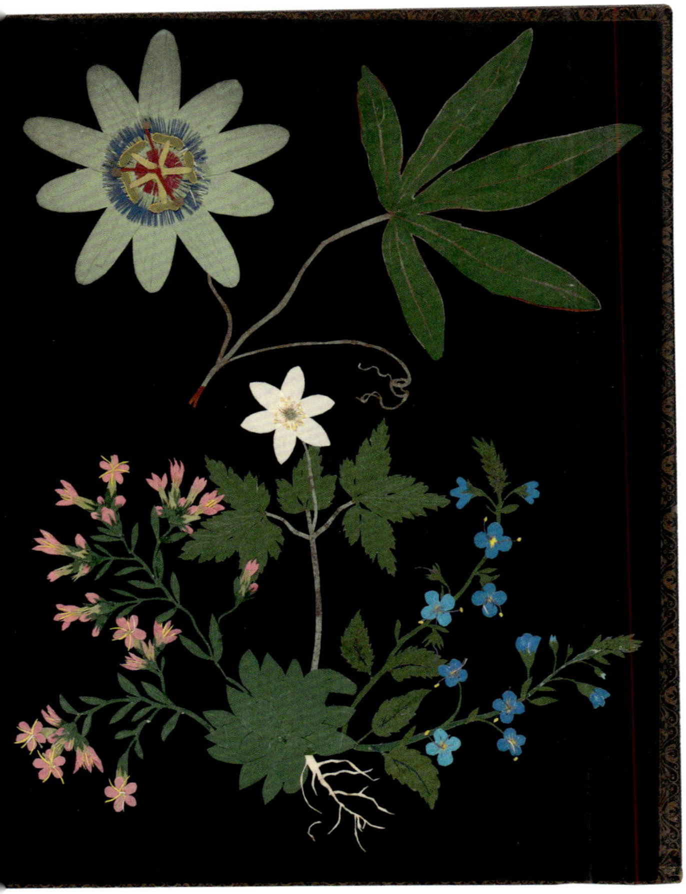

Figure 42. Ellen W., "Passionflower" from *Album of Cut-Paper Flowers*, ca. 1835, hand-cut paper collage. Yale Center for British Art, Paul Mellon Fund

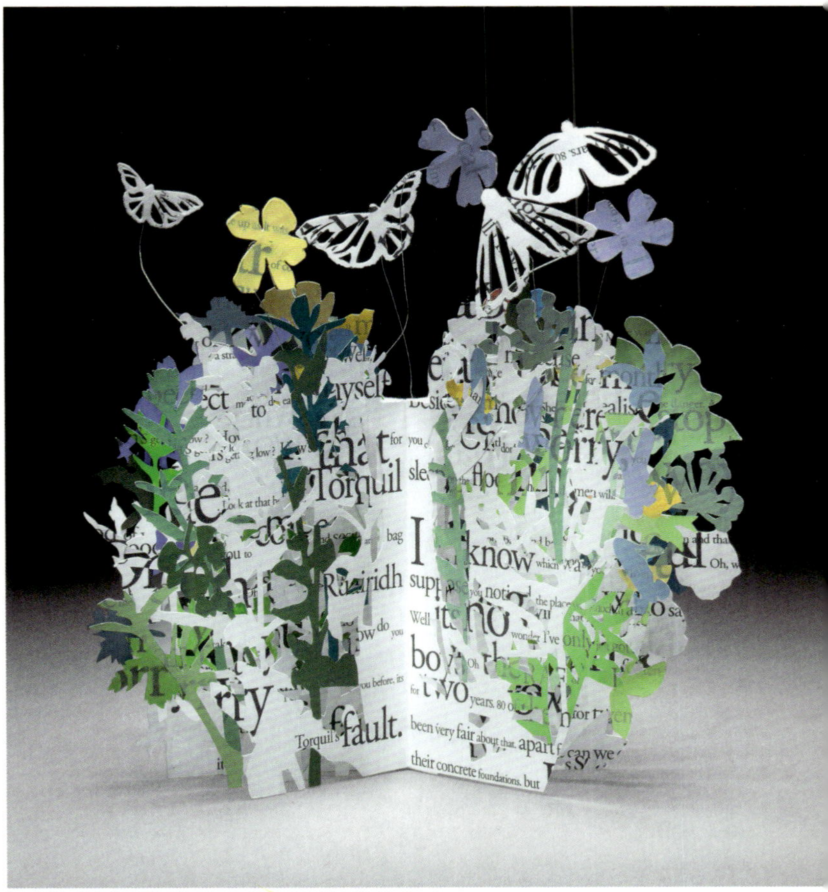

Figure 43. Sarah Morpeth, *A Garden Book*, Elsdon, Northumberland, 2011, hand-cut paper, watercolor, and wire. Yale Center for British Art, Friends of British Art Fund

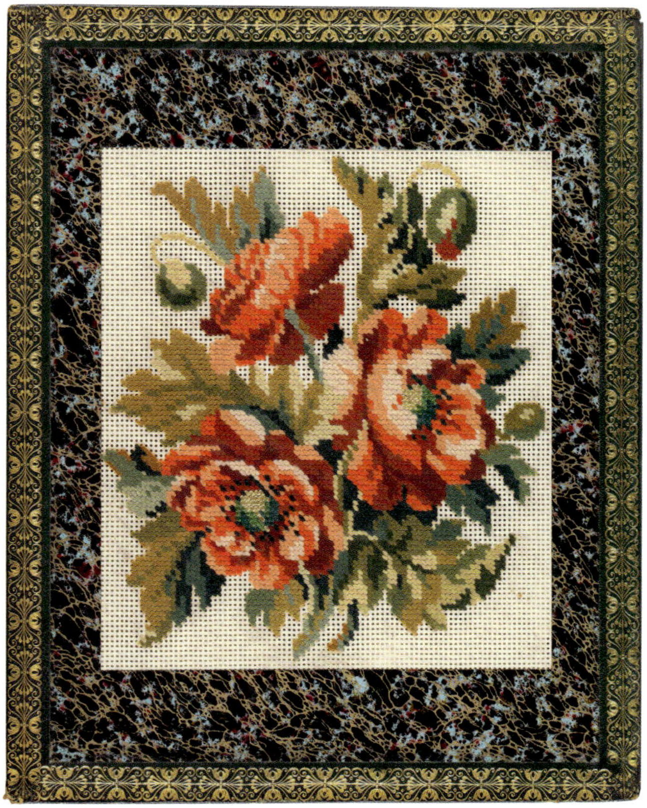

Figure 44. Ellen W., Berlin woolwork from *Album of Cut-Paper Flowers*, ca. 1835. Yale Center for British Art, Paul Mellon Fund

Beginning in the nineteenth century, the craft uses of paper shifted dramatically from the bespoke to the industrial, which, along with the introduction of mass-produced commercial art supplies and illustrated art manuals, led to the democratization of handicrafts that were formerly the domain of privileged leisure.[11] This new craft sensibility emerges in Ellen W.'s album in a Berlin woolwork picture of a spray of roses mounted onto each inner board, which transforms her album into a drawing room decoration in its own right (fig. 44). A mass-produced form of needlework that replaced embroidery as one of the most pervasive mid-Victorian crafts, Berlin woolwork allowed for the fast and effective copying of pictures through their direct application to a coarse canvas filled through with inexpensive, thick strands of colored wool.[12] Designs were appropriated from paintings and illustrated books, and circulated through women's periodicals, while paper patterns, complete with hole punches for the needle, and prepared canvases were available for purchase. Embracing the mass-produced commodity, shifting fashion, and the cheap, ephemeral, and disposable, Victorian handicraft was transformed into a medium that valorized the trappings of middle-class modernity.[13]

A paper basket, circa 1820, marks this modernizing trajectory (fig. 45). Lattice-cut, silk-stitched, and hand-painted with flowers, it invokes a combination of the amateur arts to recreate a ceramic fruit bowl in ephemeral form, down to the imitation of gilt slip through gold paper applied to its scalloped edges. Intended for the collection of visiting cards, the basket references the role of craft as a performance of status, while advertising the thrift and skill of its maker. In the Victorian handicraft revolution, such artistic replication was simplified through decalcomania (fig. 46). In this transfer process, printed designs of flowers, fruits, shells, grasses, birds, and butterflies, available in sheets from stationers and fancy goods repositories, were applied to plain surfaces and varnished, "it being the aim of the operator," according to *Artistic Amusements: Being Instructions for a Variety of Art Work for Home Employment* (ca. 1879), "to imitate hand-painting upon small objects."[14]

In this era industrialization and urban flux as well as imperial expansion allowed nature to acquire new agency as an inspiration for handicraft production through the

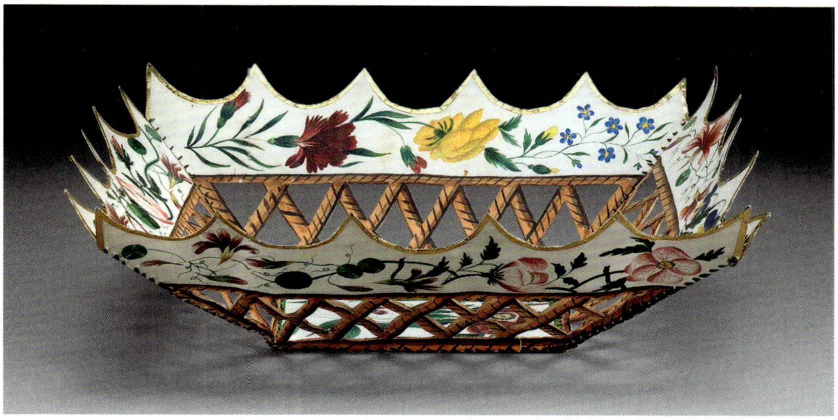

Figure 45. Cut-paper basket, England, ca. 1820, hand-cut paper with watercolor, stitched with silk thread. Yale Center for British Art, Paul Mellon Fund

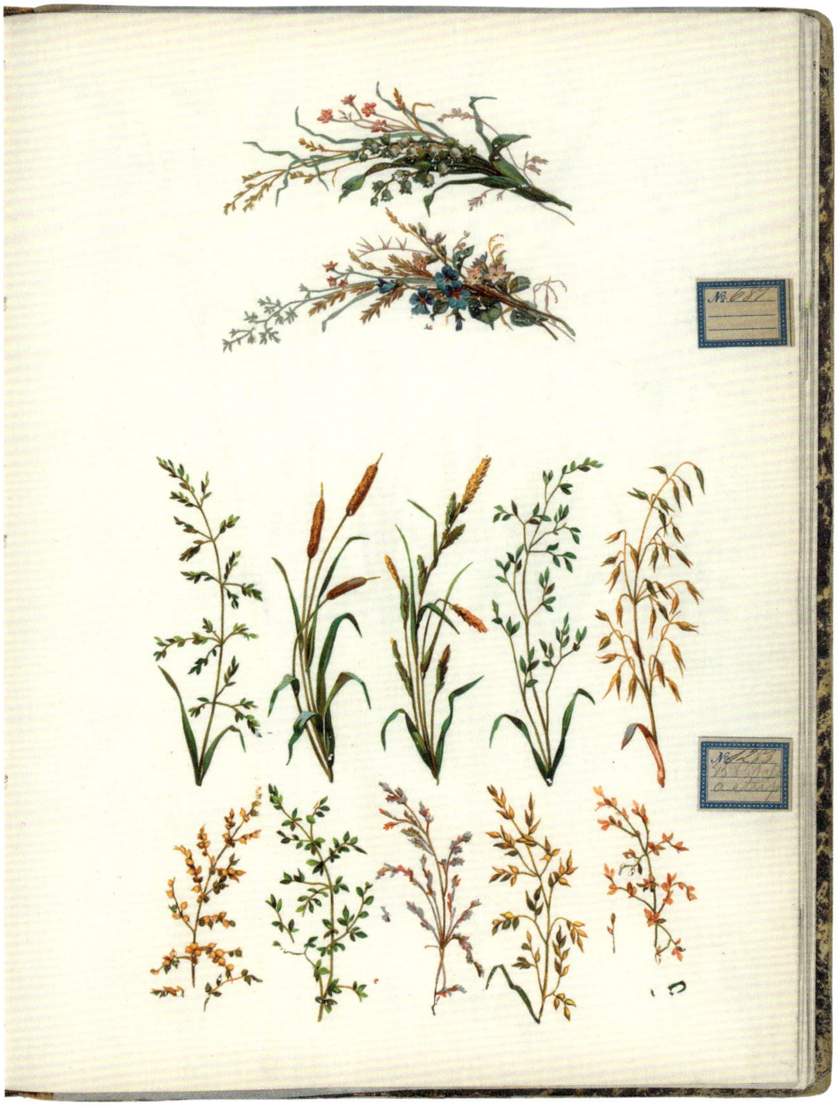

Figure 46. *Decalcomania Sample Book*, ca. 1870, chromolithographs with pen and ink labels. Yale Center for British Art, Paul Mellon Fund

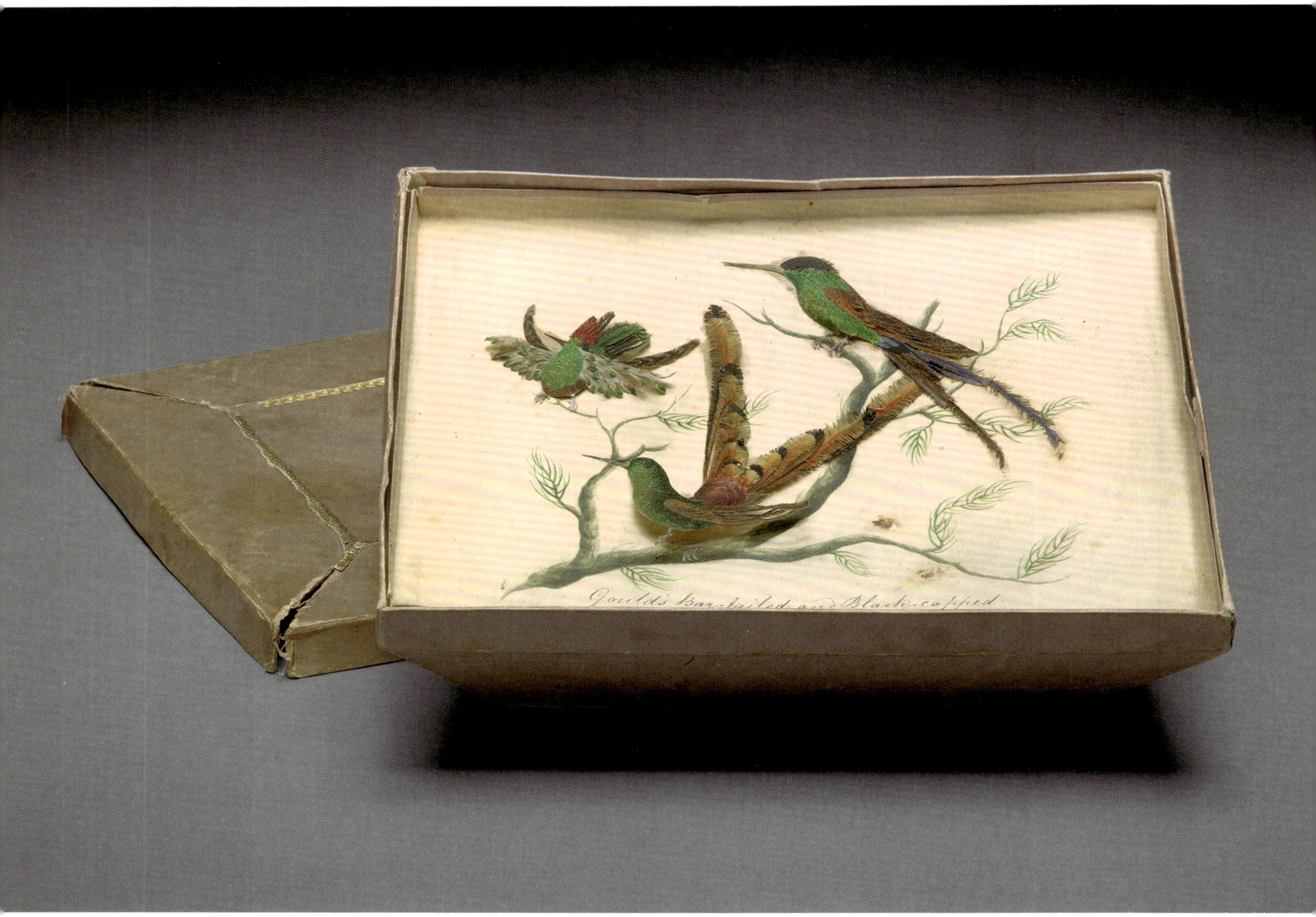

Figure 47. "Gould's Bar-tailed and Black-capped Hummingbirds" and box, *Cut-Paper Featherwork Illustrations of Gould's Birds*, ca. 1865, cut paper, watercolor, and pen and black ink. Yale Center for British Art, Paul Mellon Fund

proliferation of amateur natural history pursuits invigorated by a stream of specimens from around the globe. A set of six cut-paper "featherwork" birds based on illustrations in the ornithological publications of John Gould unites the precepts of modern Victorian handicraft with such scientific endeavor (fig. 47). Meticulously cut, textured, hand-colored, and padded to evoke a three-dimensional, almost taxidermic form, these compositions employ paper, the cost of which had been reduced after the repeal of the paper tax in 1861, to replicate featherwork pictures. Instructions for their production were included in *Elegant Arts for Ladies* (1861).[15] A cheaper alternative to the rare or exotic feathers employed in earlier examples of featherwork, such as James Forbes's late eighteenth-century painting of a highland muirfowl embellished with its autumn plumage (fig. 48), this cut-paper series adapts the aristocratic trophy bird into a form of democratic decoration.

A mélange of domestic and foreign species, the box features a crane, buzzard, ostrich, painted bunting, Pennantian broad-tailed parrot, and "Gould's Bar-tailed and Black-capped Humming Birds," the latter group inspired by the colored lithographs in John Gould's lavish *Monograph of the Trochilidae* (1849–1861), drawn from tropical specimens in his own collection. Painted in delicately modulated shades of green to simulate the iridescent emerald plumage of these exotic specimens that so captivated the Victorian public, the hummingbirds are mounted on a paper tray, which nests within a handmade box along with the other bird compositions.[16] Covered in moiré-patterned paper decorated with a border of gilt paper strips available in packets from specialized emporia such as Ackermann's Repository of Arts (fig. 49), this paper box of birds merges mass production and cheap material with painstaking handicraft. It serves as an ephemeral cabinet of curiosities for the modern era, much like Julie Cockburn's vivid *Beetles Book*, in which exotic jewels of nature from far-flung areas of the globe are contained and replicated with startling illusion to delight an urbanized audience (fig. 50).

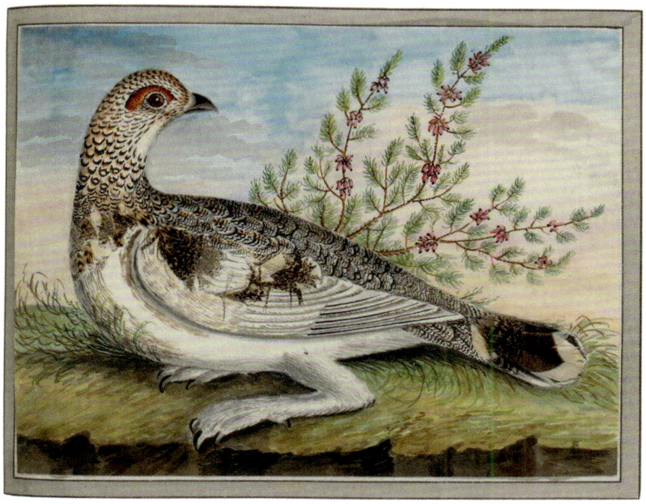

Figure 48. James Forbes, "The Ptarmakin or Ptarmakan" (detail) from *Views in England, Wales, and Scotland* (vol. 3), 1799, watercolor, gouache, and pen and ink, with natural specimens. Yale Center for British Art, Paul Mellon Collection

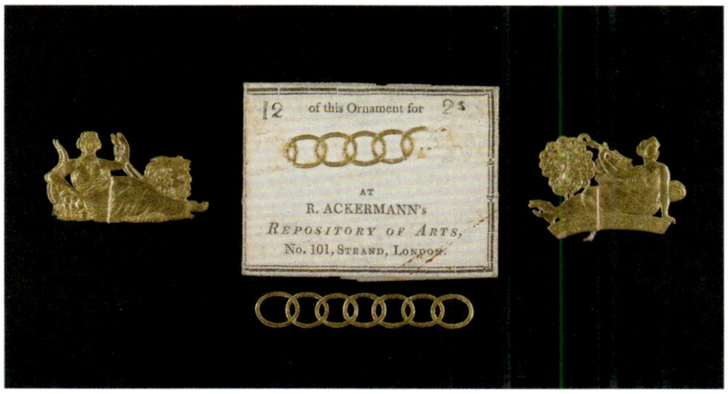

Figure 49. *Packet of Gilt Ornaments for Decorating Albums* (detail), London: Ackermann's Repository of Arts, ca. 1820. Yale Center for British Art, Paul Mellon Fund

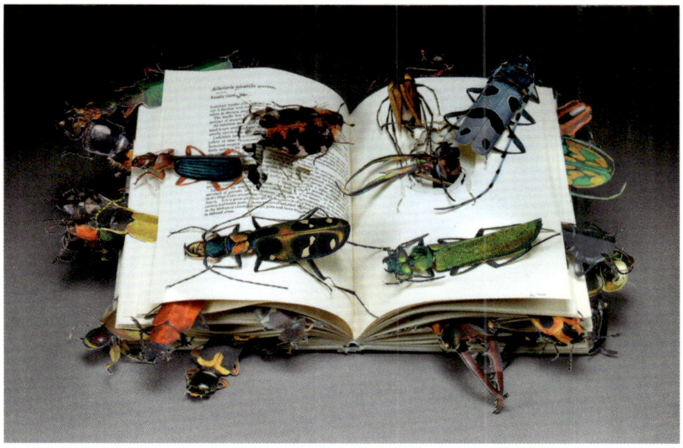

Figure 50. Julie Cockburn, *Beetles Book*, 2002, altered book. Yale Center for British Art, Friends of British Art Fund

Nature Preserved

While such handicraft was seen to enhance nature through artifice, the sheer materiality of the unadorned natural specimen was regarded as the zenith of singular beauty.[17] Scrawled across the top of a page devoted to the *Ranunculus bulbosus*, or buttercup, in Anne Pratt's *Wild Flowers* (1852–1853), an inscription by a former owner of this popular field guide finds art deficient to nature adulterated:

> Reader – take the finest gilded & burnished picture in Looking glass frame you ever saw . . . and place on it a fresh Buttercup – & oh what a miserable gilding is that of the artist to the gilding of Nature![18]

Next to an image of the flower printed by William Dickes using the Baxter process, a form of color relief printing used in the production of inexpensive books, this fervent directive asserts the artistic primacy of raw nature at a time when the introduction of new representational technologies in natural illustration were transforming the specimen into an evocative medium of drawing, from photography to nature printing (fig. 51).[19] One of a number of inscriptions on the page, it also points to the tangible character of the field guide as an interactive portal to a natural world linked in the Victorian imagination with a disappearing bucolic lifestyle in which nature's cycles were intimately observed, as epitomized in Gilbert White's *The Natural History and Antiquities of Selborne* (1789). Through self-guided botanizing in local meadows, woodlands, and coasts, as well as further afield through expanding transport systems and imperial networks, the amateur naturalist, weekend excursionist, and intrepid traveler could experience the material wonder of natural history through the collection and dis-

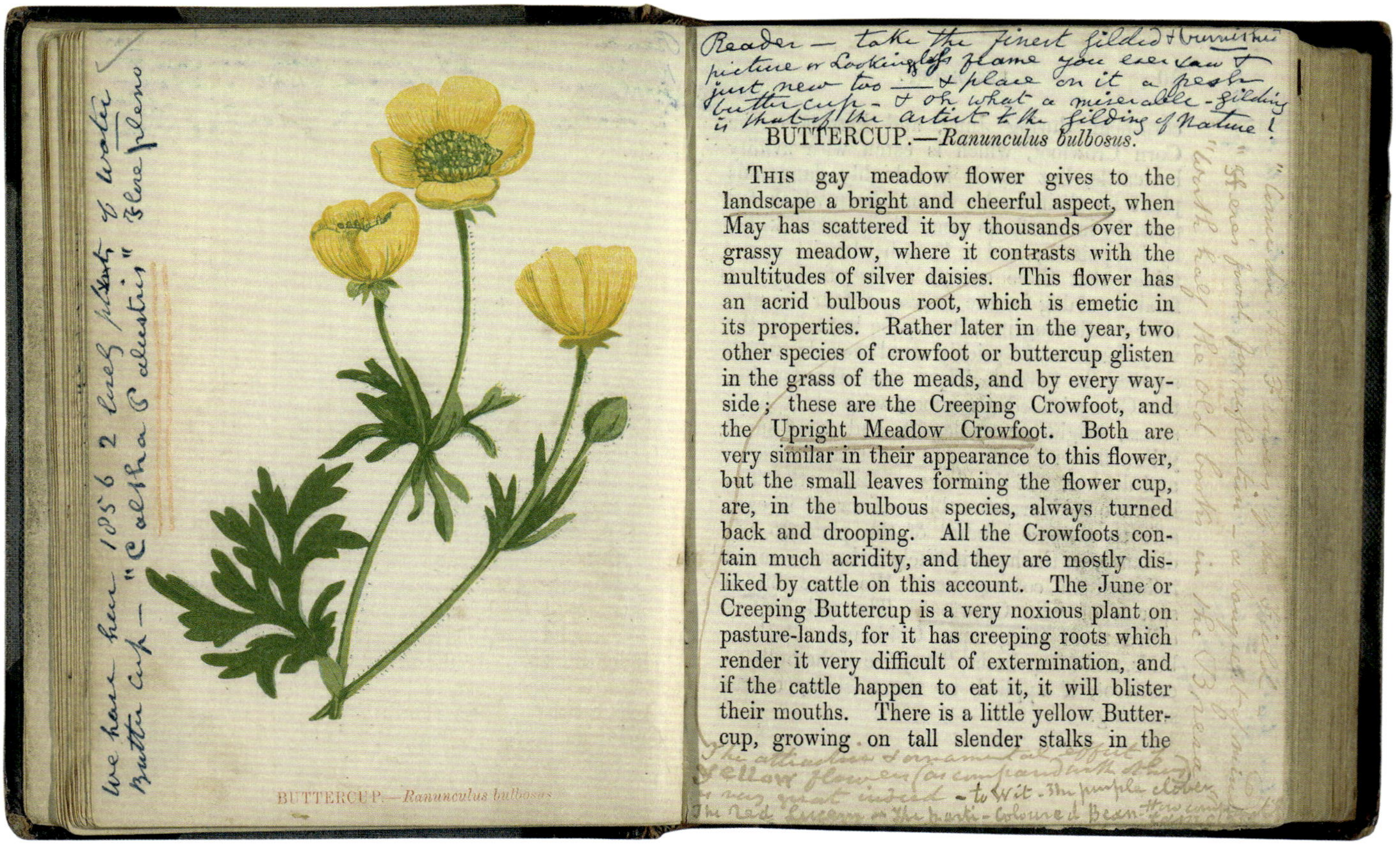

Figure 51. Anne Pratt, "Buttercup—Ranunculus bulbosus" from *Wild Flowers*, London: Society for Promoting Christian Knowledge, 1852–1853, chromolithograph with pen and ink annotations. Yale Center for British Art, Paul Mellon Collection

play of specimens from nature's emporium, the artistic arrangement of which was a veritable craft process.

An interclass, intergender, intergenerational, and international phenomenon, Victorian natural history was propagated through pictures generated by a newly disposable visual culture of industrial arts and media.[20] Illustrations were promoted as a means to sell science through their decorative appeal and flooded the market in the 1840s in cheap guides and manuals that featured mass-produced chromolithographs of specimen portraits, composite bouquets, and diagrams as an aid to identification.[21] Such representational conventions elicited pleasure through decontextualization, transforming the specimen into an object of desire, a singular jewel plucked from obscurity to dazzle on the page. Painted in flat shades of lurid pigments, these illustrations, as exemplified in Pratt's buttercup, provided a simple and direct translation of the specimen that was synthetically perfect, yet lacked the marvelous idiosyncrasies of plant matter.[22] Retaining subtle nuances of color and texture, the preserved specimen displayed a seemingly infinite level of detail that extended beyond its representational capacity to stand as a testament to the ineffable experience of communing with nature.

These two forms of illustration, painted and preserved, coalesce in the herbarium of Miss Rowe, a member of the Liverpool Naturalists' Field Club who compiled an herbarium of 101 genera of local flora in 1861, which she named and ordered according to Joseph Dickinson's *The Flora of Liverpool* (1851).[23] Paying homage to the elusive figure of the nineteenth-century female amateur, Mark Dion and J. Morgan Puett give form to Miss Rowe in their neo-Victorian photographic series, *The Ladies' Field Club of York* (1999), in which the personification of botany embodies the matrix of science, art, and fashion that shaped such recreational collection and display (fig. 52). Presenting an illustrative supplement to Dickinson's guide, Miss Rowe's collection demonstrates the sustained material intercourse that existed between botanical literature and amateur practice to create an intimate knowledge of the landscape, the lack of which informs Tracey Bush's *Nine Wild Plants* (2006), a set of collaged fragments of logos in the silhouettes of herbarium specimens. Bush's work combines Delany's "paper mosaicks" with decalcomania to criticize

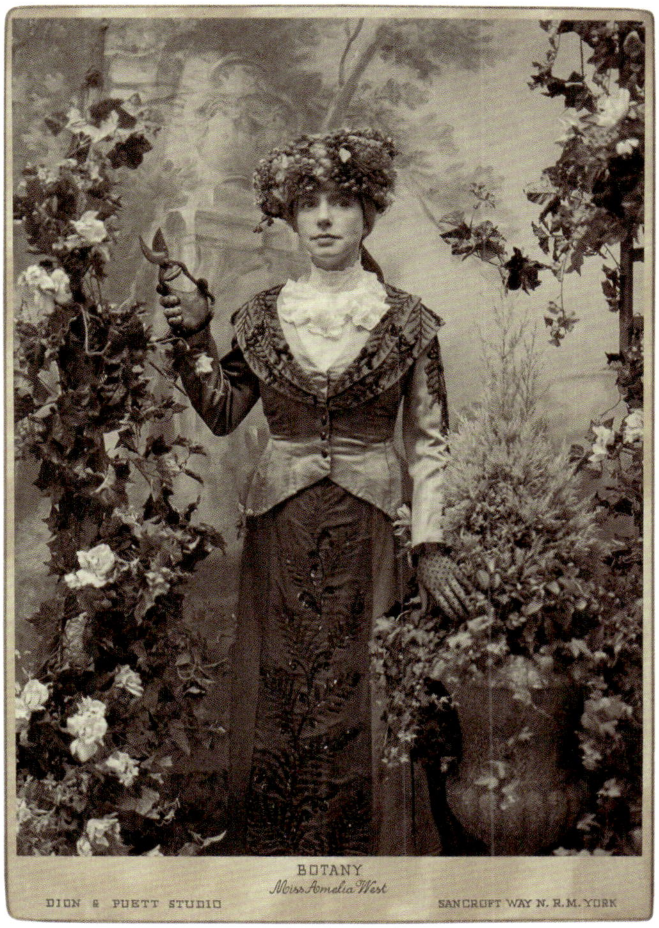

Figure 52. Mark Dion and J. Morgan Puett, "Botany" from *The Ladies' Field Club of York*, York: Impressions Gallery, 1999, photographic card. Yale Center for British Art, Friends of British Art Fund

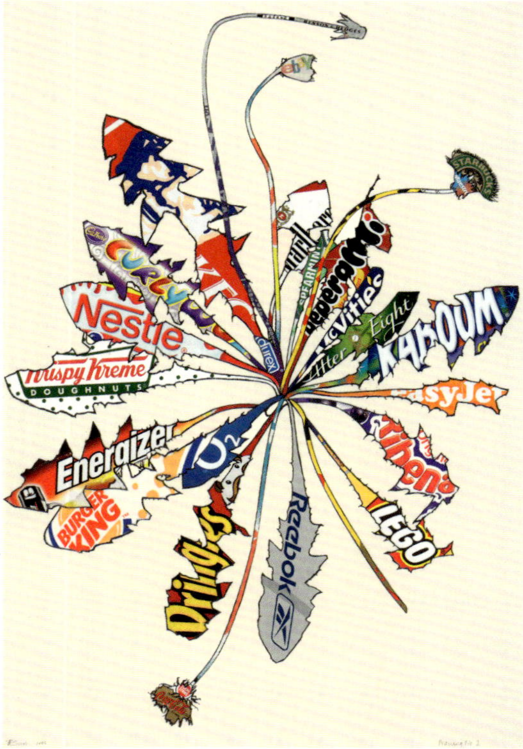

Figure 53. Tracey Bush, Study for herbarium sheet for *Nine Wild Plants: Dandelion*, 2006, hand-cut paper collage with pen and ink. Yale Center for British Art, Friends of British Art Fund

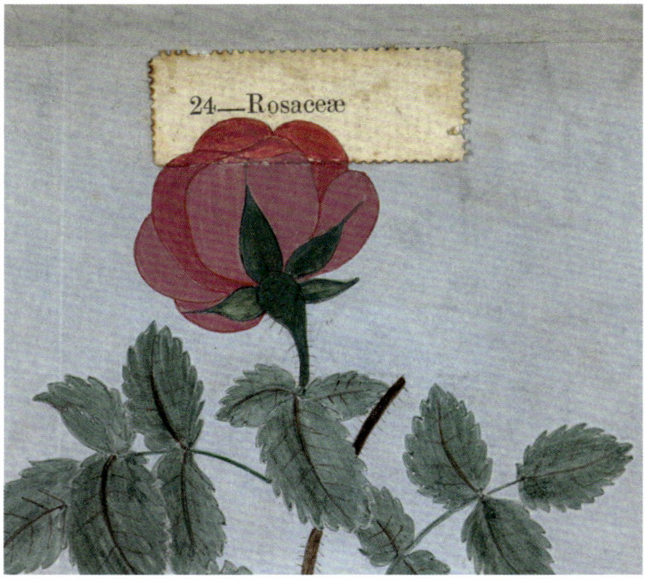

Figure 54. Miss Rowe, "24. *Rosaceae*" (detail) from *Collection of Botanical Specimens*, Liverpool, 1861, envelope with watercolor and printed label. Yale Center for British Art, Paul Mellon Fund

an overfamiliarity with commercial brands at the expense of an awareness of local flora (fig. 53).

Miss Rowe's herbarium consists of folders of pressed specimens, inscribed with their date and location of collection, grouped according to genus in envelopes decorated with painted flora. A number of these appear to bloom beyond their borders, partially obscuring printed botanical labels (fig. 54), the "repulsive technicalities" of which, as Shirley Hibberd lamented, so often disfigured the beauty of the cabinet.[24] Engaging with the intimate tactility of Miss Rowe's envelopes, Mandy Bonnell's *Lichen* (2012) presents a curated meditation on the poetic minutiae of Victorian natural history collection steeped in personal aesthetic experience, a phenomenon also revisited in Paula MacGregor's whimsical archival experiment, *Hollyhock* (2009).

The combination of romance and empiricism that defined such a collection was extended through the language of flowers, a veiled discourse on the symbolic reading of plants. It was invoked in Miss Rowe's engraved name card, which was included in her herbarium and decorated with a painted forget-me-not, described by Twamley as "the sentimental favourite of modern Poetasters" through its association with remembrance.[25] Employed to foster bonds of friendship and attachment, flirtation and courtship, the exchange of flowers, bouquets, and such floral ephemera was seen by Robert Tyas, author of *The Language of Flowers, or, Floral Emblems of Thoughts, Feelings and Sentiments* (1869), as "the mark of the politest attention."[26] Valentines, in particular, provided a popular means of floral intercourse from the 1830s, developing in tandem with the commodification of craft to frame the language of flowers in machine-made paper lace adorned with embossed scraps, fabric, netting, and pressed specimens (fig. 55).[27]

Amateur natural history was similarly ensconced in a growing genre of commercial paraphernalia, suggested by Miss Rowe's use of printed botanical labels, which intersected with the mass production of art materials to transform natural illustration into a handicraft industry. Framed through the conventions of the rectangular page into a work of art, pressed specimens were mounted on drawing paper, secured with gum arabic, strips of binding paper, or stitches of thread, and arranged with camel hair brushes or knitting needles.[28] Seaweed specimens above all, "by Lady's

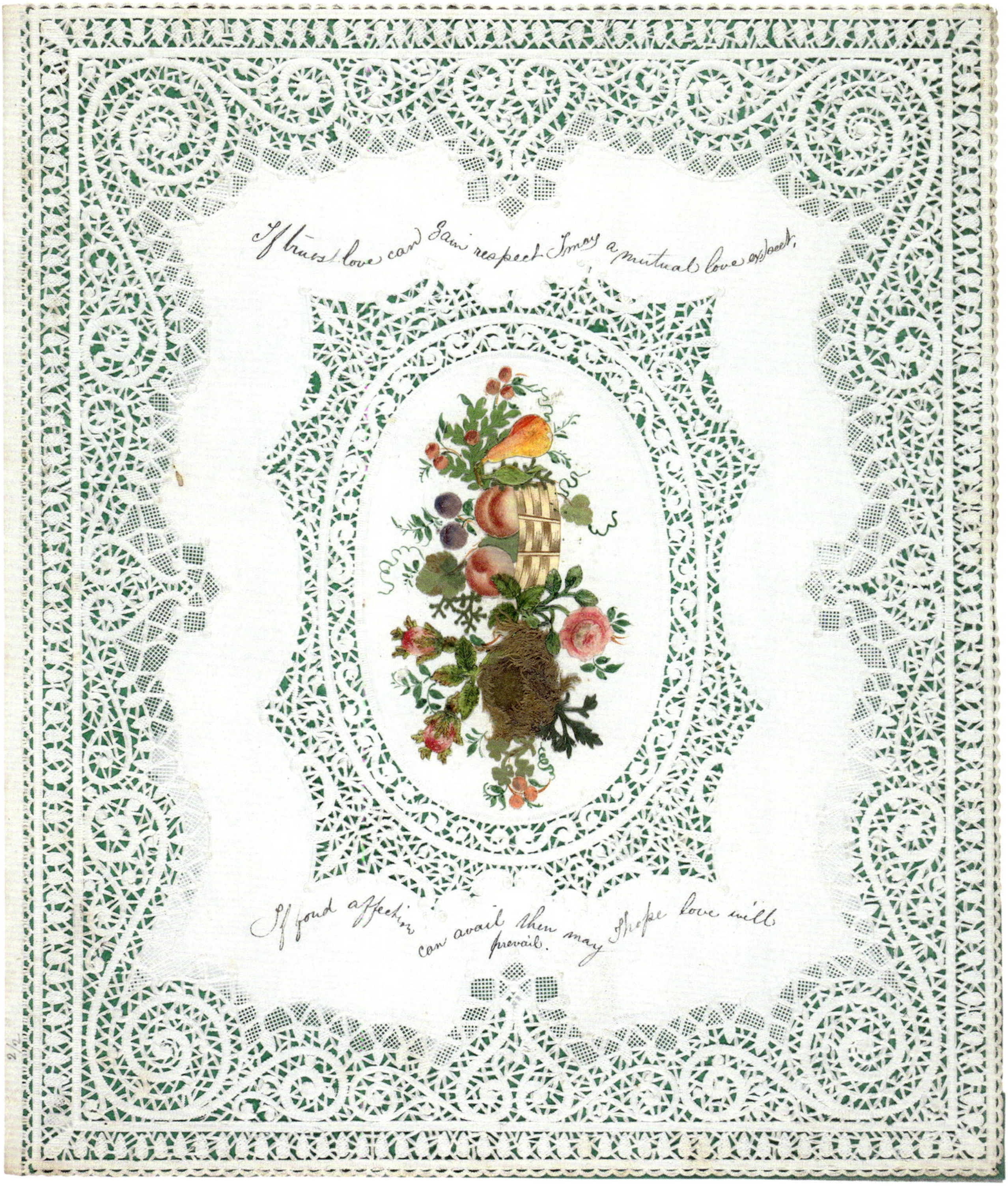

Figure 55. *If truest love can gain respect I may a mutual love expect*, London: De La Rue & Co., ca. 1866, lace-paper valentine with hand-coloring, pen and ink, and natural specimen. Yale Center for British Art, Paul Mellon Collection

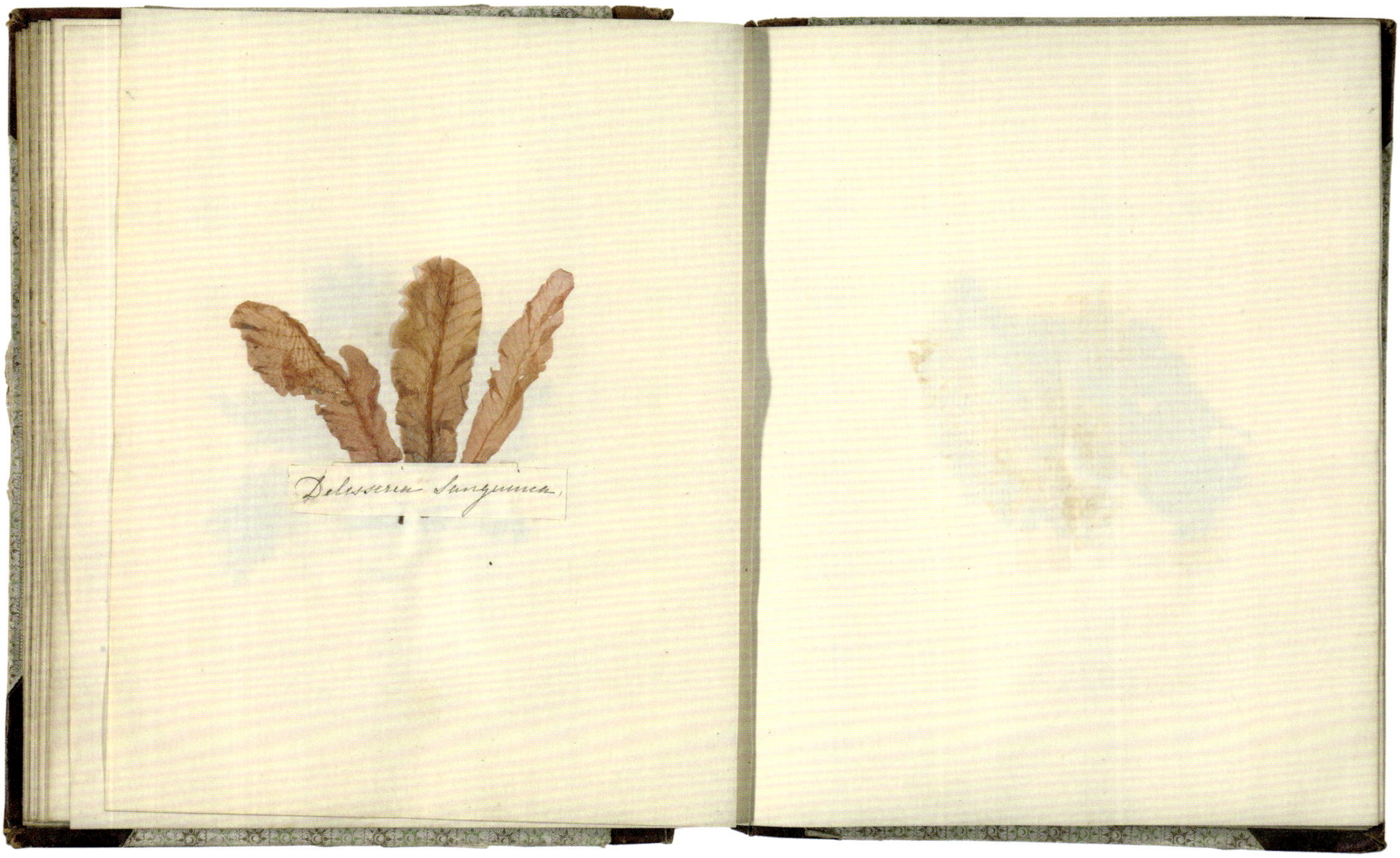

Figure 56. "*Delesseria sanguinea*" from *Specimens of Sea Weed*, ca. 1840, natural specimen with pen and ink label. Yale Center for British Art, Paul Mellon Fund

gentle fingering displayed," were described by Reverend David Landsborough in *A Popular History of British Sea-Weeds* (1851), as "exquisite paintings."[29] A process that was both disciplined and disciplining, the preservation of seaweed had much in common with tactile craft practices that revolved around the acts of cutting and pasting, shaping and assembling (fig. 56).[30]

Once gathered, specimens were kept in a container of water until laid out for mounting through the floating process, in which paper was slipped under the submerged fronds and lifted to the surface to simulate their submarine buoyancy. Remaining clumps were separated by hand with a fine brush or needle and blotting paper or linen was laid on top. Multiple sheets were prepared and layered in this manner and placed under pressure using weights until the specimens dried out and adhered to the paper through the natural sugars or carrageen they emitted during pressing. Flexing the natural specimen into an artistic format akin to watercolor or embroidery, the amateur naturalist extended the sensory experience of nature through such preservation to collaborate in Nature's process of self-representation.[31]

Envisioned through such material manipulation, marine specimens were also equated with decorative commodities. In *The Marine Botanist* (1853), Isabella Gifford compared seaweeds to fine tracery and fragments of velvet, while they were promoted as objects of design in their own right by Samuel Joseph Mackie in *Art-Studies from Nature* (1872).[32] Their relationship to ornamental manufactures was more

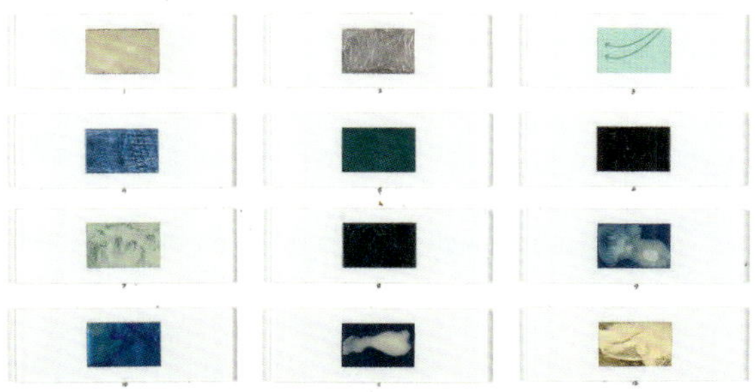

Figure 57. Nicholas Phillips, *You're Never Alone in the Ocean: Life of Plankton* (detail), London, 1981, pen and ink with relief etchings and glass slides. Yale Center for British Art, Friends of British art Fund

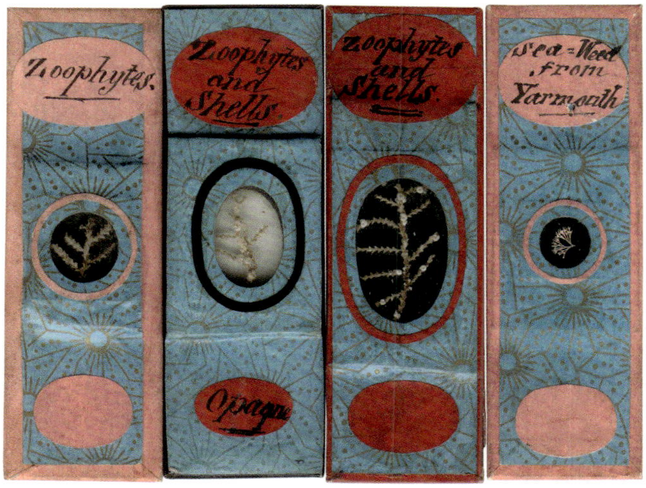

Figure 58. *Microscope slides (zoophytes, shells, and seaweed from Yarmouth)*, ca. 1860, decorated paper, glass, pen and ink, with natural specimens. Division of Historical Scientific Instruments, Peabody Museum of Natural History, Yale University

than just complementary, however. Hibberd noted that marine algae were "rich in soda, potash, iodine, bromine, and other substances . . . largely used in the arts," from photographic production to papermaking.[33]

The aesthetic dimensions of seaweeds were further expanded through the use of the microscope, which unveiled internal formations of the marvelous, the fluid worlds of which are imagined in Nicholas Phillips's *You're Never Alone in the Ocean: Life of Plankton* (1981; fig. 57). Instructions on how to prepare algae for the microscope are included in guides such as William Grosart Johnstone's *The Nature-Printed British Sea-Weeds* (1859), which advised on the "accessories requisite" of glass slips, forceps, scissors, Japan varnish, gold size, hair pencils, needles with handles, pieces of chamois leather and linen, gutta-percha rings, paste, gum, and paper covers.[34] For the latter, Thomas Davies, author of *The Preparation and Mounting of Microscopic Objects* (1873), suggested that ornamental papers "of innumerable patterns and colors" could be easily procured from most opticians "at a cost little more than nominal."[35]

A homemade set of slides housed in a cabinet at the Yale Peabody Museum of Natural History, mounted with seaweeds, zoophytes, and shells collected from the seaside resort of Yarmouth, demonstrates that even the scopic lens of Victorian natural history was enriched through such adornment. Encased in glass slips embellished with red and pink labels, these marine fragments float against an oceanic paper ground covered with a gilt pattern that evokes the "mathematical designs" of magnified specimens (fig. 58).[36] The minute structures of such scientific specks dazzle with their neo-rococo abundance of microscopic twistings and contortions, inscribed like sentimental portraits in oval frames that temper the hypervisuality of the disembodied eye through the intimate tactility of craft.

Nature Memorialized

Through such curation, the natural specimen was transformed into an artifact of human construction capable of conveying its maker's familial, domestic, and national affiliations.[37] Shells, feathers, porcupine quills, fish scales, and beetle wings were arranged in geometric designs on boxes and frames, while pressed plant specimens were incorporated into collages and pictures that turned the drawing room into a three-dimensional natural history album.[38] Crafted, gifted, traded, and exhibited, natural artifacts provided a personalized counterpoint to the circulation

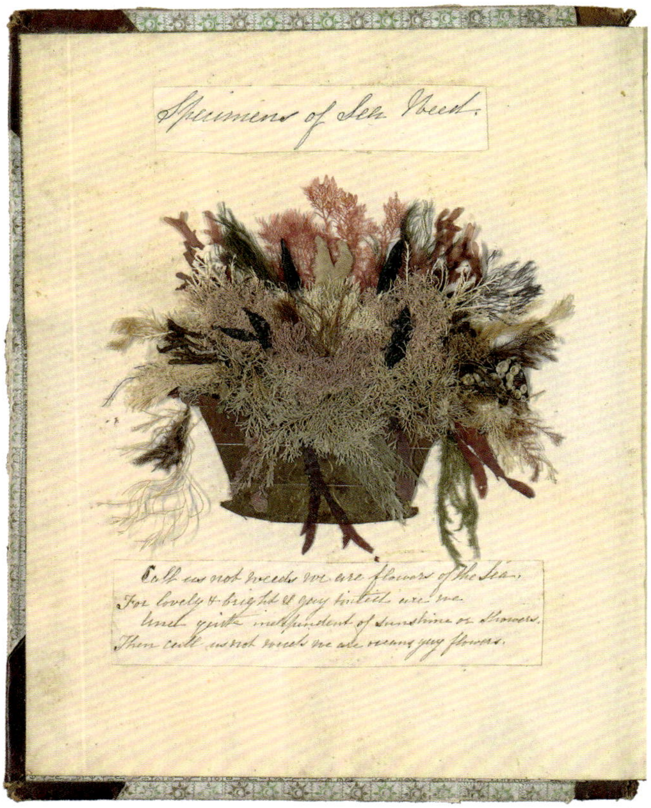

Figure 59. Title page from *Specimens of Sea Weed*, ca. 1840, natural specimens with decorative paper and pen and ink labels. Yale Center for British Art, Paul Mellon Fund

of mass-produced commodities in a progressively global Victorian economy.[39] As "still lifes" of nature, they also functioned as poignant *memento mori*, juxtaposing the fashioned material remains of specimens against their symbolic connection to impermanence.

Within the genre of the natural artifact, the transnational craft motif of a basket of ocean flowers encapsulated the taste, tactility, and exchange value that defined the handicraft marketplace. Designs for these baskets composed of preserved marine specimens were disseminated through artistic manuals such as *Leaf and Flower Pictures and How to Make Them* (1860) and natural history gift books, including Mary Matilda Howard's *Ocean Flowers and their Teachings* (1846), the frontispiece of which displays a paper basket overflowing with multicolored pressed seaweeds.[40] Subscribing to these stylistic conventions, albums compiled during the mid-to-late Victorian era in Britain, the colonies, and the United States contain similar arrangements often accompanied by a widely reproduced stanza of poetry attributed to E. L. Aveline and dedicated to the flowers of the sea (fig. 59).[41] Judged according to "the combination of colors and the delicacy and taste displayed in their arrangement," seaweed baskets were produced as seaside mementos, sold at bazaars, and exhibited in women's courts at international exhibitions as emblems of feminine industry that flowed between the public and private spheres and circulated via modern systems of media and transport through formal and informal imperial networks.[42]

As an affordable, portable repository for handicraft display, the Victorian album played a significant role in the transmission of such aesthetic and cultural traditions associated with popular natural history pursuits. Indeed, the album ultimately hinged upon its transmission value, maintaining a tenuous balance between memory, display, and exchange. Composed of assembled fragments of past experience preserved and transported, the natural artifact transformed the album into a secular reliquary.[43] Memory bouquets, urns, and wreaths figured significantly in this nostalgic material landscape, the terrain of which was tangibly traversed through textured collages of dried flowers and mosses fashioned in the form of such classical commemorative devices (fig. 60). A cut-paper urn with scroll handles and garland drapery in the Ellen W. album suggests that templates for these motifs were drawn from neoclassical prints and pattern books supplemented by instructions in craft manuals. Both men and women contributed to such memorial arrangements, as revealed in an album wreath made from ferns and flowers gathered at the Valley of Death in the Crimea by Richard William B. Crowther, a member of the Sixty-Third Regiment who took part in the Battle of Inkerman in 1854 (fig. 61).

Embossed with similar designs, the front covers of albums often reflect these material memories contained within, suggesting that the album format was an intrinsic part of its experience.[44] Set in motion through the turning of pages, the album revolved around revelation and tactile engagement, an experience challenged in Ronald King's immobile *Termite Grid* (1998), a "log book" read through the immaterial hollow woodworm tracks on the log's surface.

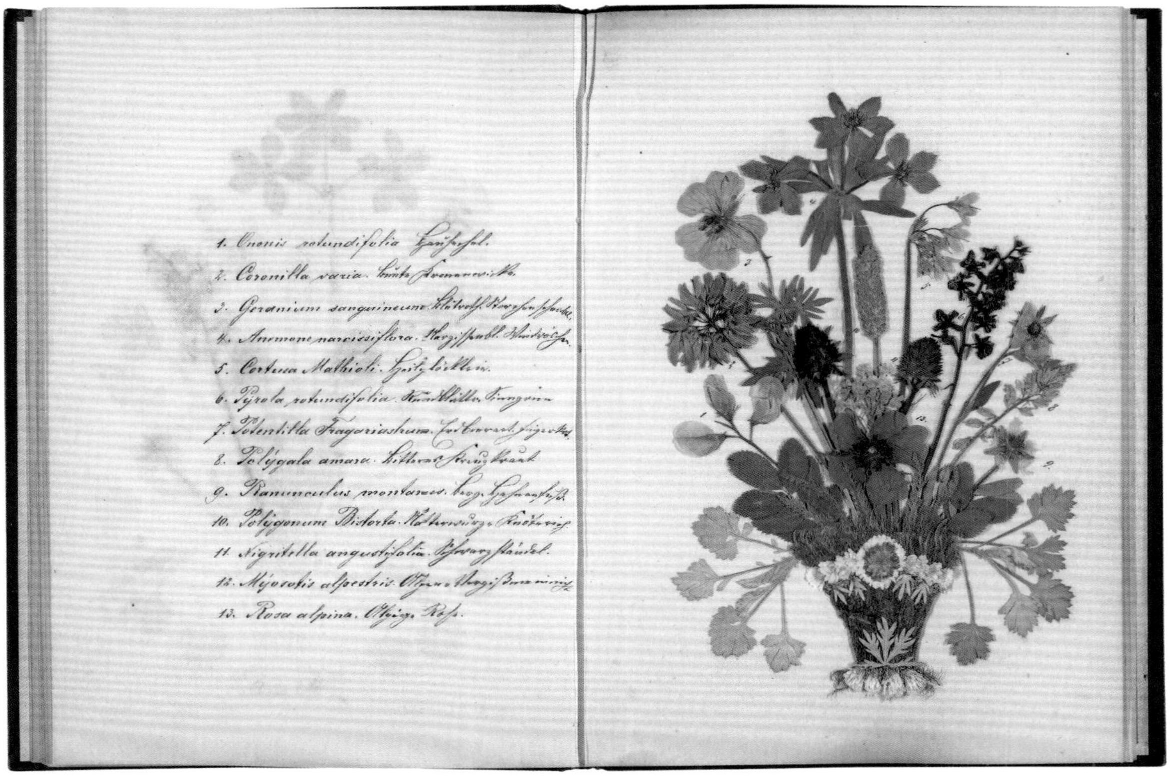

Figure 60. Bouquet of floral specimens from *Flora, an album of pressed floral arrangements*, ca. 1860, natural specimens with pen and ink. Yale Center for British Art, Paul Mellon Fund

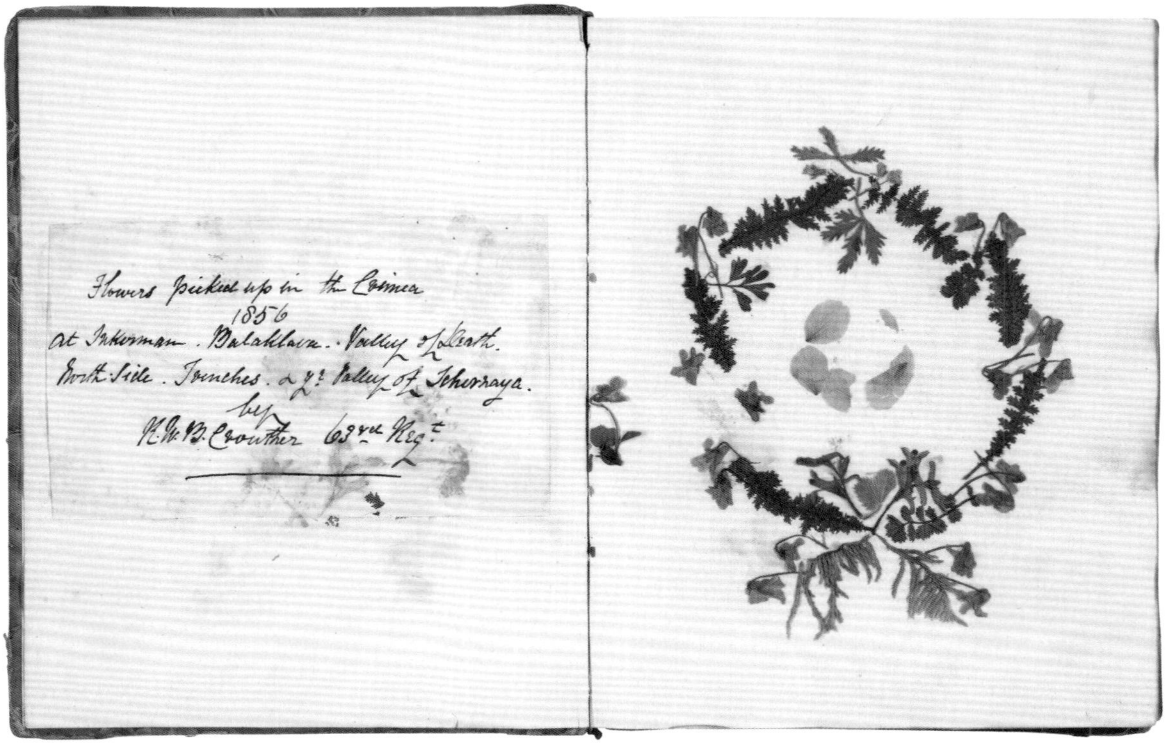

Figure 61. Wreath of floral specimens from *Album of Dried Flower and Seaweed Specimens*, 1856–1863, natural specimens with pen and ink. Yale Center for British Art, Paul Mellon Fund

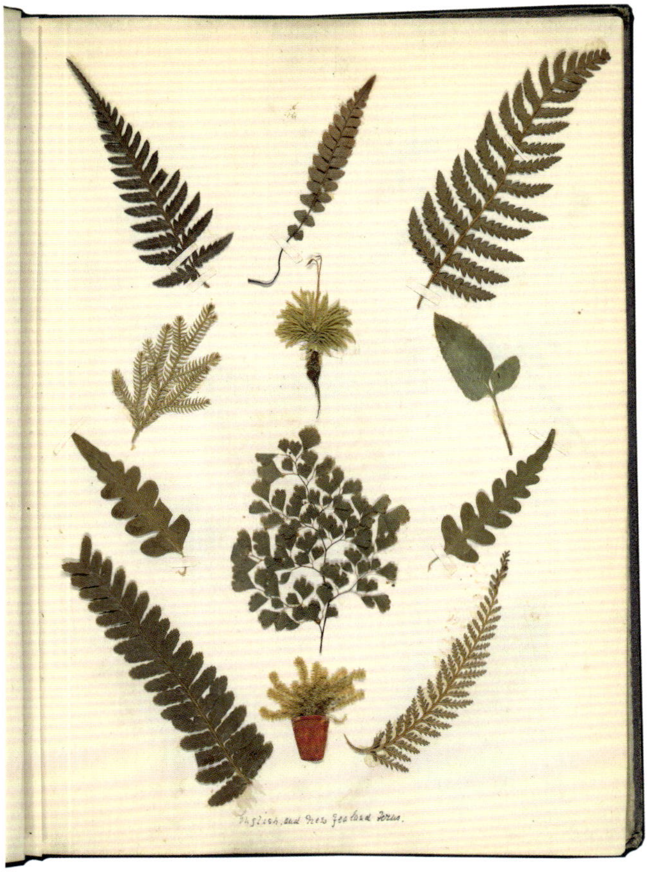

Figure 62. Ann Paley, "English, and New Zealand Ferns" from *Fern and Flower Album*, 1873–1878, natural specimens with pen and ink labels. Yale Center for British Art, Paul Mellon Fund

Bridging the gap between object and self, the physical communion of touch embodied the Victorian viewer.[45] The tangible nature of such album arrangements, in terms of both production and consumption, provided a medium through which to connect with the artifacts of empire, a familiarizing process enacted in the fern and flower album of Ann Paley (fig. 62).[46]

Incorporating ferns from New Zealand in addition to pressed specimens from her native Yorkshire, Paley's album taps into the prevailing pteridomania, or fern fever, that infected the Victorian imagination, developing into a commercial, aesthetic, and social phenomenon in the second half of the nineteenth century. Described by Philip Henry Gosse in *The Romance of Natural History* (1860) as "twinkling verdure, through whose filigree work the sunbeams are sparkling," ferns were admired for a modesty and gracefulness that offset the profusion of color and pattern that defined the mid-Victorian interior.[47] Embraced in the home by women, who according to the naturalist Charles Kingsley, were "more active, more cheerful, more self-forgetful over it, than they would have been over novels and gossip, crochet and Berlin-wool," the fern motif was applied to furniture, cast ironwork, textiles, ceramics, glass, and souvenirs embellished with pressed specimens, fern transfers, or spatterwork, which often featured foreign species.[48] Ferns from Australasia were especially valued on the international market for their size, abundance, botanical diversity, and connection to an exoticized primeval antipodes. In Paley's album, the ornamental combination of domestic and colonial specimens translates scientific colonization into an exploratory parlor pastime.

Taking on added value over spatial in addition to temporal distances, the album of specimens wrought into natural artifacts represented, according to *Beeton's Book of Home Pets* (ca. 1867), the ultimately compelling souvenir:

> How many memories of the vanished past would such a collection serve to conjure up! . . . And if those memories should be tinged with a sweet sadness, when through their magic power the eye can almost see some once familiar scene that we may never again re-visit, and the ear almost hear the welcome accents of some well-known voices whose tones are hushed forever, they would be nonetheless pleasant; but the book, that had spoken to the heart of things that were, would become a "Home Pet" that we should prize more highly and treasure more closely, after each communion with the witnessed fronds that lie between its pages.[49]

This book, and the exhibition that it accompanies, opens a portal to "the heart of things that were," reinvigorating the intimate, palpable, and wondrous communion with nature that was crafted in the Victorian book and album amid a backdrop of an industrial culture catapulting into modernity.

1. *The Hand-Book of Useful and Ornamental Amusements and Accomplishments* (London: Smith, Elder, and Co., 1845), 1.
2. Constance Classen, "Feminine Tactics: Crafting an Alternative Aesthetics in the Eighteenth and Nineteenth Centuries," in *The Book of Touch,* ed. Constance Classen (Oxford and New York: Berg, 2005), 229; Talia Schaffer, *Novel Craft: Victorian Domestic Handicraft and Nineteenth-Century Fiction* (Oxford and New York: Oxford Univ. Press, 2011), 33–36.
3. Shirley Hibberd, *Rustic Adornments for Homes of Taste, and Recreations for Town Folk in the Study and Imitation of Nature*, 2nd ed. (London: Groombridge and Sons, 1857), 2.
4. Celeste Olalquiaga, *The Artificial Kingdom: A Treasury of the Kitsch Experience* (New York: Bloomsbury, 1998).
5. Louisa Anne Twamley, *The Romance of Nature, or, The Flower Seasons Illustrated* (London: Charles Tilt, 1836), vii.
6. James Sowerby, *A Botanical Drawing-Book, or, An Easy Introduction to Drawing Flowers According to Nature*, 2nd ed. (London: Richard Taylor & Co., ca. 1807), title page.
7. Ann B. Shteir, "'Fac-similes of Nature': Victorian Wax Flower Modelling," *Victorian Literature and Culture* 35 (2007): 660; Nancy Dunlap Bercaw, "Solid Objects/Mutable Meanings: Fancywork and the Construction of Bourgeois Culture, 1840–1880," *Winterthur Portfolio* 26, no. 4 (Winter 1991): 246.
8. Samantha George, *Botany, Sexuality and Women's Writing: 1760–1830* (Manchester and New York: Manchester Univ. Press, 2007), 31.
9. Mark Laird and Alicia Weisberg-Roberts, eds., *Mrs. Delany and Her Circle* (New Haven and London: Yale Univ. Press, 2009), 35.
10. Schaffer, *Novel Craft*, 69.
11. Ann Bermingham, *Learning to Draw: Studies in the Cultural History of a Polite and Useful Art* (New Haven and London: Yale Univ. Press, 2000).
12. As Talia Schaffer has suggested, it aligned the Victorian craftsperson with the factory worker, who was valued for her reliable, economical, and efficient production. See Schaffer, *Novel Craft*, 42–43.
13. Schaffer, *Novel Craft*, 8.
14. *Artistic Amusements: Being Instructions for a Variety of Art Work for Home Employment, and Suggestions for a Number of Novel and Saleable Articles for Fancy Bazaars* (London: "The Bazaar" Office, 170, Strand, W.C., ca. 1879), 14.
15. *Elegant Arts for Ladies* (London: Ward and Lock, 1861), 82.
16. In addition to featherwork, these shimmering flying jewels were incorporated through handicraft into taxidermic earrings and brooches in which their preserved carapaces were set in gold and silver.
17. Lynn L. Merrill, *The Romance of Victorian Natural History* (Oxford and New York: Oxford Univ. Press, 1989), 51.
18. Anne Pratt, *Wild Flowers* (London: Society for Promoting Christian Knowledge, 1852–1853), 20.
19. Carol Armstrong and Catherine de Zegher, eds., *Ocean Flowers: Impressions from Nature* (Princeton: Princeton Univ. Press, 2004).
20. Anne Secord, "Botany on a Plate: Pleasure and the Power of Pictures in Promoting Early Nineteenth-Century Scientific Knowledge," *ISIS* 93 (2002): 29.
21. Gil Saunders, *Picturing Plants: An Analytical History of Botanical Illustration* (Berkeley and Los Angeles: Univ. of California Press, 1995), 106.
22. Saunders, *Picturing Plants*, 102.
23. Joseph Dickinson, *The Flora of Liverpool* (Liverpool: Deighton and Laughton, 1851).
24. Shirley Hibberd, *The Seaweed Collector, A Handy Guide to the Marine Botanist* (London: Groombridge & Sons, 1872), 4.
25. Twamley, *The Romance of Nature*, 157.
26. Robert Tyas, *The Language of Flowers; or, Floral Emblems of Thoughts, Feelings, and Sentiments* (London: G. Routledge and Sons, 1869), 35.
27. Leigh Eric Schmidt, "The Fashioning of a Modern Holiday: St. Valentine's Day, 1840–1870," *Winterthur Portfolio* 28, no. 4 (Winter 1993): 229.
28. Andrea DiNoto and David Winter, *The Pressed Plant: The Art of Botanical Specimens, Nature Prints, and Sun Pictures* (New York: Stewart, Tabori & Chang, 1999), 20.
29. David Landsborough, *A Popular History of British Sea-Weeds*, 2nd ed. (London: Reeve and Benham, 1851), poem opposite preface and 173.
30. Carol Armstrong, "Cameraless," in Armstrong and de Zegher, *Ocean Flowers*, 157.
31. Kathryn Tuma, "The Victorian 'Book of Nature': Modeling, Mimesis, and the Community of Forms," in Armstrong and de Zegher, *Ocean Flowers*, 221.
32. Isabella Gifford, *The Marine Botanist: An Introduction to the Study of Algology*, 3rd ed. (London: Darton and Co., 1853), 164, 248, 289, 338; Samuel Joseph Mackie, "Sea-Weeds as Objects of Design," in *Art-Studies from Nature, as Applied to Design*, ed. Frederick Edward Hulme (London: Virtue & Co., 1872), 94–95.
33. Hibberd, *The Seaweed Collector*, 19; Robyn Stacey and Ashley Hay, *Herbarium* (Cambridge: Cambridge Univ. Press, 2004), 143. Iodine was employed in the production of daguerreotypes and calotypes, while agar extracted from boiling seaweeds was combined with paint to create the marbleized endpapers of books.
34. William Grosart Johnstone, *The Nature-Printed British Seaweeds* (London: Bradbury and Evans, 1859–1860), 246.
35. Thomas Davies, *The Preparation and Mounting of Microscopic Objects* (London: Robert Hardwicke, 1873), 32.
36. Mackie, "Sea-Weeds," in Hulme, *Art-Studies from Nature*, 94–5.
37. Susan Pearce, *On Collecting: An Investigation into Collecting in the European Tradition* (London and New York: Routledge, 1995), 14; Schaffer, *Novel Craft*, 14.
38. Thad Logan, *The Victorian Parlour* (Cambridge: Cambridge Univ. Press, 2001), 135.
39. Schaffer, *Novel Craft*, 14.
40. *Leaf and Flower Pictures and How to Make Them*, 2nd ed. (New York: A. D. F. Randolph, 1860), 44–45; Mary Howard, *Ocean Flowers and Their Teachings, by the Author of Wild Flowers and Their Teachings* (Bath: Binns & Goodwin, 1846), 41.
41. "Oh! Call us not weeds, but flowers of the sea,/ For lovely, and gay, and bright-tinted are we! . . ."
42. Henry Day, *Art Recreations* (Boston: J.E. Tilton and Company, 1863), 292; Schaffer, *Novel Craft*, 8.
43. Elizabeth Edwards, "Photographs as Objects of Memory," in *Material Memories*, eds. Marius Kwint, Christopher Breward, and Jeremy Aynsley (Oxford: Berg, 1999), 228.
44. Elizabeth Edwards, "Grasping the Image: How Photographs are Handled," in Classen, *The Book of Touch*, 421–25.
45. Geoffrey Batchen, *Forget Me Not: Photography and Remembrance* (Amsterdam: Van Gogh Museum, 2004), 49; Susan Stewart, "From the Museum of Touch," in Kwint, Breward, and Aynsley, *Material Memories*, 35.
46. Ariane Fennetaux, "Female Crafts: Women and *Bricolage* in Late Georgian Britain, 1750–1820," in *Women and Things, 1750–1950*, eds. Maureen Daly Goggin and Beth Fowkes Tobin (Burlington, VT, and Farnham, England: Ashgate, 2009), 104.
47. Philip Henry Gosse, *The Romance of Natural History*, 2nd ed. (London: James Nisbet & Co., 1861), 339; Sarah Whittingham, *Fern Fever: The Story of Pteridomania* (London: Frances Lincoln, 2012), 33.
48. Charles Kingsley, *Glaucus; or, The Wonders of the Shore* (London: Macmillan and Company, 1855), 4; Whittingham, *Fern Fever*, 71–72.
49. *Beeton's Book of Home Pets* (London: Ward, Lock and Tyler, ca. 1867), 832.

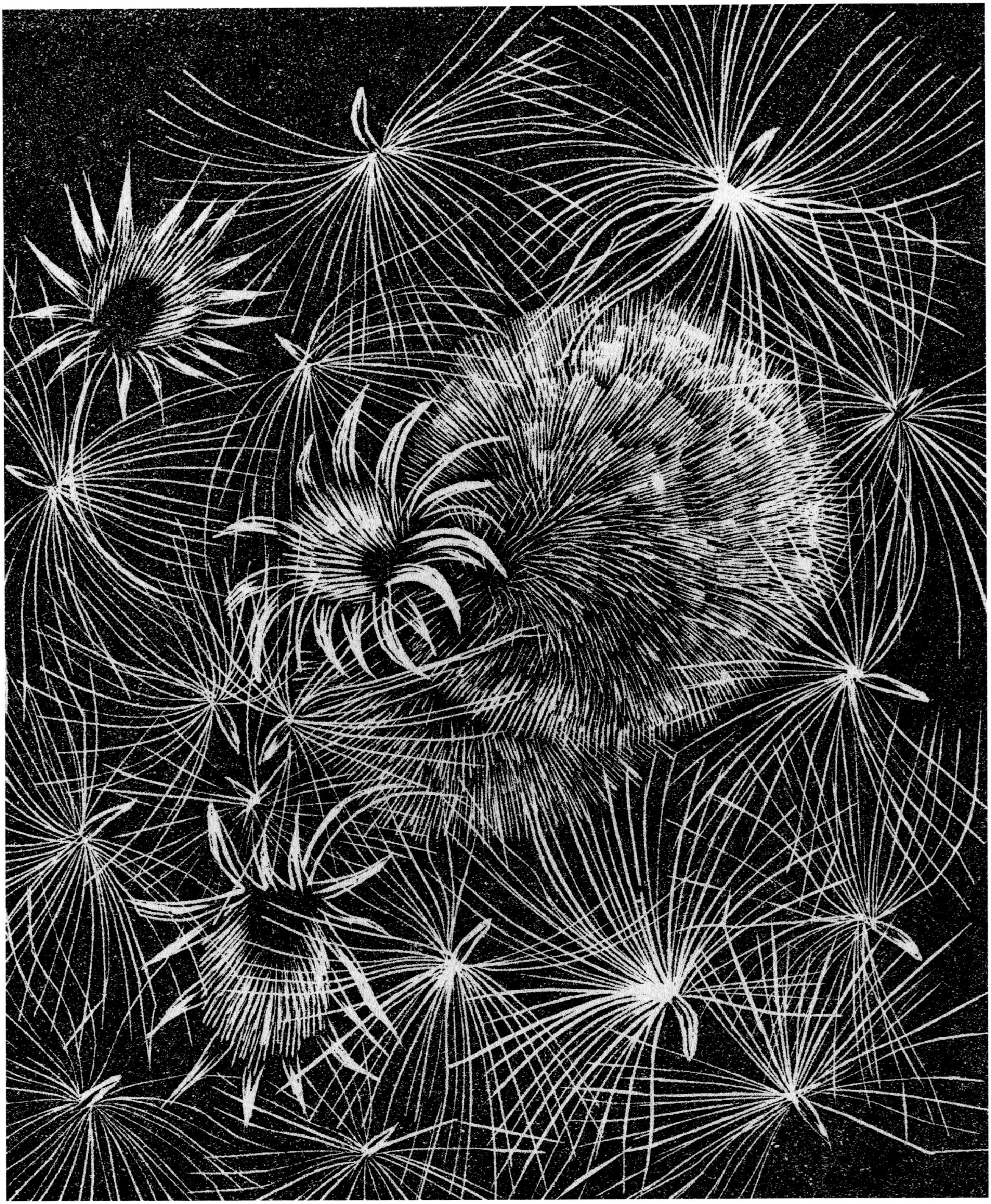

"A still, small voice"
Sister Margaret Tournour, Wood Engraver and Naturalist

David Burnett

Many collectors do not or cannot meet the artists whose work they collect. However, as a collector of contemporary British wood engravings, I have been fortunate over several decades to correspond with and often meet artists in this medium as well as to collaborate with a dozen in publications that have united text and image (fig. 63).[1] In addition, I have been glad to be friends with several. Such friendships are not only valuable in themselves but also, through conversation and more particular study, foster a better understanding of the images that these artists create.

Sister Margaret Tournour (1921–2003) was both a nun and an artist. In her later life, during her retirement from teaching for her order, her primary religious vocation sustained and informed her second calling as an artist. There were two streams but a single source.

A Londoner, she studied at Croydon School of Art from 1938 to 1940 and subsequently worked as a freelance illustrator of children's books, principally for Oxford University Press. She was also a painter in oil and watercolor. Like two other notable wood engravers, Eric Gill and David Jones, her near-contemporaries, she was a Catholic convert. In 1951, on her thirtieth birthday, she entered the Society of the Sacred Heart and devoted herself to God. At that time this order was enclosed, and much of her day, apart from teaching, was spent in prayer and silence. She undertook various duties and was also a class mistress in schools of her order, teaching children in all subjects, including the natural world. In her later years, principally in the northeast of England from 1972 to 1998, she was permitted to engrave again, and from the midseventies until her death she cut some five hundred blocks. In ascending the gyre, or winding stair, of life she returned to the art of her youth, though at a higher level both as an artist and as an individual. It was a remarkable achievement, particularly as she had never enjoyed good health and in later life was incapacitated and confined to a wheelchair. She also suffered from osteoporosis and had to wear a surgical collar to support her neck.

Retirement was a misnomer: there was to be change and a new direction, but no rest. She remained fully engaged with the life and discipline of her community and with worship and prayer, but she now had leisure to devote herself to art and, in particular, to wood engraving. After a life of service to others there was to be a second harvest. A talent which had been sacrificed to others was not lost and did not atrophy, and latterly bore much fruit. Her art was no doubt a consolation in infirmity, but it was also a spirited and positive response to it, a pearl seeded and ripened

OPPOSITE: Sister Margaret Tournour, *Thistledown* (enlarged detail), 1985, wood engraving. Collection of David Burnett

David Burnett is a poet and collector of books and wood engravings living in Edinburgh, Scotland.

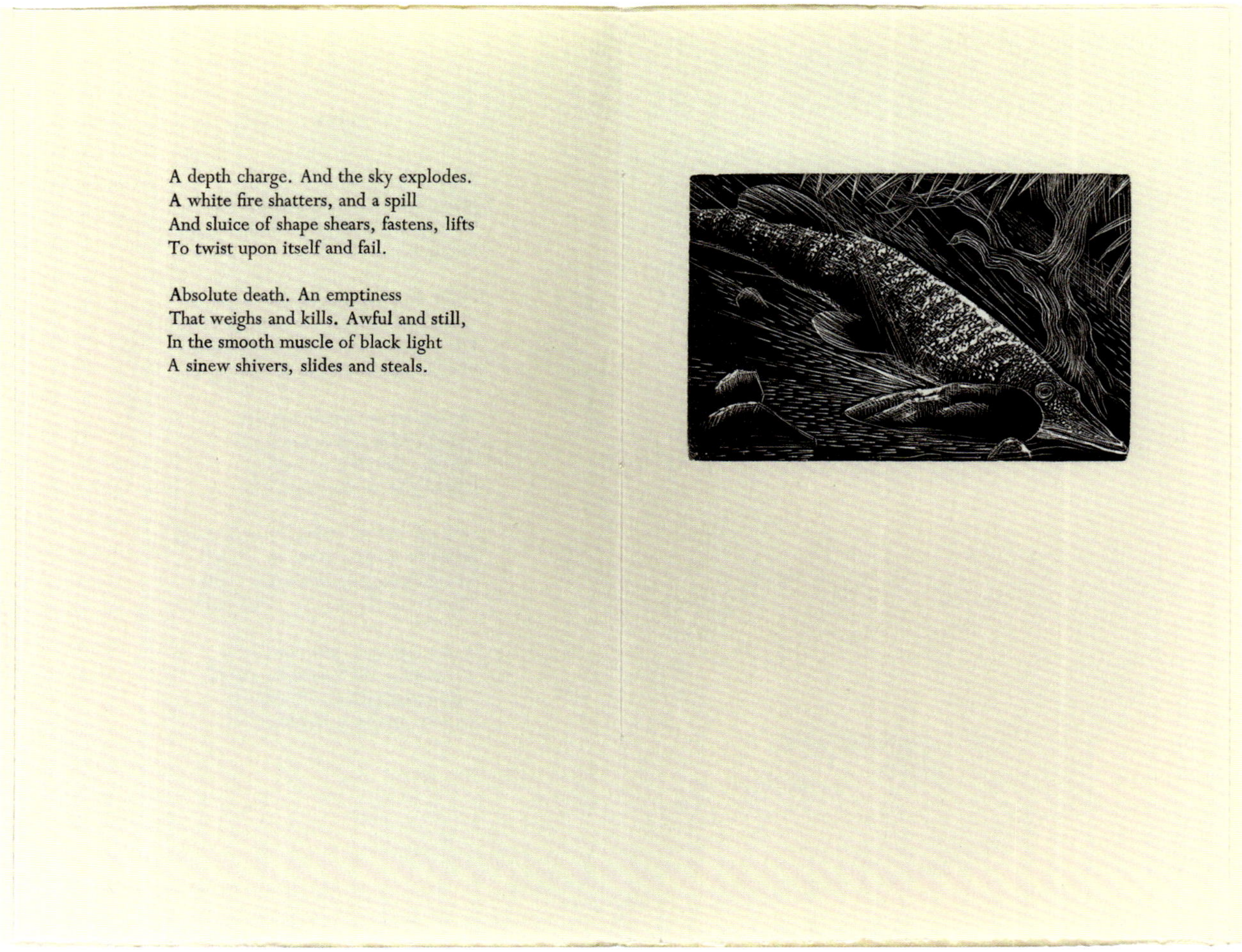

A depth charge. And the sky explodes.
A white fire shatters, and a spill
And sluice of shape shears, fastens, lifts
To twist upon itself and fail.

Absolute death. An emptiness
That weighs and kills. Awful and still,
In the smooth muscle of black light
A sinew shivers, slides and steals.

Figure 63. David Burnett, "Pike" (conclusion), with wood engraving by Sister Margaret Tournour, Durham, England, 1997. Collection of David Burnett

by its grit. The nun and the artist were rooted in a single life of faith and prayer, and it was this common tree that nourished and sustained each branch. Her achievement as a wood engraver in her later years was enriched and informed by her vocation as a nun. To understand and to value her one needs to know both.

Her early work, in so far as it survives, is competent, but relatively slight and insubstantial. Significantly, much of this is in pen and ink or scraperboard: her forte was line and not color. She was to accept and work with this in wood engraving, even though in retirement she still occasionally worked in oil, watercolor, and pastel. Her maternal grandfather was believed to have been an engraver, but she herself was effectually self-taught, like many other wood engravers. She studied for a day with John Farleigh, a pupil of Noel Rooke, at the Central School of Art in London, but she did not consider that this tuition had been very helpful. However, as Clare Leighton has noted, this is an art in which there is least to teach and most to learn. As a nun, Sister Margaret was vowed to poverty, and in retirement she did not have a studio, so the economies of this medium were also advantageous. In later life she worked in both the morning and the afternoon in her room.

Her tools, like those of most British wood engravers, were supplied by Stanley Lawrence of London; and she engraved not only upon expensive boxwood but also upon lemon, lime, pear, and sycamore. For economy she occasionally would engrave upon each side of a block, and she also

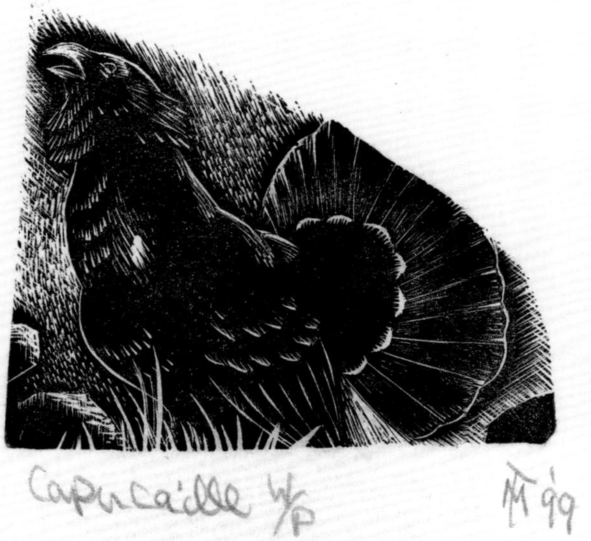

Figure 64. Sister Margaret Tournour, *Capercaillie*, 1999, wood engraving. Collection of David Burnett

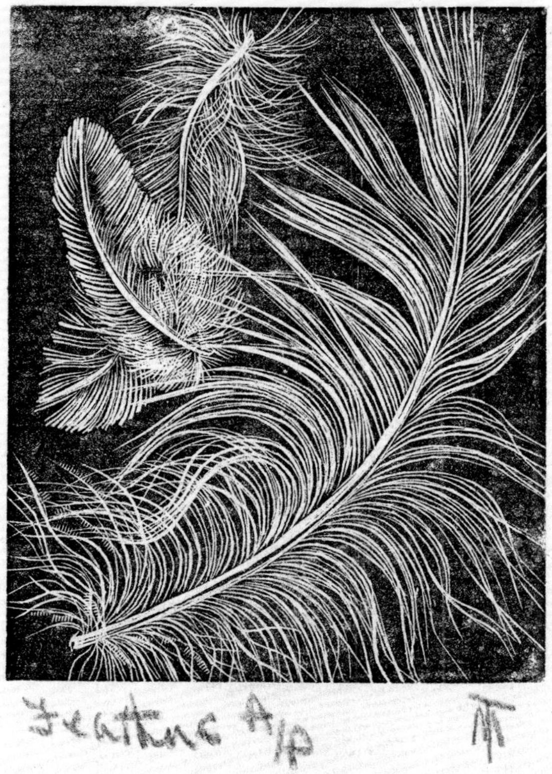

Figure 65. Sister Margaret Tournour, *Feathers*, ca. 1995, wood engraving. Collection of David Burnett

made use of misshaped trial blocks (fig. 64). Unusually, but of necessity, she worked with the sandbag and block upon her lap and with the further aid of a lens slung from her neck and of an angled lamp upon a table in poor light. She was ambidextrous, but engraved with her right hand and wrote with her left. She made few preparatory drawings, preferring to work occasionally from a light pencil sketch on the wood and then from progressive prints pulled from it as the image evolved. None of her blocks was editioned by her since she burnished prints from each with a silver spoon, so impressions of most of her blocks are rare. Such impressions realize fully the fine incisions and subtle tones of her work. The delicate lineaments and textures of fur, feathers, and foliage, as well as each nuance of light, whisper upon the wood (fig. 65). Each impression, moreover, is individual and hence differs from the uniform prints taken at a press: each offers a unique interpretation of the image. A significant number of her blocks, however, were also printed with text in limited editions, often to a high standard despite the problems of working blocks in the press and of rendering exactly her exquisite and ethereal incisions.

As an artist she made a virtue of her infirmity. She was unable with ease to cut deeply and extensively in the wood, but the frailty and almost miraculous delicacy of her lines, which became more marked in her later years, are congruent to and necessary for her chosen effects. Much of the block is not gouged with the scorper, but the consequent and characteristically dark palette of her images reinforces by

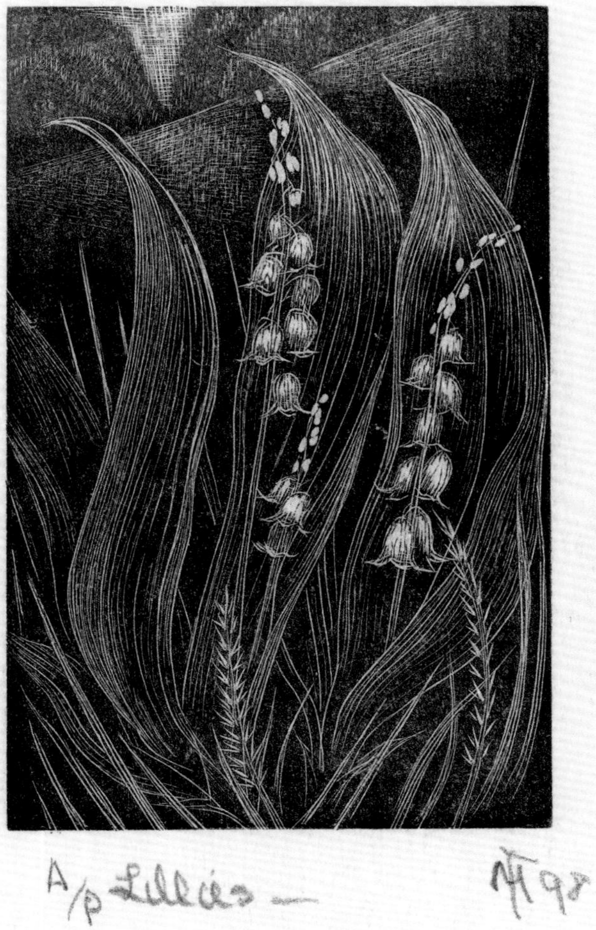

Figure 66. Sister Margaret Tournour, *Lilies*, 1998, wood engraving. Collection of David Burnett

contrast the intricate and elfin lights of the incisions (fig. 66). Similarly, her confinement did not constrain but on the contrary focused and sharpened her observation of the natural world, which was also informed by lifelong study and teaching. She noticed how much we can observe about us every day, often phenomena which might at first sight appear slight or insignificant. Her work rewards close attention.

As with many other earlier and contemporary wood engravers, she also engaged with the printed word. A significant, though not a very large, proportion of her work was also printed with text. As a young woman she had illustrated books and had in some cases also designed their *mise-en-page.* She did, therefore, welcome and respond to the marriage of image with text, which the wood engraving favors and letterpress facilitates, so in later life she was happy that her work was associated with printed texts, in particular passages from the Bible chosen by her as well as texts created independently of her images. In the last decade of her life some twenty books, pamphlets, and keepsakes were issued with illustrations by her.

Just as Thomas Bewick had illustrated and written about quadrupeds and birds two centuries before, so she chose to illustrate beasts, birds, fishes, and plants from the Bible with appropriate biblical texts (fig. 67). Each engraving alludes and suggests but does not comment, so that, as with a parable, the reader must actively engage with each in order to grasp its meaning independently and thereby also more fully. Her purpose is not to close and settle but to open and illuminate. In addition to such biblical selections in several books, she also provided images for other texts and poems, which were issued as small collections, keepsakes, and Christmas cards. Such collaborations exemplify the symbiosis and synergy of image with text and need to be understood in this context: one must see to read but also read to see.

For her, as for Bewick, the natural world is central, though her close observation is rooted in the spiritual, not the secular. She is, indeed, a Saint Francis of the graver who bears witness to and celebrates God's world. For her to see is also to pray since here, too, is God's anchor-hold. Each image is not a copy but an equivalence and a transcendence. Her visionary engagement with the world is, moreover, sharpened by acute and meditated perception and realized by excellent draftsmanship, for the hand also sees. With

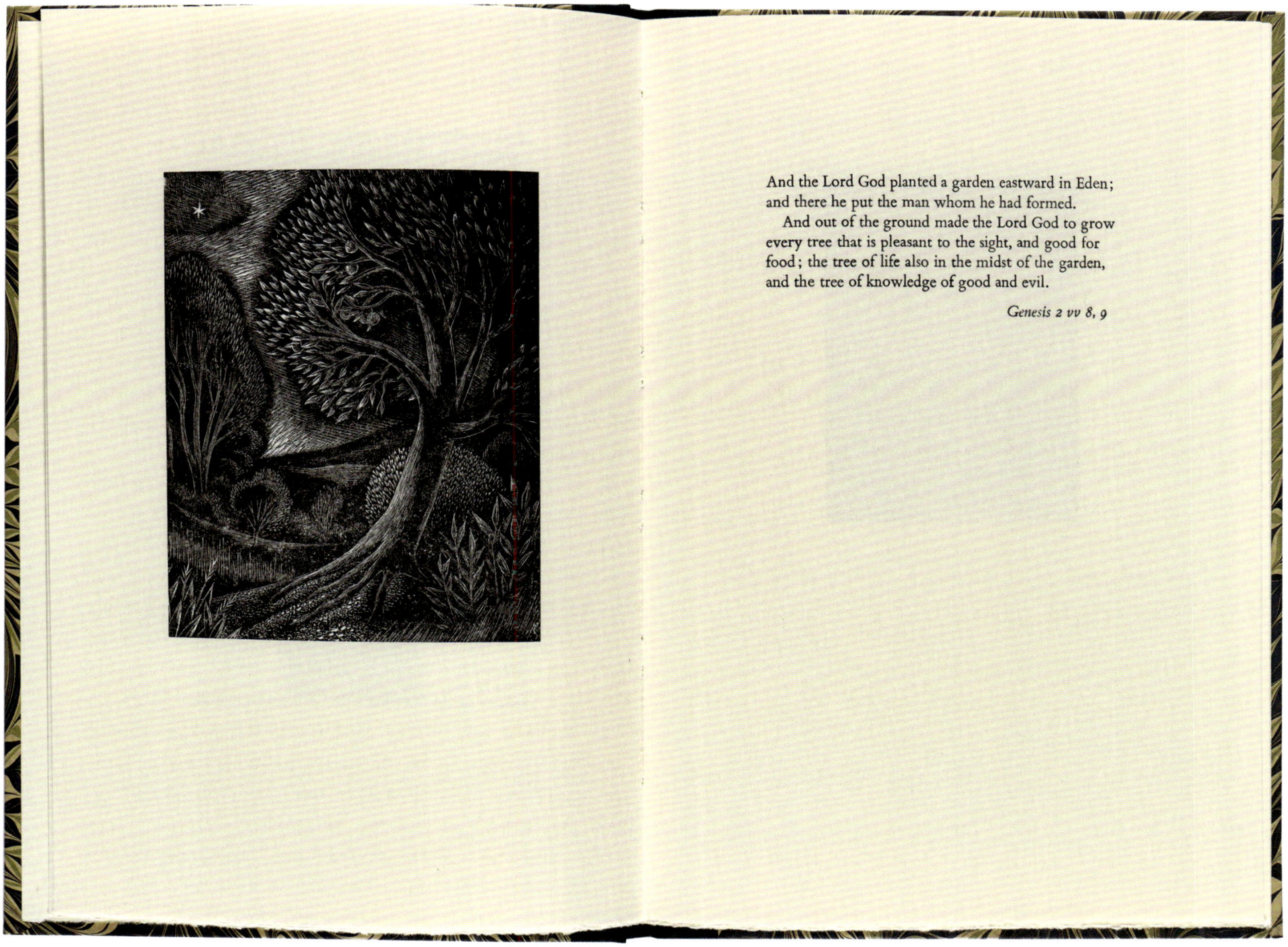

And the Lord God planted a garden eastward in Eden; and there he put the man whom he had formed.

And out of the ground made the Lord God to grow every tree that is pleasant to the sight, and good for food; the tree of life also in the midst of the garden, and the tree of knowledge of good and evil.

Genesis 2 vv 8, 9

Figure 67. Sister Margaret Tournour, "Paradise Tree" from *Bible Trees*, Durham: Black Cygnet Press, 1997, wood engraving. Collection of David Burnett

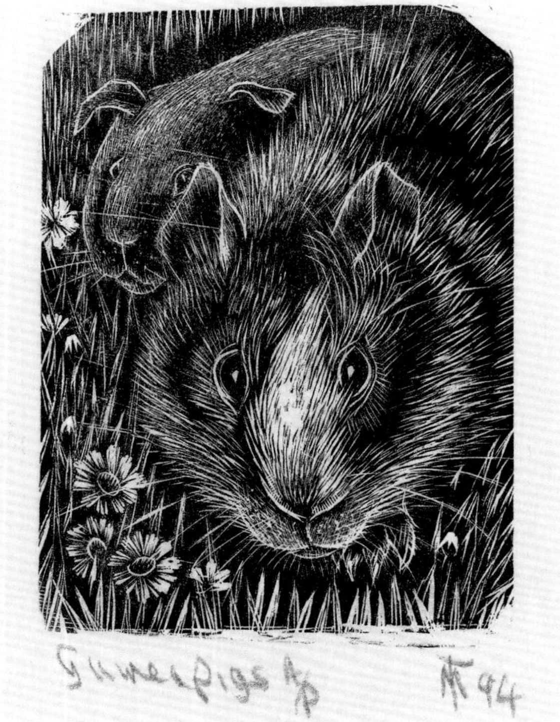

Figure 68. Sister Margaret Tournour, *Guinea Pigs*, 1994, wood engraving. Collection of David Burnett

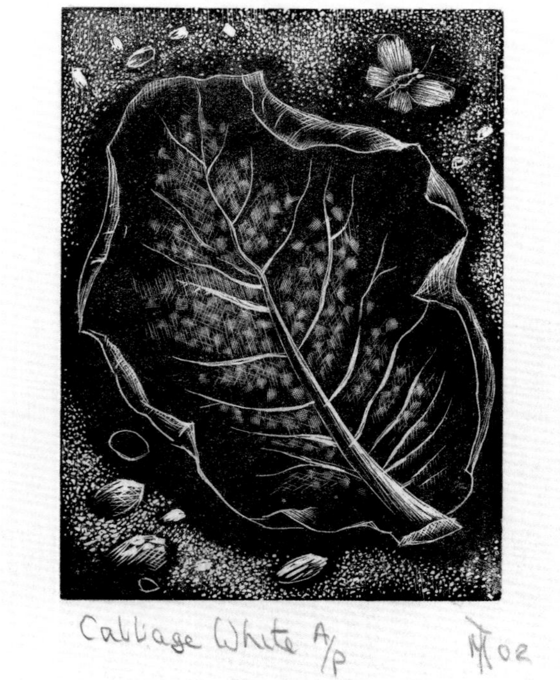

Figure 69. Sister Margaret Tournour, *Cabbage White*, 2002, wood engraving. Collection of David Burnett

affection there is, too, a deep empathy, the "negative capability" of John Keats, with its insight and inwardness. In the alert gaze of a guinea pig or a mouse, the outstretched paw of a resting cat, or the strut and panoply of a capercaillie, the character of each creature is inwardly perceived and surely realized, both affectionately and also occasionally with her puckish humor (fig. 68). Each is instinct with life.

Much of her work is explicitly inspired, informed, and reinforced by consonant passages from the Bible and should always be understood in this context. Lilies, for example, can and should be cherished for their intrinsic beauty, but in admiring them we should also ponder Solomon and the lilies of the field as well as their association with the Virgin. Similarly, a passion flower recalls the agony of Christ. In this respect her work recalls the woodcut prints of religious images prepared for pilgrims and the laity by nuns from the Low Countries in the fifteenth century. Hers, too, is a devotional art. Her reverence for and delight in nature also recall, more particularly, the early Celtic Church and, above all, Saint Cuthbert, who lived amongst, and was familiar with, the same Northumbrian landscapes as she herself. The delicate and intricate loveliness as well as the flowering and abundance of the natural world body forth for her God's fullness and blessing and presage the assurance of redemption and salvation which is His purpose for humanity. There are, as Sir Francis Bacon suggests, not only the book of revelation of the Scriptures, but also in the natural world a second book, and in it a further revelation. In her work each image is alight and burns with a splendor not wholly its own. She sees, but also sees beneath and beyond.

Her vision is Adamic, and through her eyes we see the world afresh, new minted from its creation. With Louis MacNeice she cherishes and celebrates "the drunkenness of things being various," the immeasurable variety and complexity of the world.[2] She also emphasizes with Duns Scotus the distinctive and inescapably individual *haecceitas,* the quiddity, of the creation. A hedgehog or a flower or even a leaf is always itself (fig. 69). The general is never sufficient, and it is the particular that matters, as with William Blake or the inscape of Gerard Manley Hopkins.

This is an art that has affinities with that of Blake (fig. 70), Edward Calvert (fig. 71), Samuel Palmer, and early Graham Sutherland, as well as with writers such as

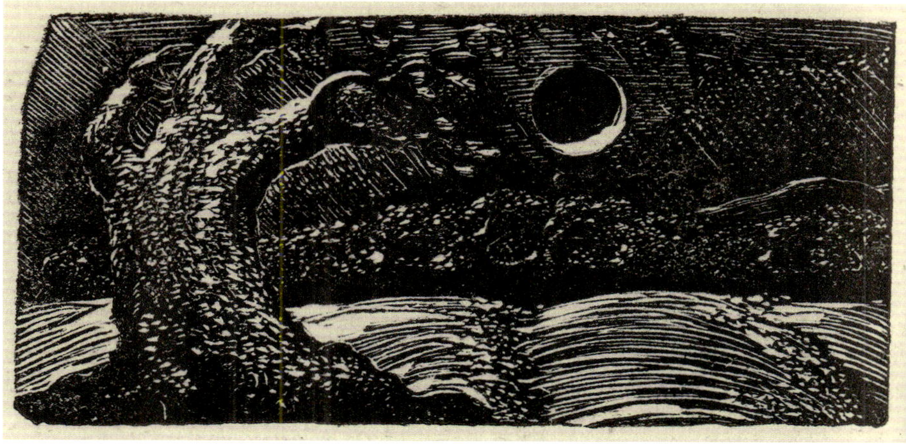

Figure 70. William Blake, "Imitations of Eclogue I," in *The Pastorals of Virgil, with a Course of English Reading, Adapted for Schools* (vol. 1, p. 15), ed. Robert John Thornton, London: F. C. & J. Rivingtons, 1821, wood engraving. Yale Center for British Art, Paul Mellon Collection

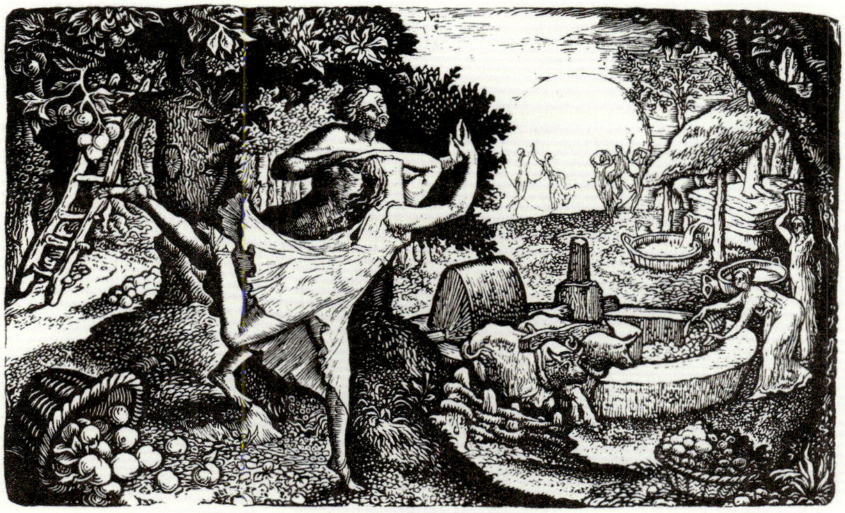

Figure 71. Edward Calvert, *Cyder Feast*, 1828, wood engraving. Collection of David Burnett

William Vaughan and Thomas Traherne. With these artists and writers she shares a pastoral tradition that intuits and affirms a numinous and healing presence in the beauty and benignity of the natural world. Eden with its promise and fulfillment lies about us still if we will look. In her work we can discover, as Palmer noted of the wood engravings of Blake, "visions of little dells, and nooks, and corners of Paradise; models of the exquisitest pitch of intense poetry."[3]

The spiritual reference and significance of the engravings, whether explicit or implicit, should not, however, lessen an appreciation of their aesthetic value. Each is also a work of art which can and should be understood and enjoyed as such. To see and to appreciate wood engraving one has to understand it on its own terms and to see it differently from art in other media.

It is, firstly, an art of miniature such as the lied in music, the epigram and haiku in poetry, and the medallion and netsuke in the plastic arts. It is intentionally small in scale, concentrating the artist's vision. Like the drop of dew in Marvell's poem, an image "Does, in its pure and circling thoughts, express / The greater Heaven in an Heaven less."[4] Economy alone, however, is never enough: the Lord's Prayer inscribed upon a grain of wheat will astonish but not inspire us. A reduction or a digest is insufficient and is not in itself a work of art. Here, however, the particular is explored both for itself and for its relation to a wider world. There is repeatedly in these images, as Emerson desiderates, a "sense of universal relation and power": as with Newton's theory of gravity, the fall of an apple is united to the movement of the earth and of the utmost stars.[5] In Blake's words, you may here "see a World in a grain of sand, / And a Heaven in a wild flower, / Hold Infinity in the palm of your hand, / And Eternity in an hour."[6]

It is, furthermore, an art of line, even though tonal gradation is often exquisitely realized in the delicately incised lines as well as in the burnishing of a print. The work is explicitly figurative, but it is not a paradox to suggest that a representational reading alone is reductive and inadequate. In responding to these images we need to appreciate and enjoy, as in calligraphy, the beauty of line in itself, to cherish its flow and openings, and the dance of line with line. With Pope's spider, she "feels in each thread and lives along the line."[7] As with all good art, it has formal virtues, of composition, of balance between dark and light, of tone and, especially important, line. Lineation is both a means to the transcription of reality and in itself an end, a breathing loveliness that quickens and shimmers in the dark. With Paul Klee she takes the line for a walk. Each line has a life and a sinuous energy of its own, and each can and should be cherished for itself as well as for what it represents (fig. 72).

Wood engraving in the West is also substantially, though by no means wholly, an art of black and white, and not of color: it is generally monochrome, and its palette is one of light and shade alone. The image is realized not only in line but also normally in one color alone. This color—to Picasso the only necessary color—is black owing in large measure to its early ancillary and later more independent association with letterpress and the book. The world itself by contrast is in color. To respond fully to wood engraving, therefore, we have to refuse the rainbow and deny the natural response and full use of our eyes. We see less fully and more narrowly, but also more intently. Just as with line, wood engraving in its emphasis upon black and white reduces and abstracts the manifold colors and tones of the world so as to concentrate our vision. Wood engravers are certainly constrained by the evolution and practice of their medium, but one can also suggest that they are drawn to it by an innate perceptual preference for black and white as well as by a cognitive predisposition to closely focused observation and analysis.

There is, moreover, a deeper significance in this preference for black. A block that has not been engraved prints wholly black. It is, as it were, the devil's visiting card, a negation and an emptiness. In the first chapter of Genesis we are told that the earth was at first without form and void, and that there was a darkness upon the face of the

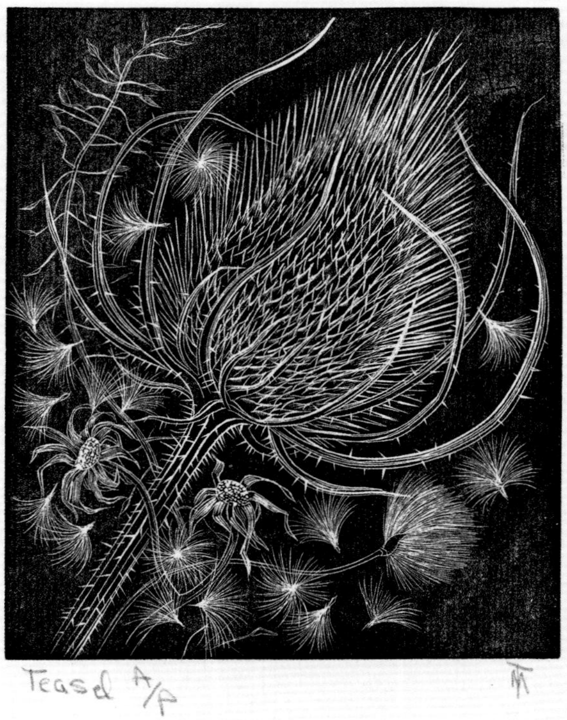

Figure 72. Sister Margaret Tournour, *Teasel*, ca. 1995, wood engraving. Collection of David Burnett

deep. This is the pure, pitch black, formless and void, which the wood engraver transforms by adding light with each incision so as to create form as God did in His creation. The artist, too, sees the light, knows that it is good, and divides it from the aboriginal dark. It is a drawing with light as well as a new creation. The wood engraver works upon a shapeless abyss and its black nothingness to form a world from light (fig. 73). In each image, therefore, we can apprehend and cherish the triumph of form yet also appreciate that this hard won creation is a balance of opposites, sustained by the tension of its contraries as are the springers of an arch. Here also are the fundamental antinomies, light and dark, day and night, summer and winter, life and death, the formed and the formless, Demeter and Hades, the poles of abundance and denial, of Carnival and Lent, between which the engraver works and to one or other of which each image tends. To perceive a wood engraving wholly is to see and enter into the cosmic dance.

The wood engraver Noel Rooke suggests that a drawing can be related to a print as speech to literature. The former is much more common, but it is also typically more provisional, tentative, and unfinished: its virtue is spontaneity, its metal quicksilver. Wood engraving, by contrast, is a meditated art and must be appreciated as such. As with a cameo in obsidian, each line has to be considered by the engraver before it is incised since more effort is required than in drawing and since correction in wood is in practice very difficult to effect. Each art, of course, is needed, and each is complementary to the other. The strength of the drawing is that it can respond swiftly and instinctively to an occasion and its circumstance, whereas the engraving is a final, deeply pondered realization of what is portrayed. Against the impression and its immediate response and improvisation there is by contrast in wood engraving a deeply considered meditation, memorable and profound.

The closely grained boxwood is nearly as hard as metal, and its art at its best is likewise perdurable. The message is more than its medium, but it is nonetheless shaped by it too.

Sister Margaret's images are founded not only upon a lifetime of patient, informed, attentive, and loving observation of the natural world but also upon a secure mastery of a demanding skill, for which persevering application and varied experience are needed as much as talent. Art is

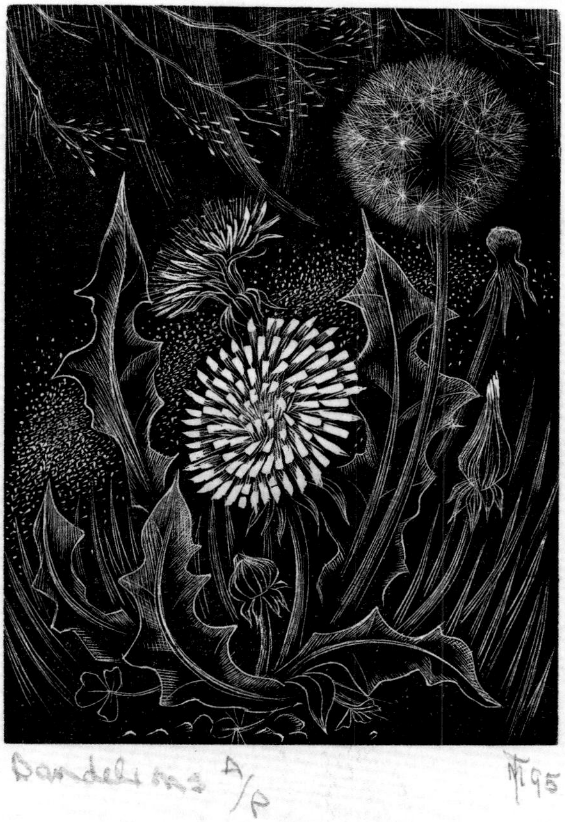

Figure 73. Sister Margaret Tournour, *Dandelions*, 1995, wood engraving. Collection of David Burnett

founded upon craft, labor, and discipline, as well as upon vision. Her vision is fully realized precisely because it is integrated with its technique and medium: the end and its means are at one. The material and the tools are actively present in the creation of each image and are thus, as it were, actors in the design. There is a truth to material and to the tools needed to realize her vision. The wood opens to the blade, and from each wound the light seeds and sows itself to root and flower in the exquisite lines. Each line is witness to her concentration as well as to the gradual evolution of the image through repeated and progressive proofs pulled from the block. Art conceals art, and the technical skill and labor that it demands are not evident, but they should be acknowledged and respected.

The exquisite and scintillating veins of an insect's wing, each strand of a bristling teasel, an airy and coruscating crown of thistledown, are realized in lines as fine as gossa-

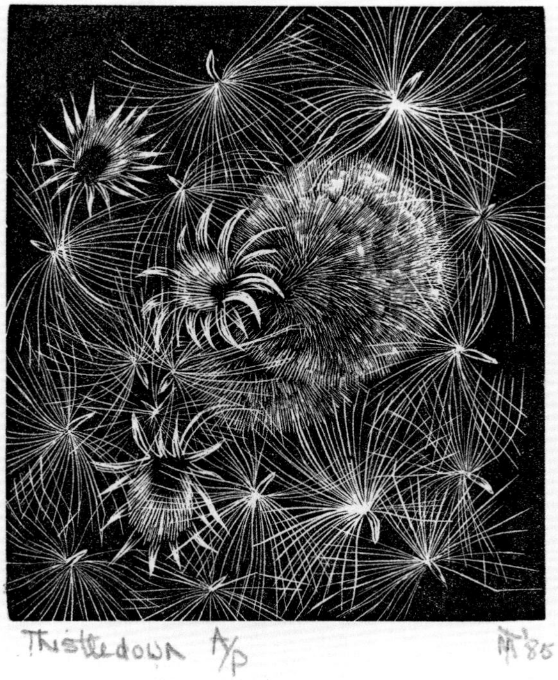

Figure 74. Sister Margaret Tournour, *Thistledown*, 1985, wood engraving. Collection of David Burnett

mer that gleam and float and twist to weave their spell (fig. 74). A tuft of grass, the braided and combed plumes of a feather, the knotted and gnarled grain of bark, the glistening sheen of a seal's skin, or the soft deliquescence of a cat's fur, are all instinct with life. The constellated globe of a dandelion shimmers with a meteor shower of light, a small rain of stars. Such images witness not only to her precise observation but also to a remarkable control of incision: in wood engraving as in acting or singing the *sotto voce* is both necessary and salutary. Hers is a still, small voice, but it commands attention as a trumpet does not.

At the same time, she is able to take advantage of the full range of her chosen instrument. There is a firm command of black as well as of white and, as a consequence, the balance and invigorating opposition of a rich, velvety darkness to pure or more subtle lights, including a shimmering dove grey. In this, likewise, there is vision but also craft.

An art founded upon a demanding craft, figurative, small in scale, rooted in a particular artistic tradition, concerned above all with our natural and not our built heritage, with beauty and not with ugliness, mere novelty or sensation, is an art contrary in many ways to contemporary critical fashion and artistic practice and their assumptions. It is, moreover, not only modest but also profoundly spiritual: in both her life and her art there is a working with light.

Her wood engravings are at their best theophanies and they should, therefore, be appreciated as such, as an altar painting or an icon, a requiem mass or a sacred oratorio, works of art that have both a human and a spiritual dimension. In a post-Christian and increasingly secular age and culture there will be many who do not share this artist's perspective. She should, however, be understood and valued as are other Christian artists such as Cecil Collins, David Jones, Stanley Spencer and, more particularly, contemporary

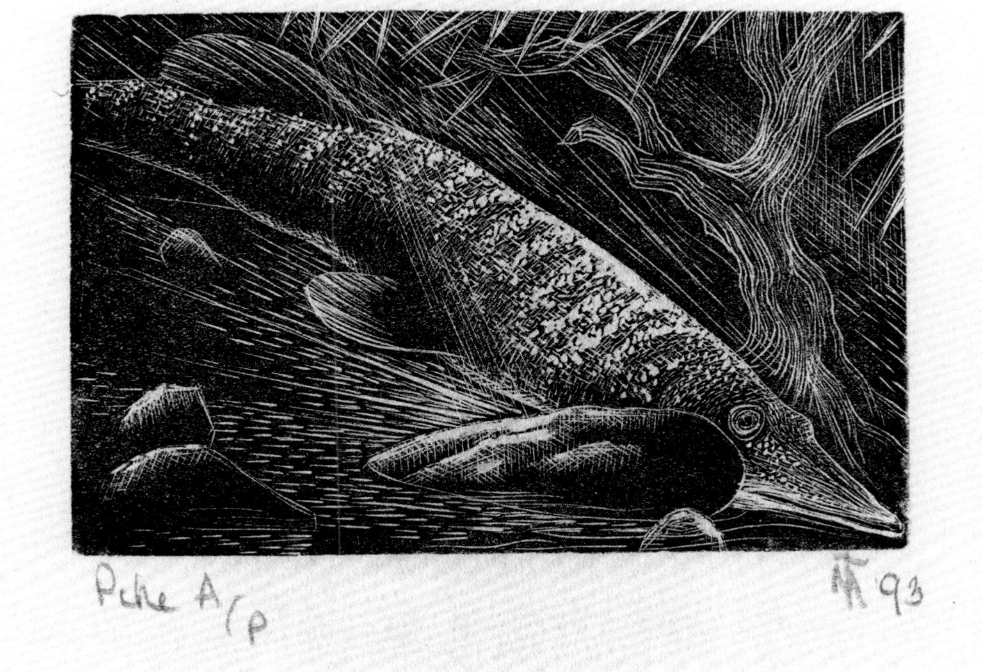

Figure 75. Sister Margaret Tournour, *Pike*, 1993, wood engraving. Collection of David Burnett

and later wood engravers such as Eric Gill and Rosemary Kilbourn. One may also note other similar artists, writers, and composers such as Georges Rouault; T. S. Eliot and Paul Claudel; and Igor Stravinsky and John Tavener. Hers is, no doubt, a still, small voice, but it is one which with these is worth listening to and which will reward attention.

For Robert Frost a poem begins in delight but ends in wisdom. A primary and desirable purpose of art is to give pleasure, but this in itself is insufficient. These wood engravings are a witness to and a celebration in images of the natural world as well as an extended meditation upon the beauty and fecundity of the creation and its creator. The deep truth is undoubtedly, as Percy Bysshe Shelley observes, imageless, but it is through images that we can approach it. One hopes that those who see and take pleasure in these images will wish to reflect and ponder as well as enjoy and to find insight with delight (fig. 75).[8]

1. Please note that the reproductions of Sister Margaret Tournour's wood engravings seen here are approximately life-sized, except where noted.
2. From MacNeice's poem "Snow," originally published in 1935.
3. Alfred Herbert Palmer, ed., *The Life and Letters of Samuel Palmer: Painter and Etcher* (London: Seeley & Co., 1892), 15.
4. Alexander B. Grosart, ed., *The Complete Works in Verse and Prose of Andrew Marvell* (London: Robson and Sons, 1872), 80.
5. From Ralph Waldo Emerson, "XII. Art," in *Essays: First Series*.
6. John Sampson, ed., *The Poetical Works of William Blake: A New and Verbatim Text from the Manuscript Engraved and Letterpress Originals* (Oxford, UK: Clarendon Press, 1905), 288.
7. From Alexander Pope's poem "An Essay on Man," first published in 1733.
8. For further biography on Sister Margaret, see: David Burnett, *Sister Margaret Tournour* (Durham: Black Cygnet Press, 2003); David Burnett, "Sister Margaret Tournour," [obituary] *The Times* (London), May 7, 2003; William Connelly, "Bibliography: Sister Margaret Tournour," *Studies in Illustration* 19 (Winter 2001): 23-32; William Connelly, "Sister Margaret Tournour," [obituary] *The Independent*, March 5, 2003.

Field Guide to the British Countryside

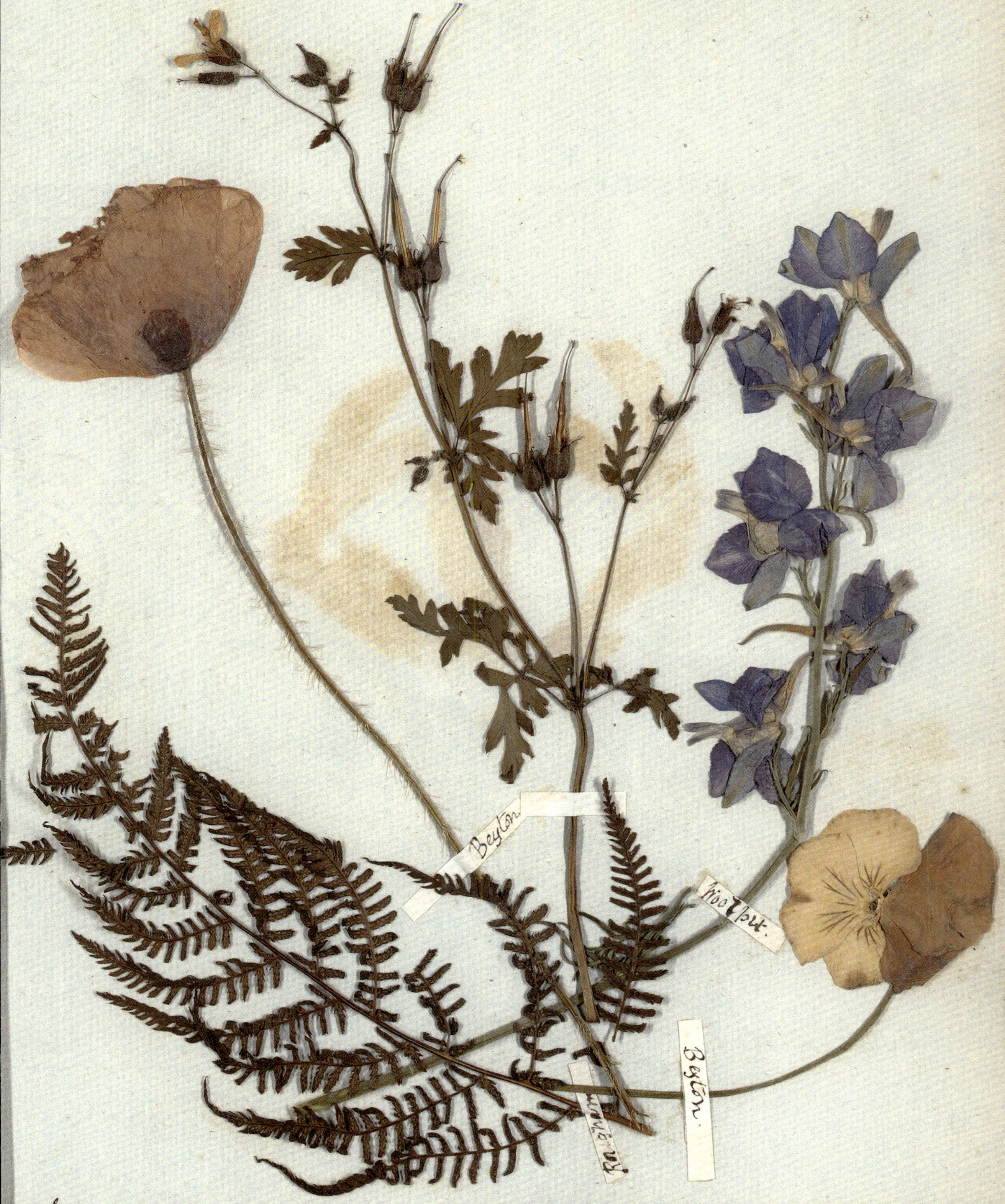

From Beyton, Rougham, Bury St. Edmunds & Woolpit. August & Sep^tr. 1859.

Wildflowers

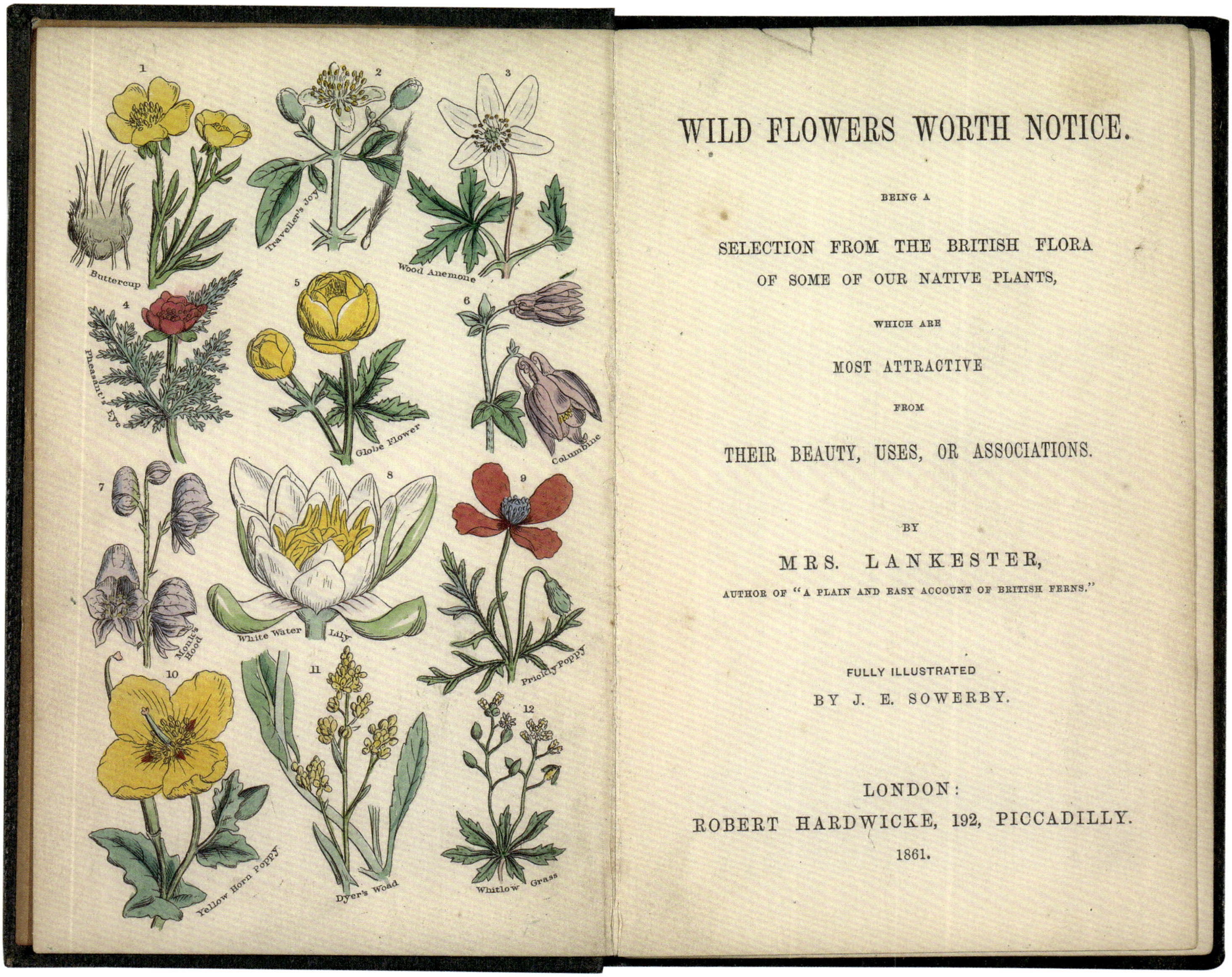

Phebe Lankester, *Wild Flowers Worth Notice*, London: Robert Hardwicke, 1861, hand-colored wood engraving. Yale Center for British Art, Paul Mellon Collection

PAGES 76–77: *Album of Dried Flower and Seaweed Specimens*, 1856–1863, natural specimens with pen and ink. Yale Center for British Art, Paul Mellon Fund

Observe, collect, and preserve the wild flowers you find; arrange and name them scientifically, if possible; but if that be not in your power, still they will always be interesting to you as a pleasing record of "times and places, and old familiar faces," which one day you will value.

Phebe Lankester

Wild Flowers Worth Notice, 1861, p. ix

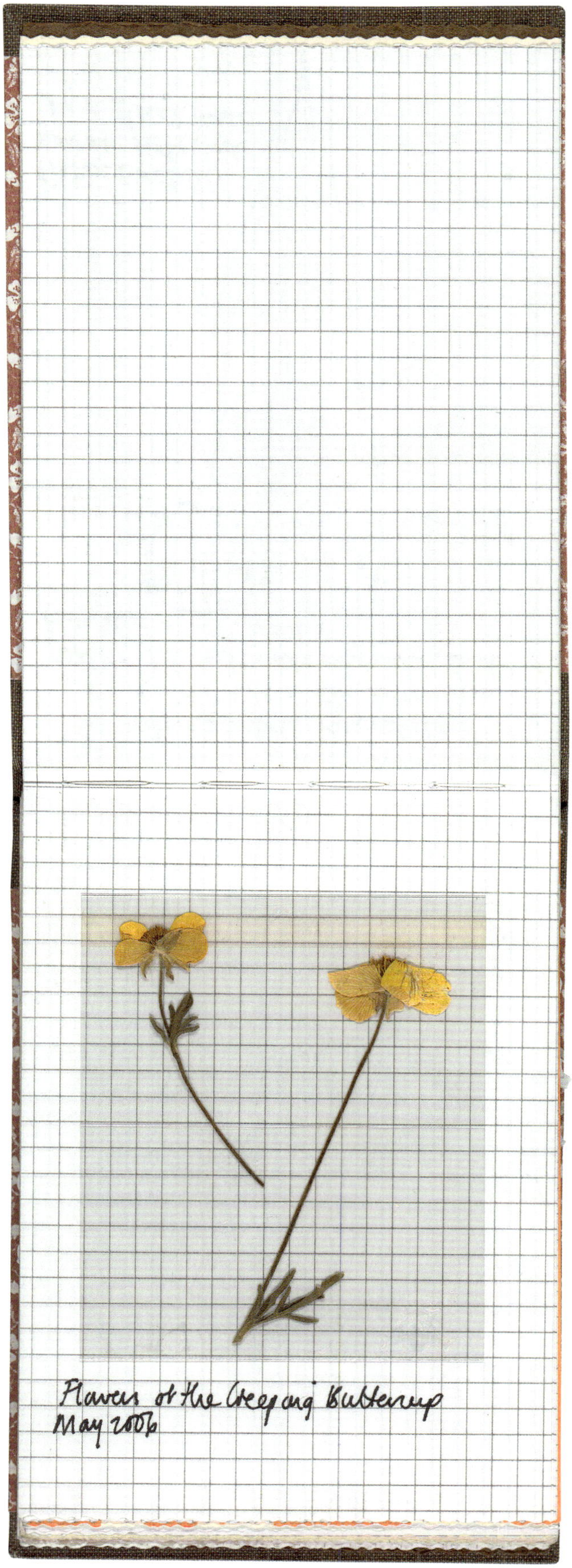

Tracey Bush, *Herbarium Notebook*, 2005, pen and ink with natural specimens and hand-cut paper collages. Collection of the artist

Buttercups and Harebells

The buttercup is one of the most common wild flowers. Everyone remembers fields of yellow in June, and anyone can recognize the flower at a glance. Yet this recognition is strangely generalized. Few people take the time to notice the remarkable individuality of each particular flower.

The drawings in my work *100 Buttercups* were made from plants taken from a field beside a church overlooking the sea in St. Monans, a small fishing village on the east coast of Scotland. The field is a mile from where I live, and the buttercups were gathered at random, with no selection process and no preference for particular flowers, on daily walks.

Pencil crayons were used to make the drawings, and as the series progressed it built an ongoing work of similarity and difference. No two flowers are ever the same, and the hundred drawings stand in for a possible infinity.

The daily act of drawing returned a quality of attention owed to this common but remarkable plant.

Buttercups are a common flower in the east of Scotland. So, too, are harebells. Together these two flowers make landscapes of yellow and blue. I wanted to immerse myself in these colors, the drawings of buttercups and harebells adding up to whole rooms of yellow and blue. This is one way of realizing an old ambition of architecture: to bring the outside indoors. I love the freshness and brightness of these colors together. To me, they make a positive statement of light and hope.

Laurie Clark
ARTIST

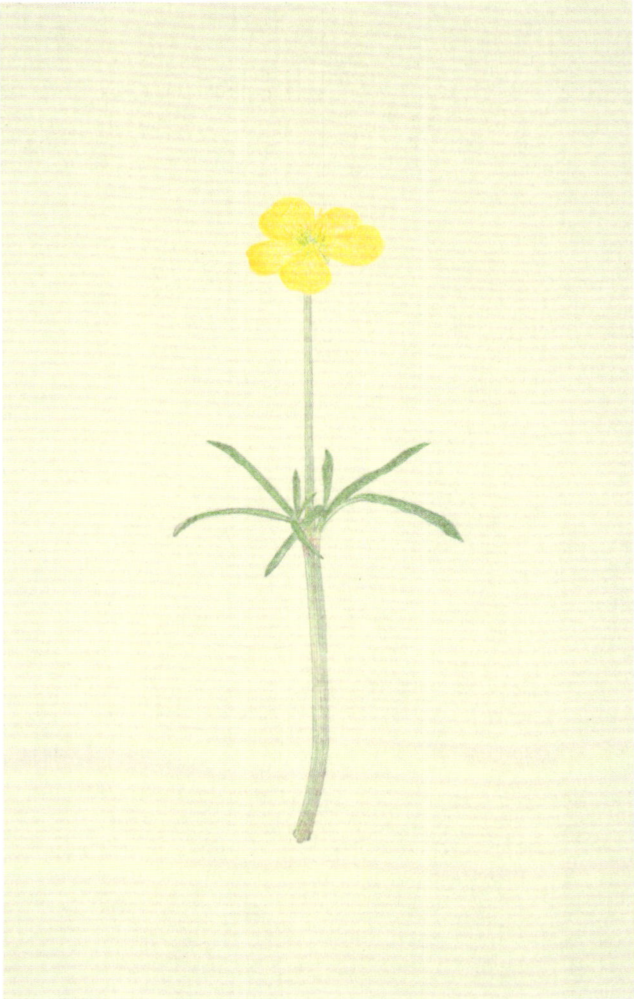

Laurie Clark, *100 Buttercups*, Fife, Scotland: WAX366, 2010, colored pencil drawing. Yale Center for British Art, Friends of British Art Fund

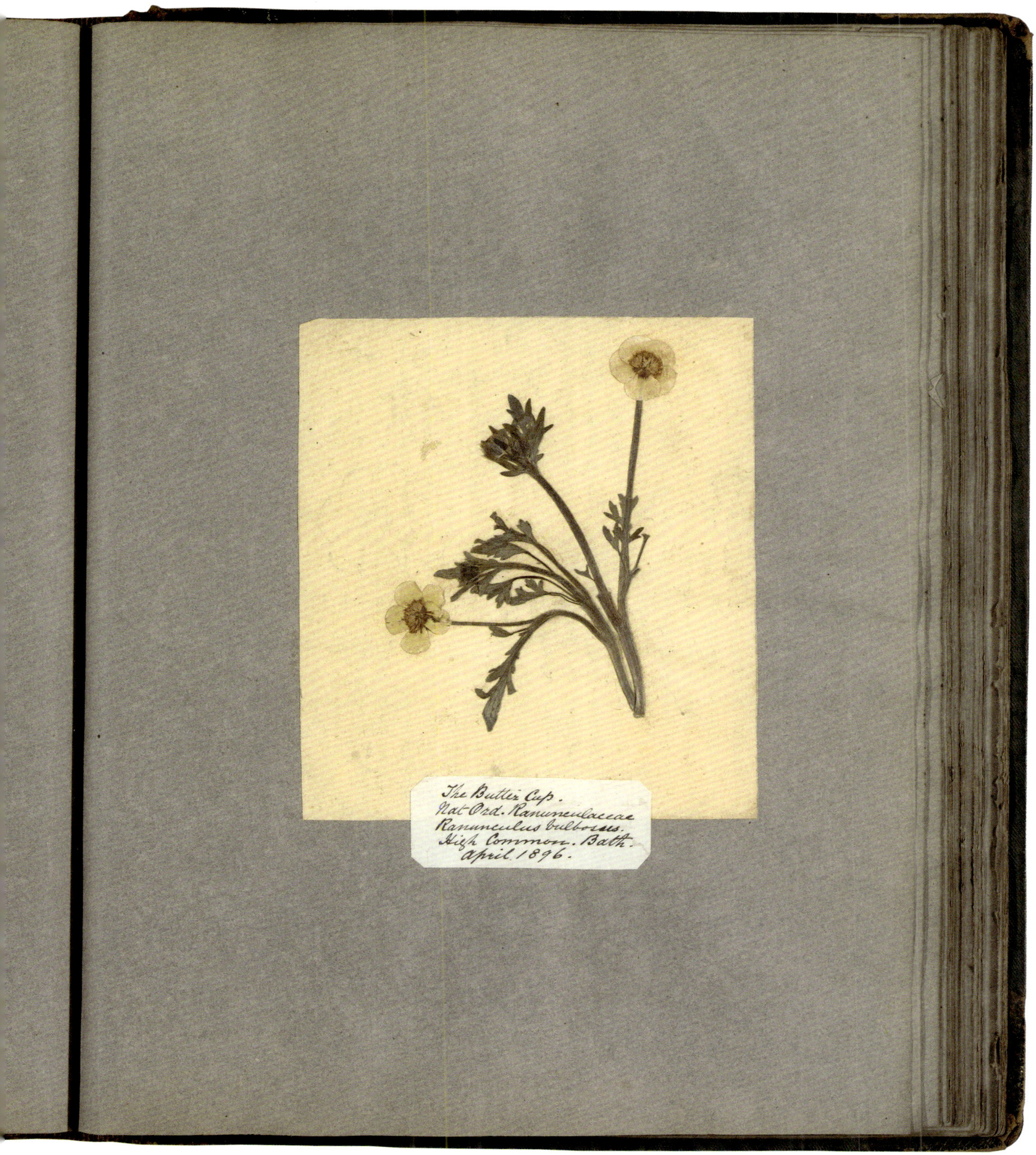

Album of Pressed Wild Flowers, 1882, natural specimens with pen and ink labels. Yale Center for British Art, Paul Mellon Fund

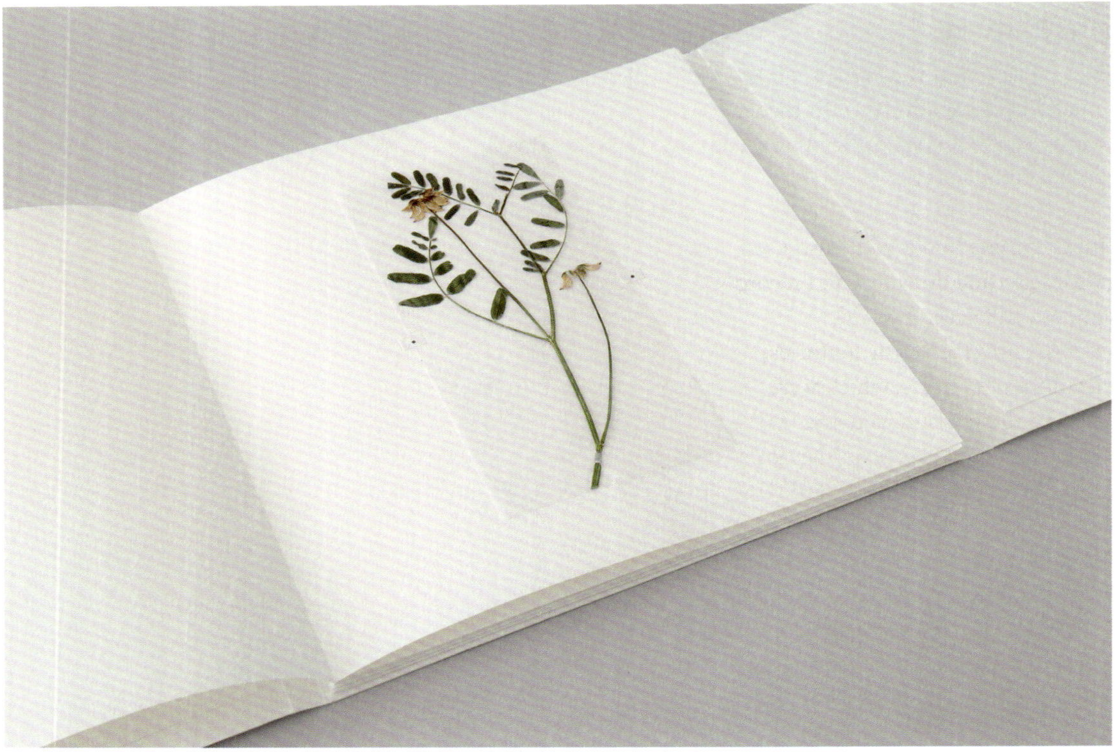

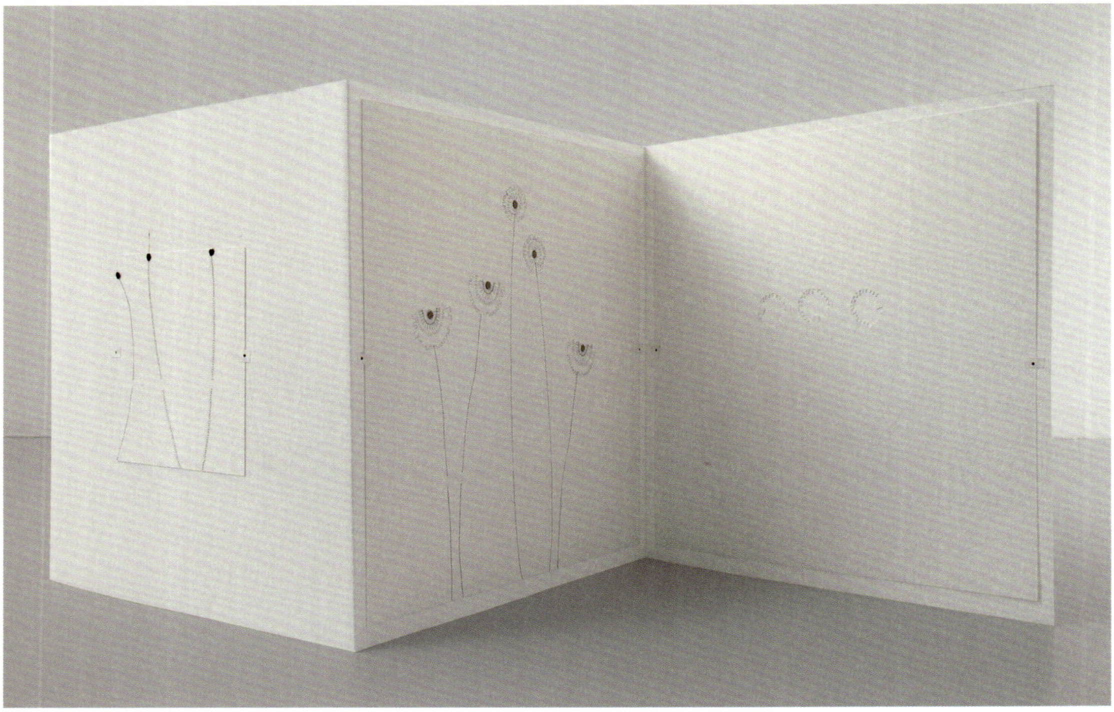

ABOVE AND DETAIL OPPOSITE: Mandy Bonnell, *Eloise Butler's Wildflower Garden, Est. 1907*, London, 2013, graphite, letterpress, hand-cut paper, and natural specimens. Collection of the artist

The Recording of Nature

I am interested in the exploration of the book as an art form, which has for me a historical affinity with the tradition of the sample or pattern book, in use for hundreds of years. Originally produced to contain explorative examples of assorted textile patterns and needlepoint often taken from plant life, these books were also used for detailed investigative scientific drawings and recordings of nature to document the natural world.

During the eighteenth and nineteenth centuries, women usually produced these samples, which were frequently considered a feminine leisure activity. However, intrepid Victorian women such as Marianne North, Constance Cummings, Maria Burges, and Miss Rowe of Liverpool took the pastime of gathering, collecting, researching, and recording very seriously and with appreciation for their subject matter. They rightly gained respect for the detailed intricacy of their drawings and watercolors from nature, and have influenced the work I do.

Pattern making is central to my imagery, and I prefer to work on an intimate scale—those small objects such as insects, shells, small plants, and seedpods that can be drawn life-sized. Drawing within the format of a pattern helps me to understand and recognize how important both nature and its growth are as a key to creativity. I have also started to draw typographic punctuation marks, transferring them into different patterns and arrangements to form their own visual narrative. I noticed that these marks related directly to botanical pattern formations as seen under a microscope.

I often have collaborated with the writer and poet Gabriel Gbadamosi, who connects my images with a poem, allowing a conversation to evolve between text and image.

My images also explore the natural history, culture, and artistic life within East Africa, focusing in particular on Lamu Island off the north Kenya coast, where I have lived and worked for many years.

Mandy Bonnell
ARTIST

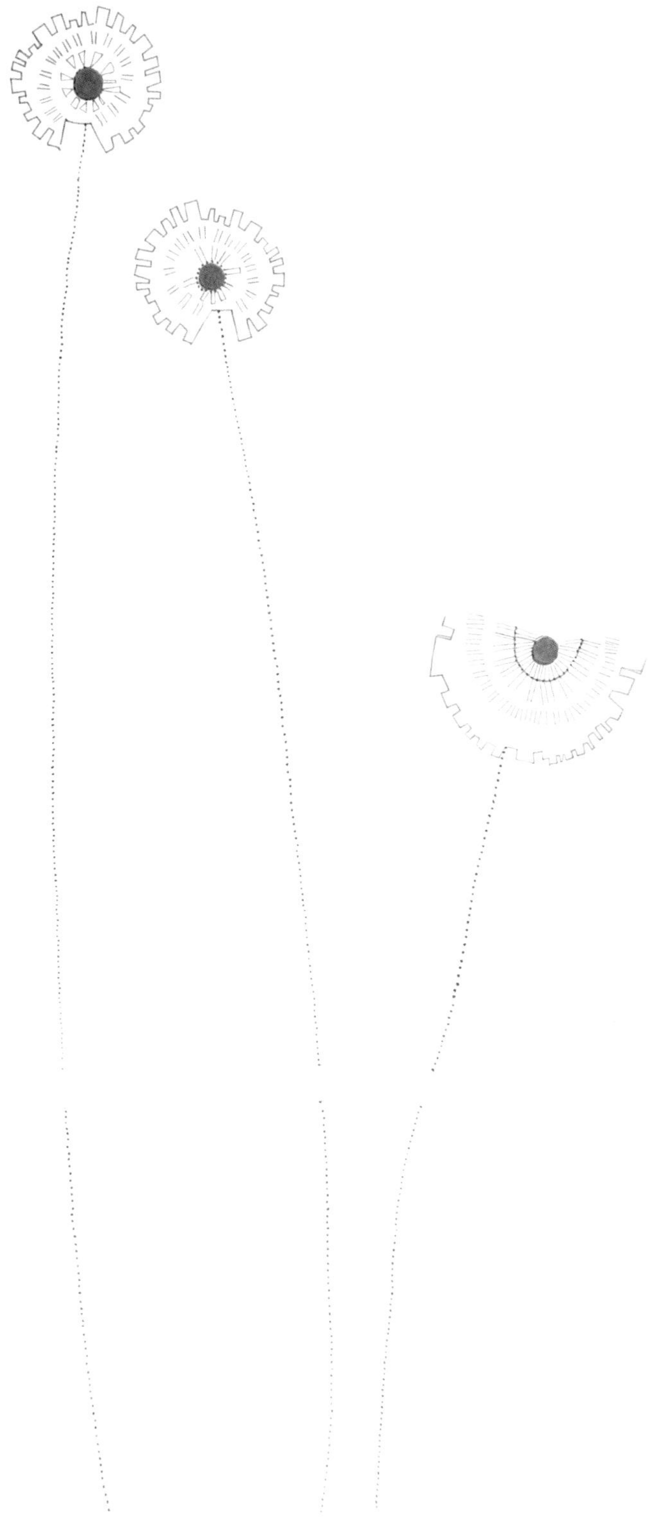

Nine Wild Plants

My work often deals with the impact that humans have on the environment. This project was inspired by the ecological thinker Paul Hawken, who noted that the average Western adult can recognize over one thousand brand names or logos, but fewer than ten local indigenous plants.

I began posing the question to people, "Which nine wild plants could you confidently identify?" Collecting responses and drawing the plants highlighted, I combined this data and visual information in unique drawings and bookworks, which formed the basis of the project. Constructed from recycled packaging, the delicate leaves and stalks of the poppy, buttercup, and dandelion were crafted from hundreds of brands and logos. The "Herbarium" drawings are botanically correct images of the nine wild plants that my quite unscientific survey revealed as the "most recognized" wild plants in Britain. The leaves, stalks, and petals are collages made from brand names and logos picked up like a bunch of wild blooms from the streets of London. These "wild plants" highlight our diminishing knowledge of indigenous flora and fauna while encouraging us to consider our growing excess of consumerism.

At the start of the project I made a list of the two hundred or so brands I hoped to find and use. I had made a rule that everything had to be genuinely discarded. I decided to dress the plants in the clothing of super brands—forming the collection of the nine most recognized. I wanted to absorb the brand recognition created from billions of pounds of global advertising into the quiet and overlooked, combining the precision of carefully observed botanical constructions with the deeply embedded brand-awareness within our culture.

Tracey Bush
ARTIST

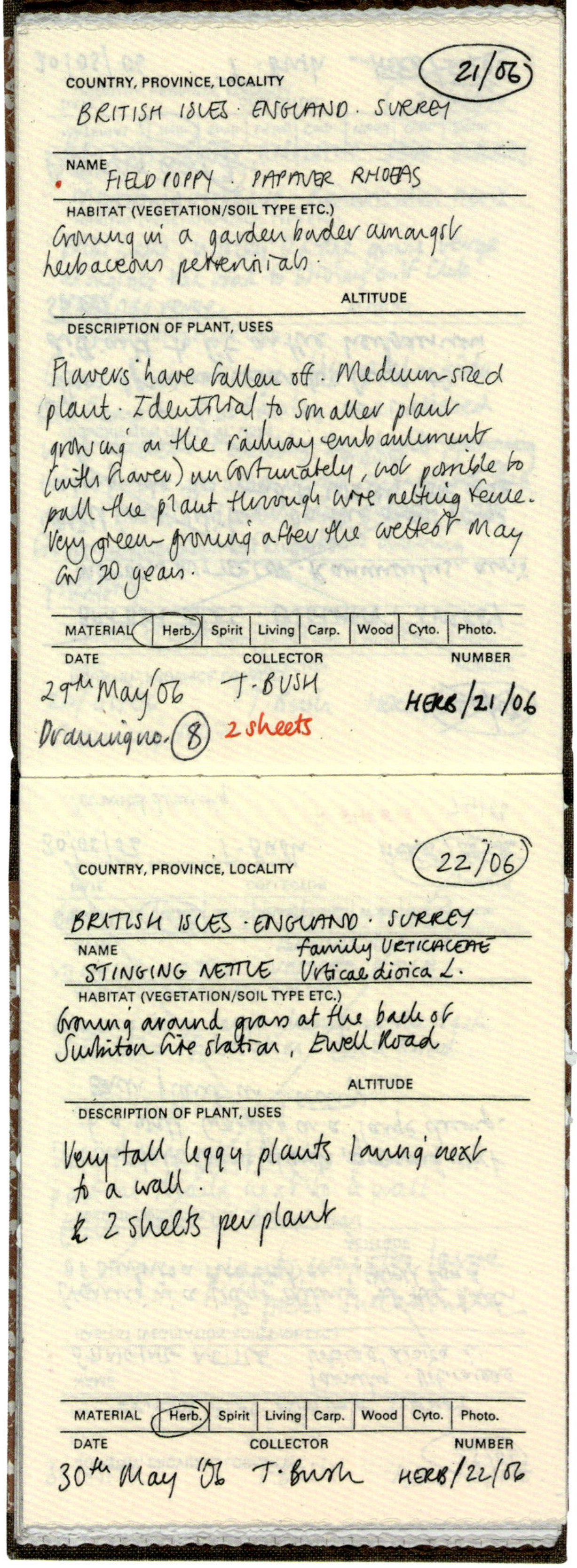

Tracey Bush, *Herbarium Notebook*, 2005, pen and ink with natural specimens and hand-cut paper collages. Collection of the artist

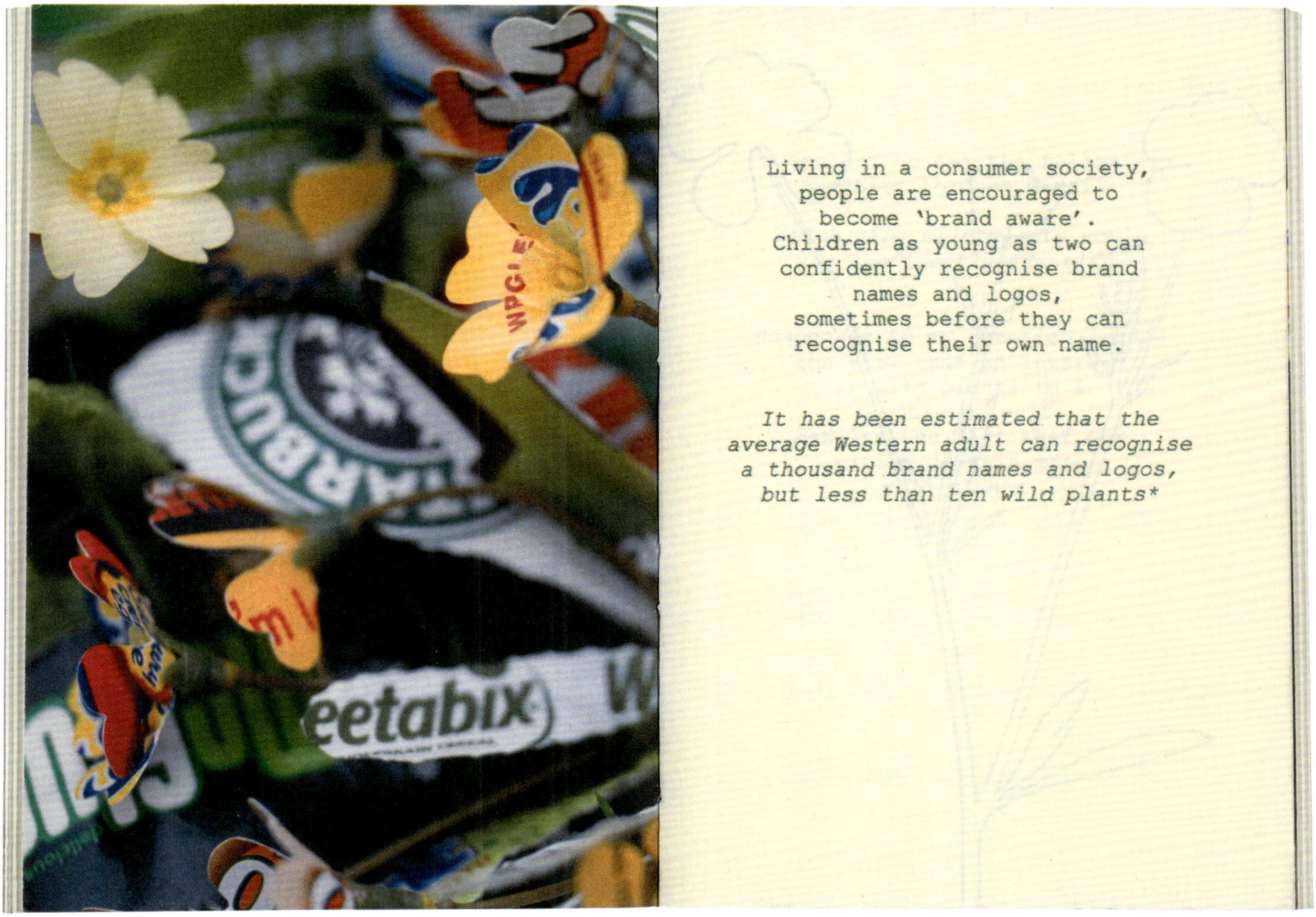

Living in a consumer society,
people are encouraged to
become 'brand aware'.
Children as young as two can
confidently recognise brand
names and logos,
sometimes before they can
recognise their own name.

*It has been estimated that the
average Western adult can recognise
a thousand brand names and logos,
but less than ten wild plants**

Tracey Bush, *Nine Wild Plants*, London, 2006, digital print, unique proof copy. Collection of the artist

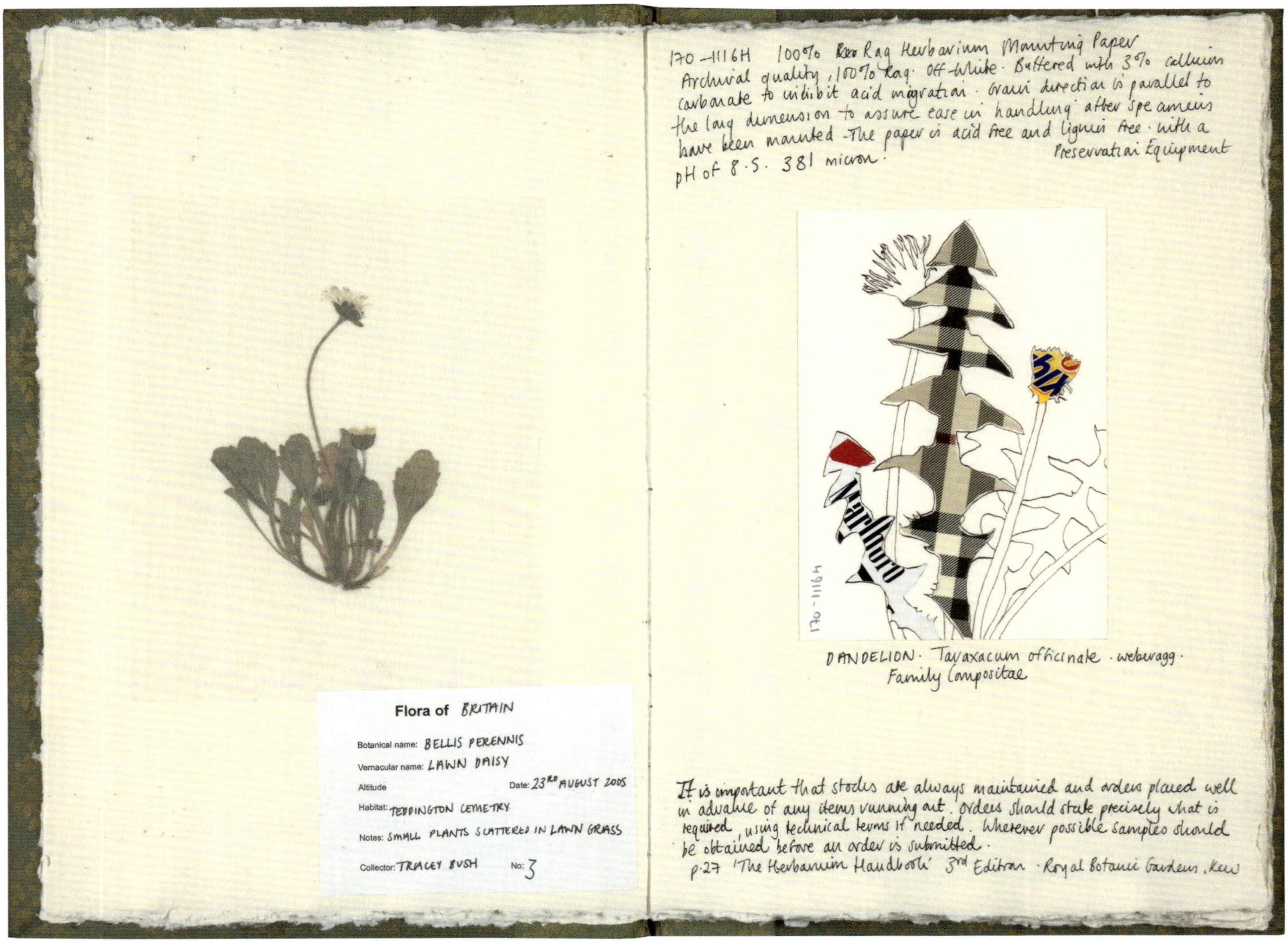

Tracey Bush, *Florilegium Sketchbook*, 2005, paper packaging collage, pen and ink with natural specimen. Collection of the artist

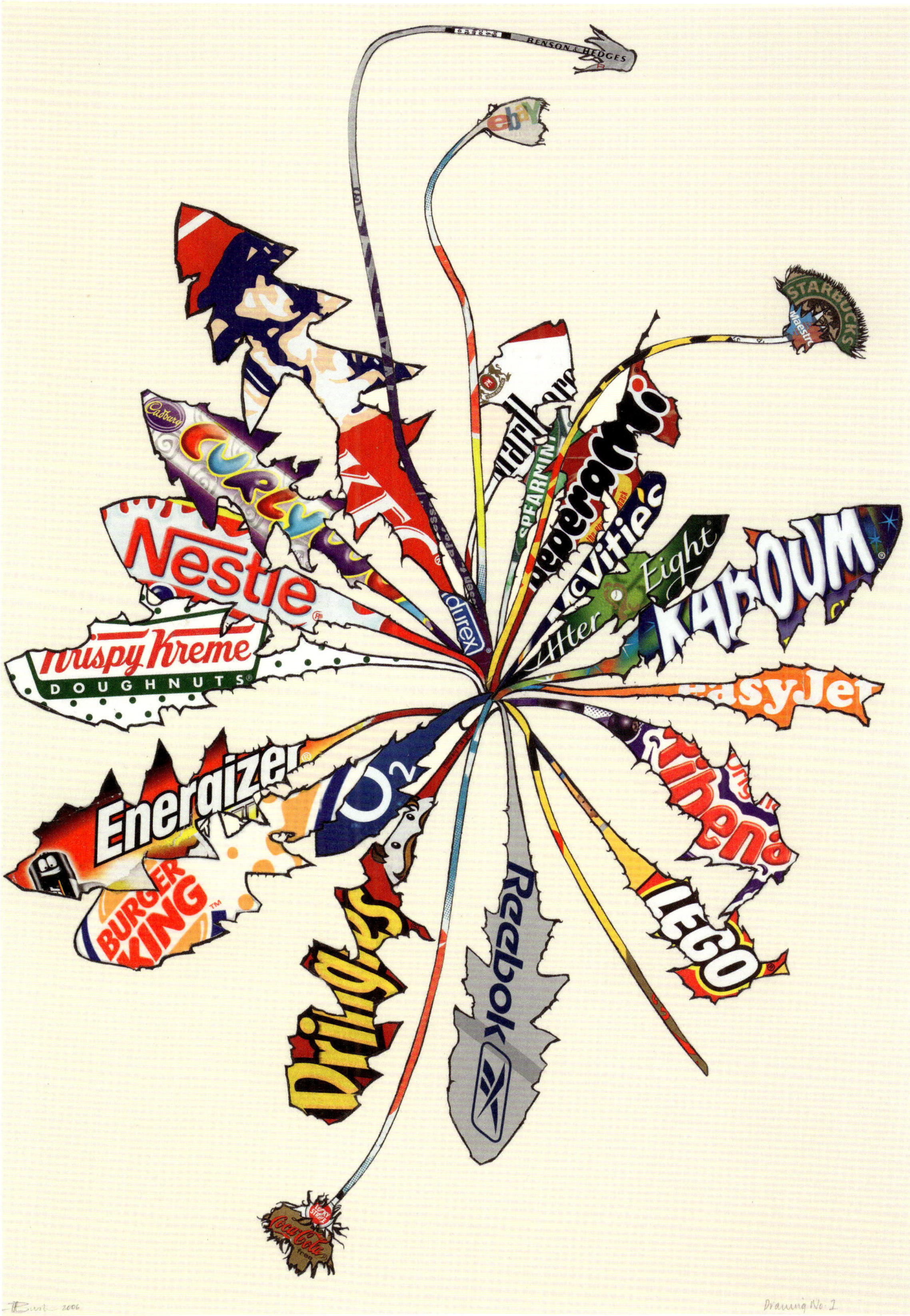

Tracey Bush, Study for herbarium sheet, *Nine Wild Plants: Dandelion*, 2006, hand-cut paper packaging collage with pen and ink. Yale Center for British Art, Friends of British Art Fund

Convergence of Art and Geography

Recently, on taking a neglected book down from a shelf, I rediscovered something that gives an emphasis to the subjects I explore within my work.

There was a label pasted inside the cover of the book stating that it was awarded to me in my penultimate year at school for receiving first prize in Art and in Geography. This gave me confidence in reminding me that these dual interests have underpinned and influenced me throughout my life.

I have always been keen to explore my immediate environment, to try to understand and explain it through making images.

Early in my career, during the 1980s, I documented the decline of the mining industry in Midlothian with a particular interest in coal bings and derelict buildings as well as disused railways and viaducts that had been the places I explored as a child. I became fascinated by nature's reclamation of these sites and also increasingly aware of the wild plants that advance to populate the places I continue to revisit.

In 2008, while carrying out research in the Herbarium at the Royal Botanic Garden Edinburgh, I was attracted by the way the herbarium sheets were made, collated, and archived. This led me to create my own paper herbarium by collecting plants from familiar sites and making etchings of them. This process was similar to that of pressing plants as the plant collector does, flattening and trapping them. After printing, the etchings were painstakingly cut out to create specimens that were then attached to my own herbarium sheets and catalogued, giving details of the date, location, and collector.

This project allowed me to make a formal record of the plants I am interested in, alongside locating them—something that is critical to my work, the convergence of Art and Geography.

Jane Hyslop
ARTIST

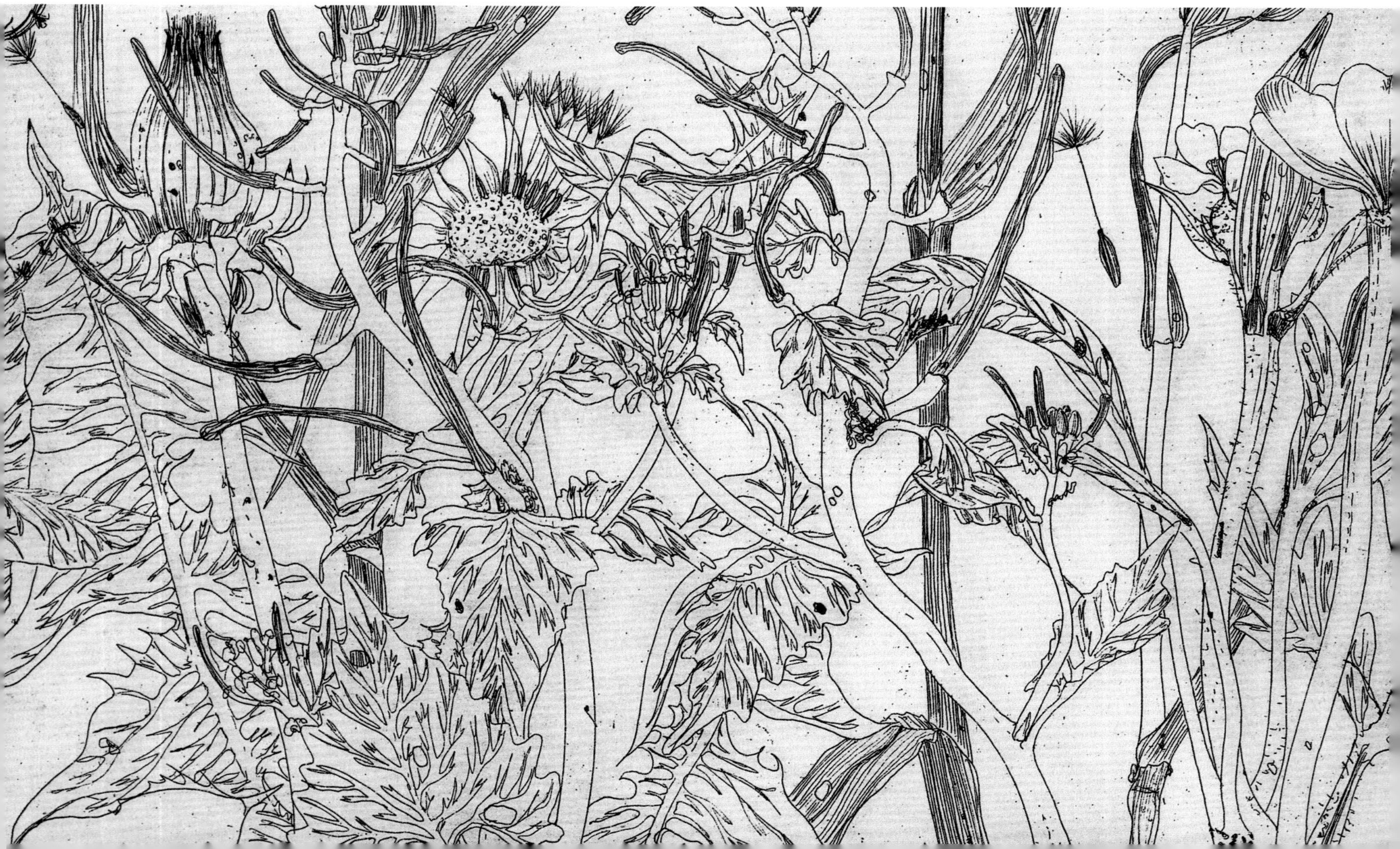

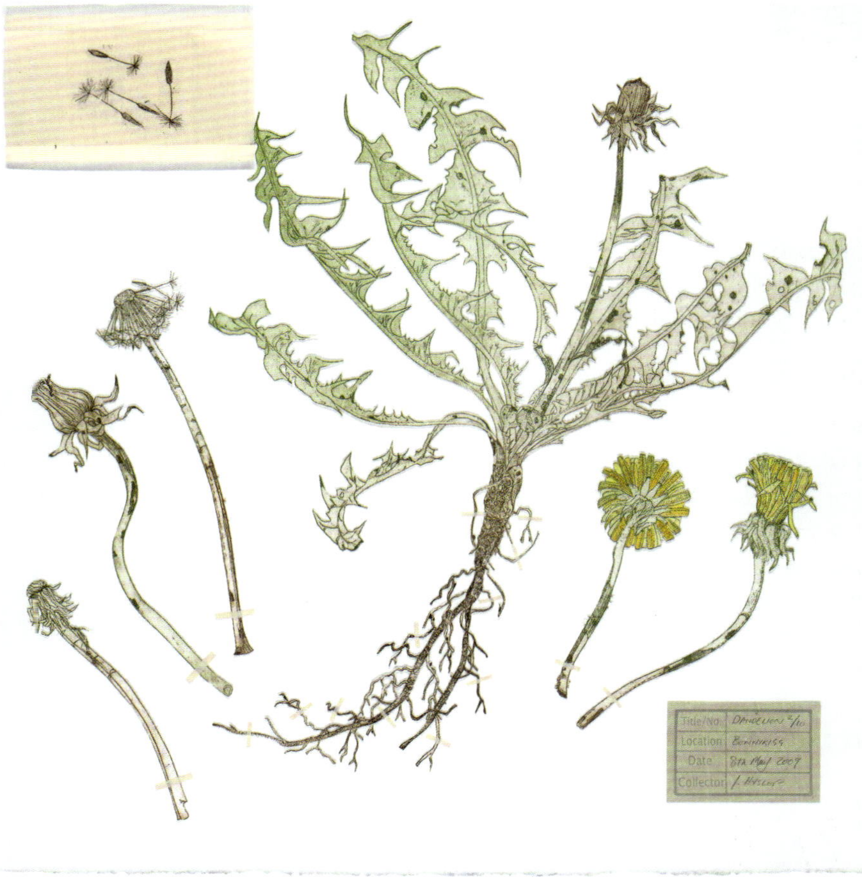

Jane Hyslop, *Herbarium series: Dandelion*, Midlothian, Scotland, 2009, hand-colored etchings and hand-cut and folded paper. Yale Center for British Art, Friends of British Art Fund

BELOW: Jane Hyslop, *Wild Plants Collected in Midlothian*, Midlothian, Scotland, 2002, etching. Yale Center for British Art, Friends of British Art Fund

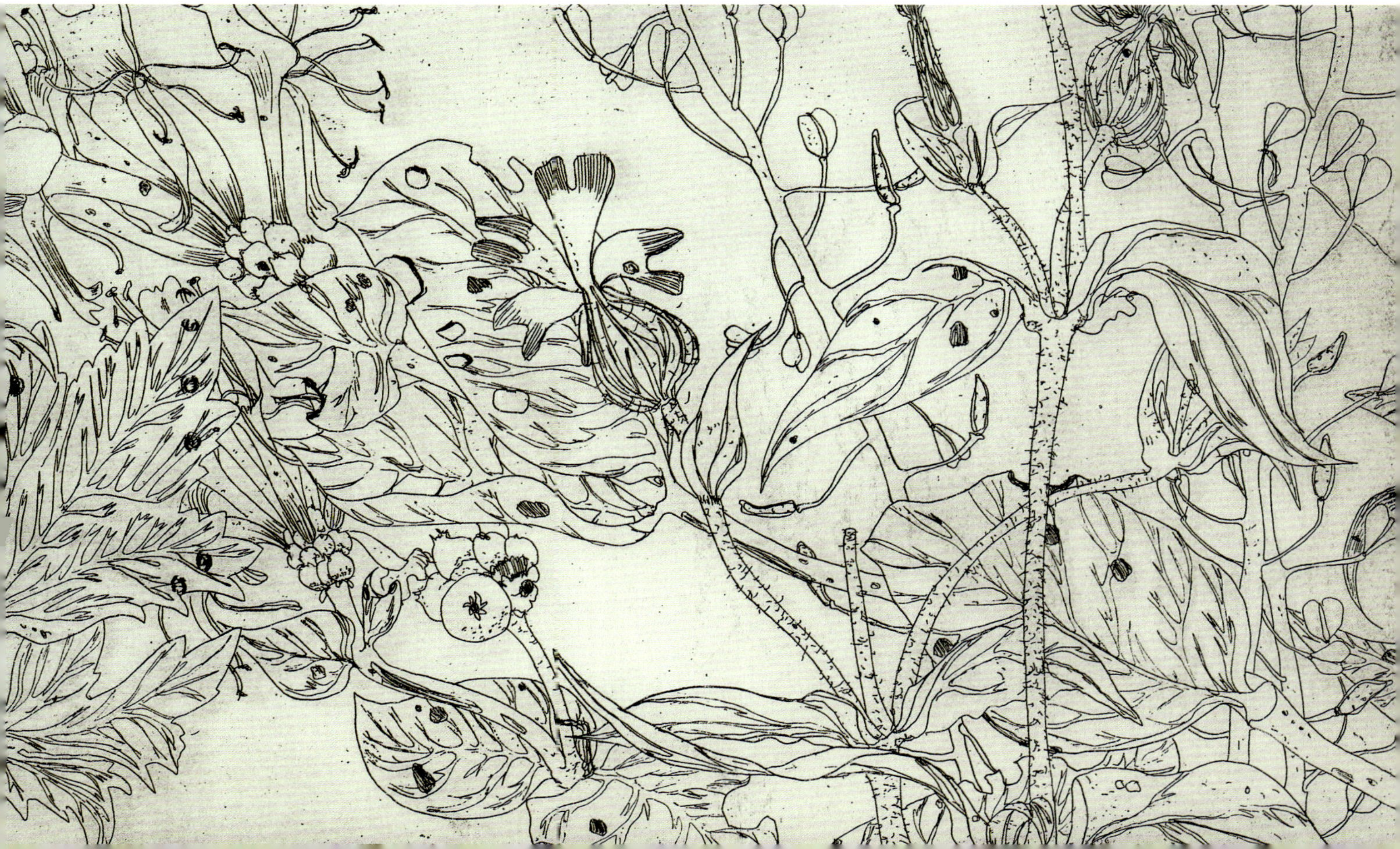

Herbarium Specimen Sheets

It is instructive to compare the herbarium sheets created by Tracey Bush and Jane Hyslop with the herbarium specimens from the Yale Peabody Museum of Natural History. While the artists have no formal botanical training, their works are remarkable in that they faithfully depict distinctive features of each plant, making it possible to identify the depicted plants to the species level.

A herbarium specimen is created by pressing, drying, and then mounting a plant to a sheet of paper, a practice that has changed little since it began in the mid-sixteenth century. Herbaria are a valuable scientific resource because herbarium specimens are important to contemporary taxonomic, evolutionary, and ecological research disciplines, to name a few.

The *Ranunculus acris* (tall buttercup) specimen was collected in 1829 in the area of Heidelberg, Germany, probably by Gottlieb Wilhelm Bischoff (1797–1854), and the *Papaver rhoeas* (corn poppy) specimen was collected somewhere in Europe in the first half of the nineteenth century, by either Bischoff or a colleague. Bischoff was a German pharmacist and botanist, a professor at the University of Heidelberg, and a director of the Botanic Garden in Heidelberg. The genera *Bischofia* and *Bischofioxylon* are named in his honor. The *Taraxacum officinale* (dandelion) specimen was collected in 1884 in Glastonbury, Connecticut, by Francis Wilson Starmer, an amateur plant collector. All of these specimen sheets were created at a time when collecting to document temperate plant diversity was among the chief botanical research pursuits.

Patrick Sweeney
BOTANIST

OPPOSITE: *Taraxacum officinale (dandelion)*, collected by Francis Wilson Starmer in Glastonbury, Connecticut, 1884, natural specimen mounted on paper. Division of Botany, Peabody Museum of Natural History, Yale University

OVERLEAF, LEFT: Tracey Bush, Herbarium sheet for *Nine Wild Plants: Common Poppy*, 2006, hand-cut paper packaging collage with pen and ink. Yale Center for British Art, Friends of British Art Fund

OVERLEAF, RIGHT: *Papaver rhoeas (Corn Poppy)*, collected by Gottlieb Wilhelm Bischoff or colleague, somewhere in Europe before 1850, natural specimen mounted on paper. Division of Botany, Peabody Museum of Natural History, Yale University

SECOND OVERLEAF, LEFT: *Ranunculus acris (Tall Buttercup)*, collected by Gottlieb Wilhelm Bischoff in the area of Heidelberg, Germany, 1829, natural specimen mounted on paper. Division of Botany, Peabody Museum of Natural History, Yale University

SECOND OVERLEAF, RIGHT: Tracey Bush, Herbarium sheet for *Nine Wild Plants: Common Buttercup*, 2006, hand-cut paper packaging collage with pen and ink. Yale Center for British Art, Friends of British Art Fund

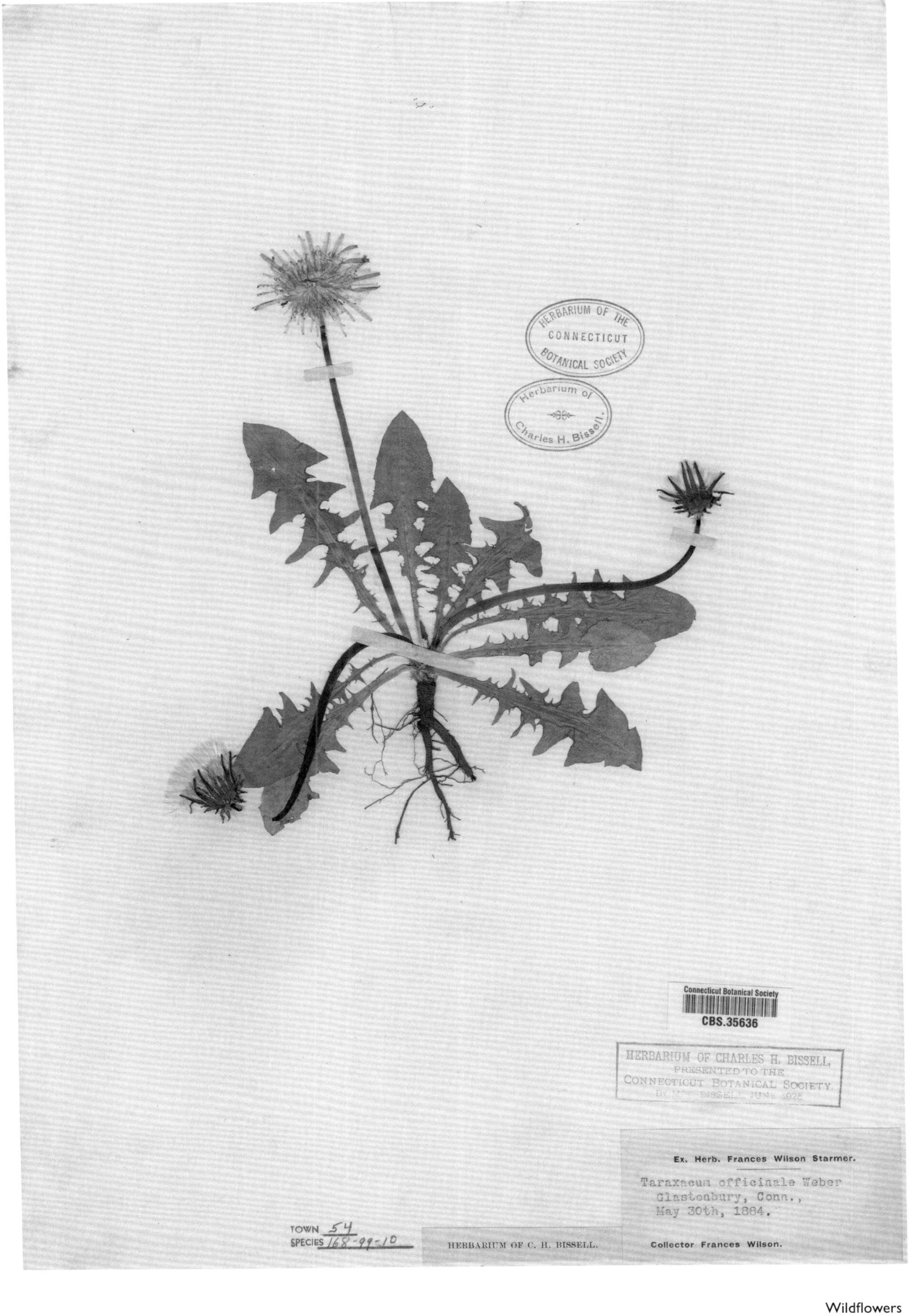
HERBARIUM OF THE CONNECTICUT BOTANICAL SOCIETY
Herbarium of Charles H. Bissell.
Connecticut Botanical Society
CBS.35636
HERBARIUM OF CHARLES H. BISSELL
PRESENTED TO THE
CONNECTICUT BOTANICAL SOCIETY
Ex. Herb. Frances Wilson Starmer.
Taraxacum officinale Weber
Glastonbury, Conn.,
May 30th, 1884.
TOWN 54
SPECIES 168-99-10
HERBARIUM OF C. H. BISSELL.
Collector Frances Wilson.

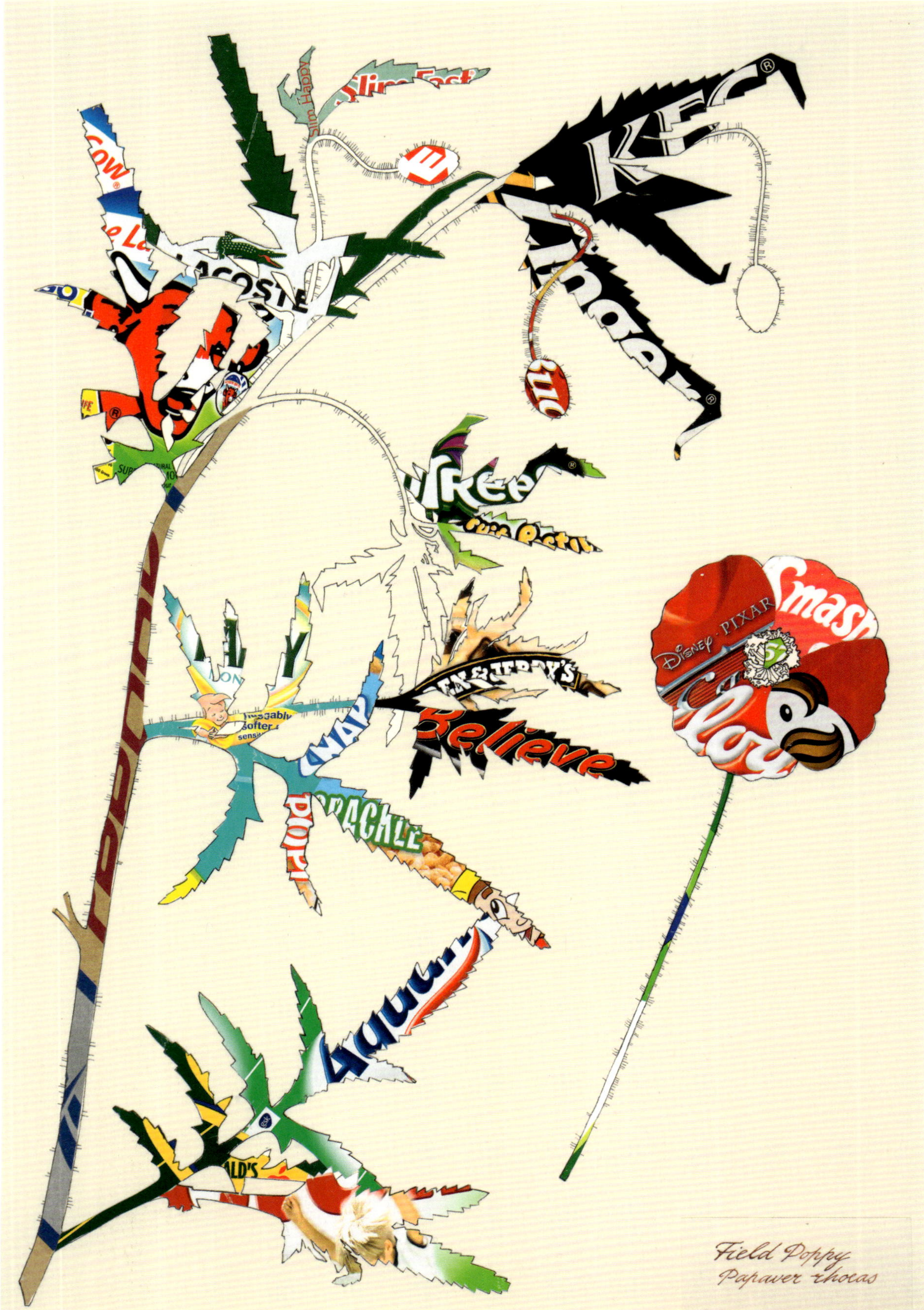
KFC
LACOSTE
Believe
DISNEY · PIXAR
Field Poppy
Papaver rhoeas

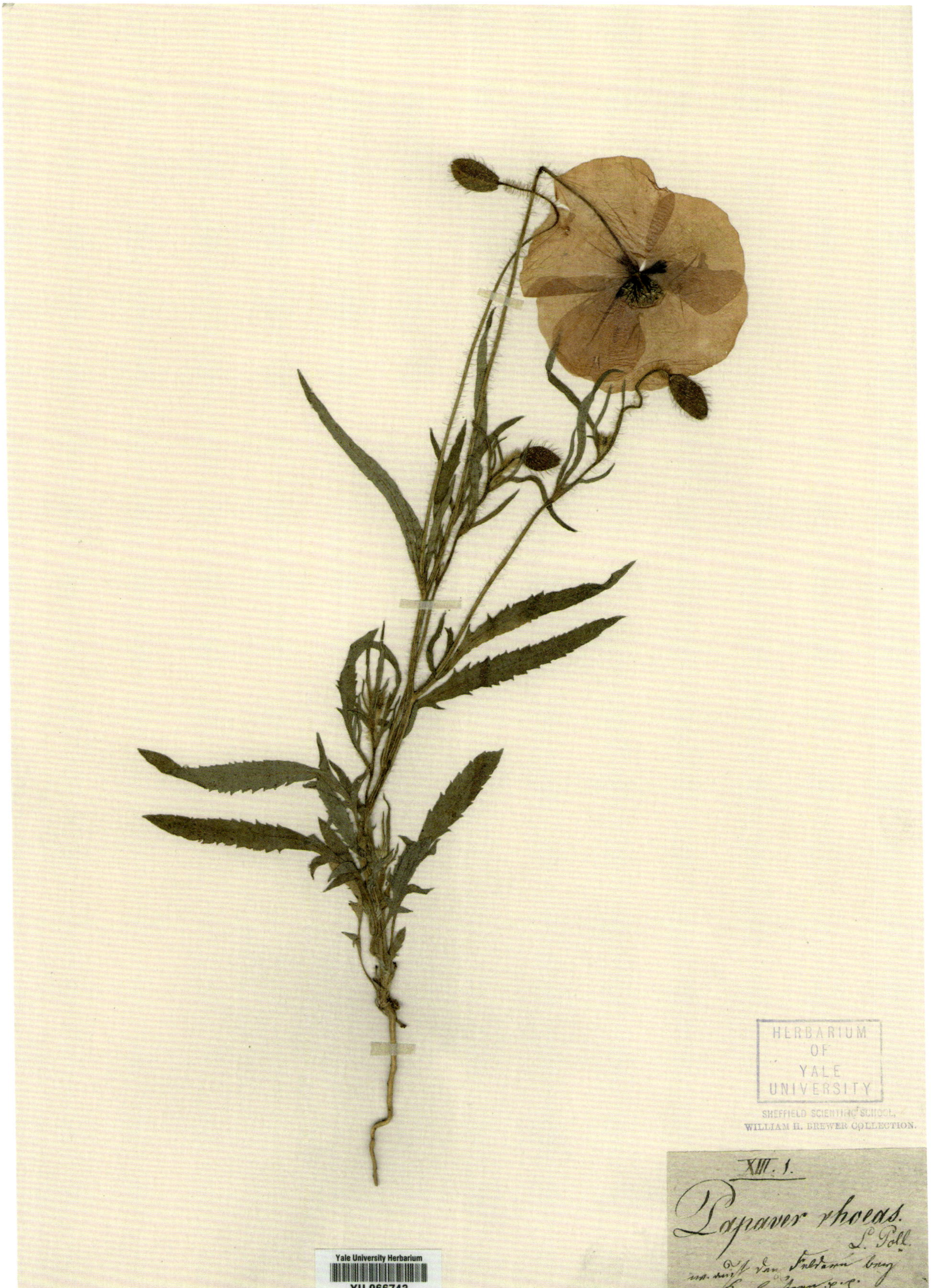
HERBARIUM
OF
YALE
UNIVERSITY
SHEFFIELD SCIENTIFIC SCHOOL,
WILLIAM H. BREWER COLLECTION.
XIII. 1.
Papaver rhoeas.
L. Poll.
Yale University Herbarium
YU.066743

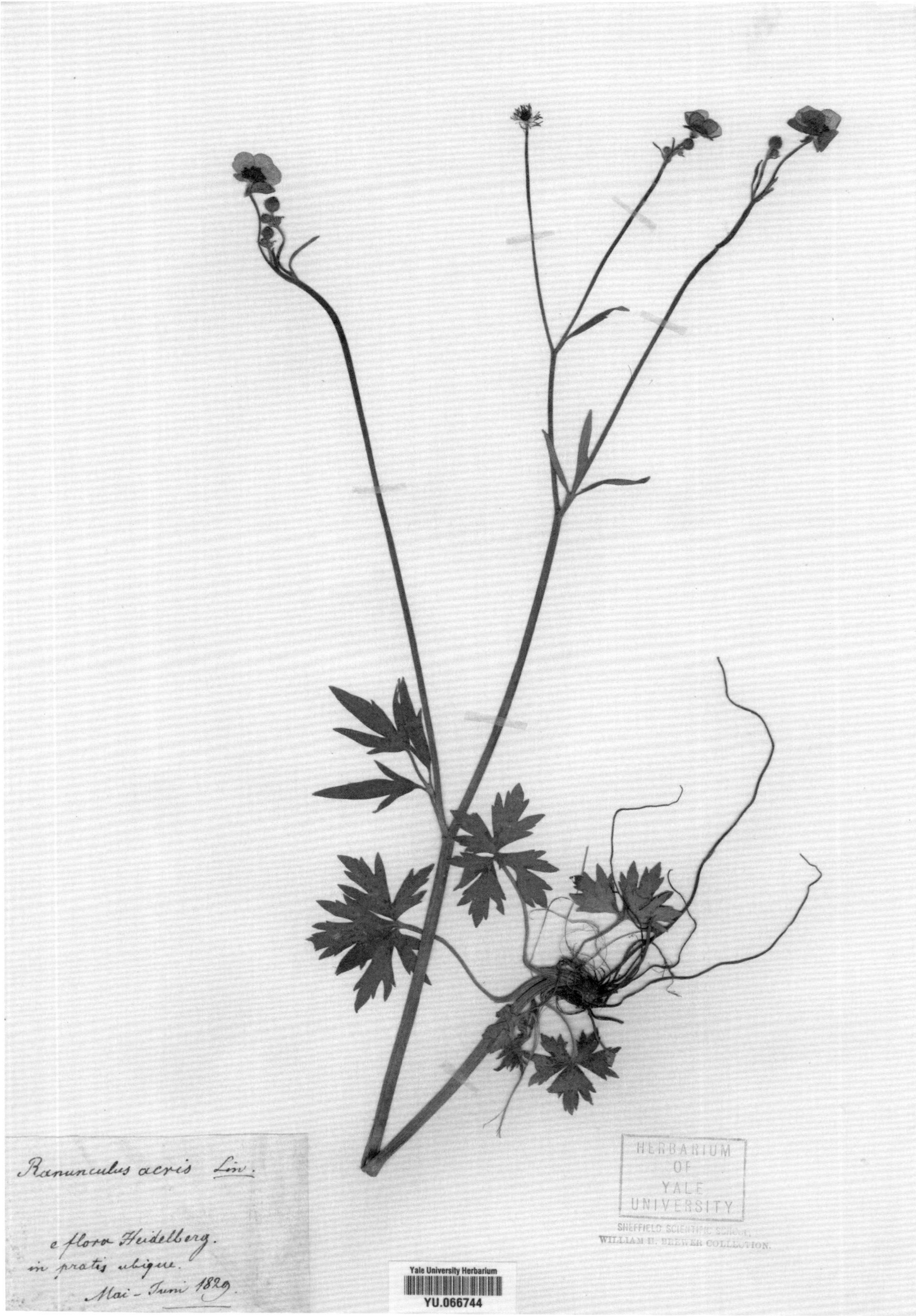
Ranunculus acris Lin.
e flora Heidelberg.
in pratis ubique.
Mai - Juni 1829.
HERBARIUM OF YALE UNIVERSITY
SHEFFIELD SCIENTIFIC SCHOOL.
WILLIAM H. BREWER COLLECTION.
Yale University Herbarium
YU.066744

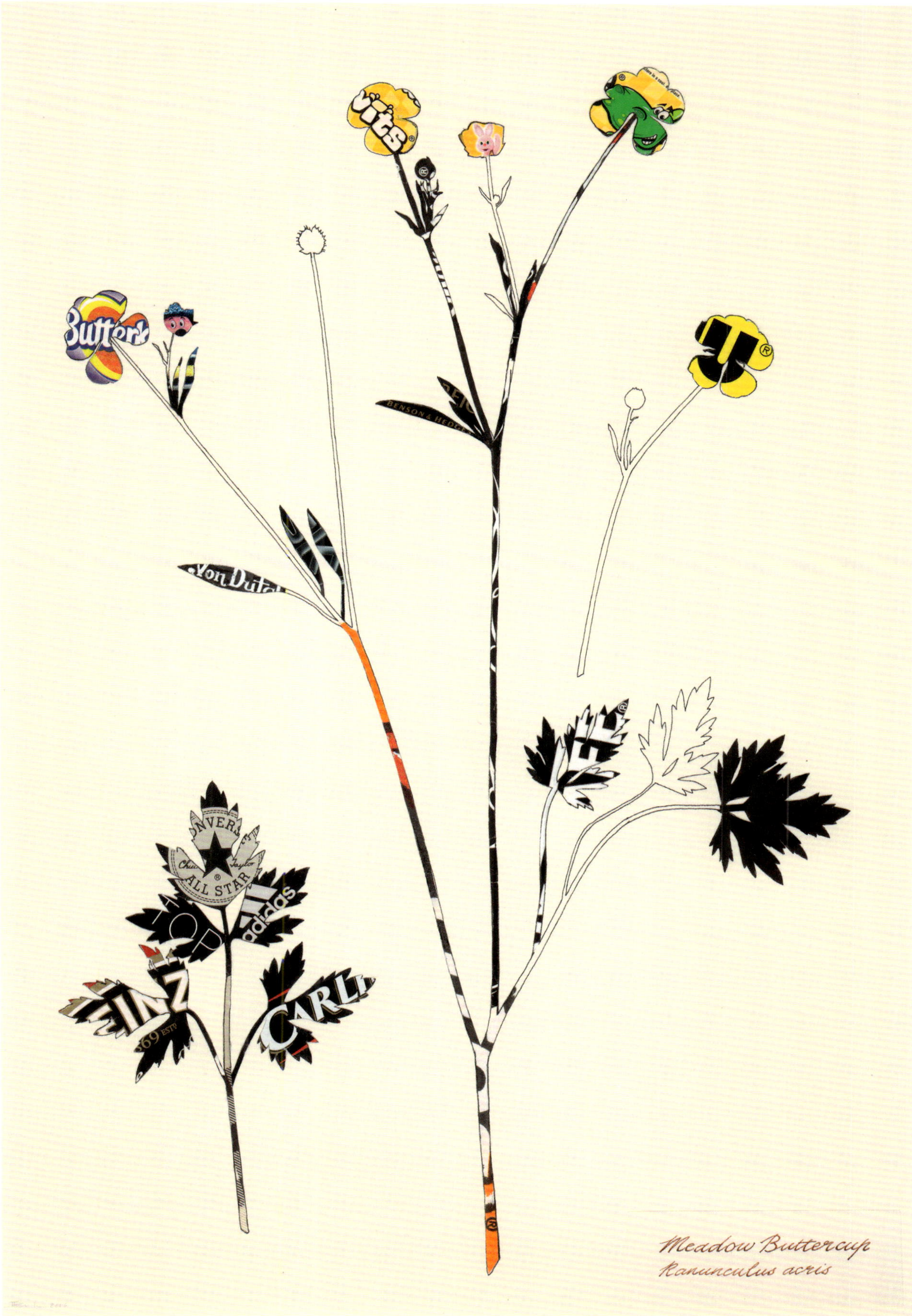
Meadow Buttercup
Ranunculus acris

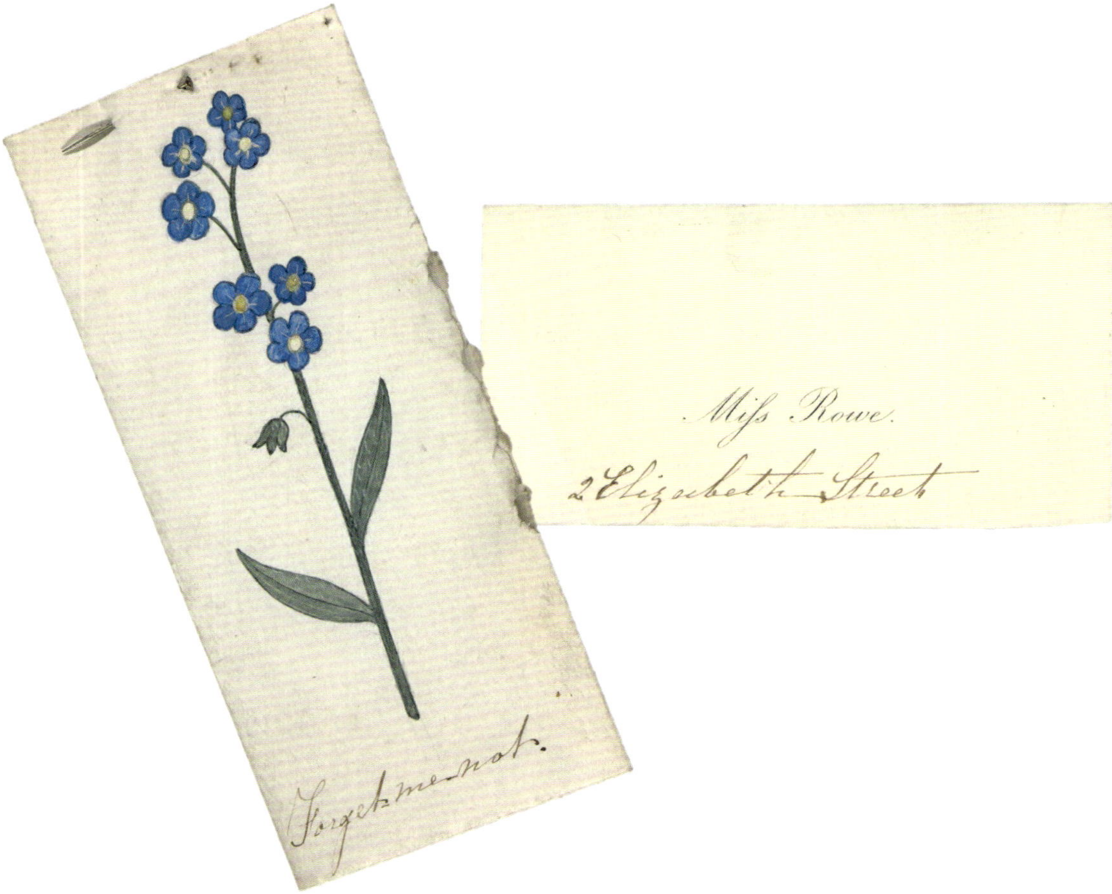

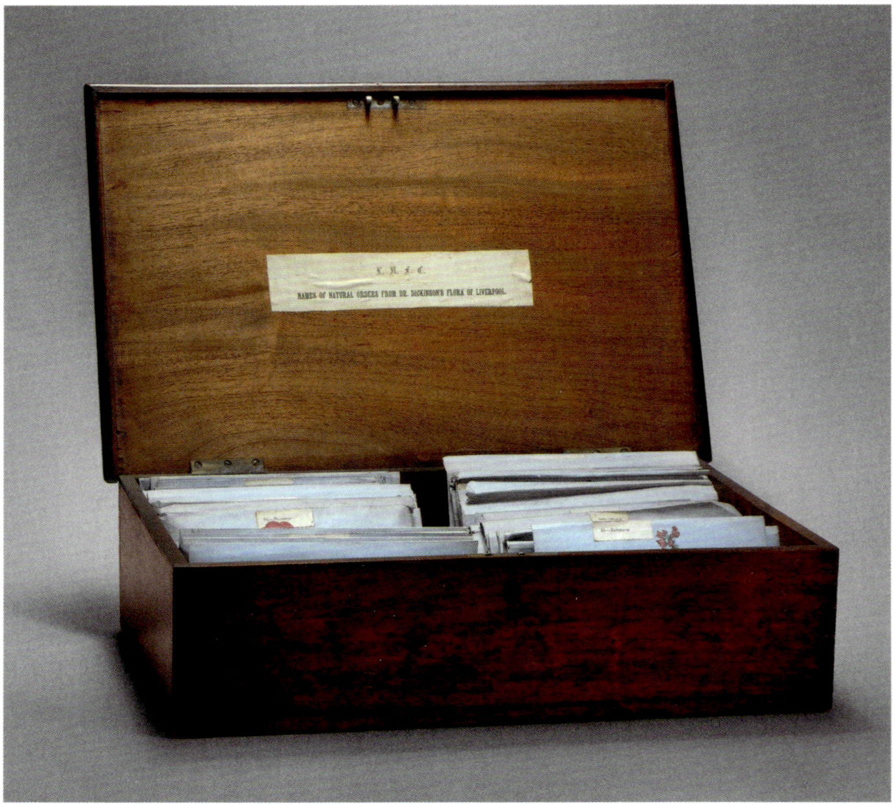

Miss Rowe, *Collection of Botanical Specimens*, Liverpool, 1861, Yale Center for British Art, Paul Mellon Fund

ABOVE: Calling card and envelope with forget-me-not drawing, watercolor and pen and ink with pin.

RIGHT: Mahogany box with envelopes of natural specimens.

OPPOSITE: "4. *Papaveraceae*" and "24. *Rosaceae*"

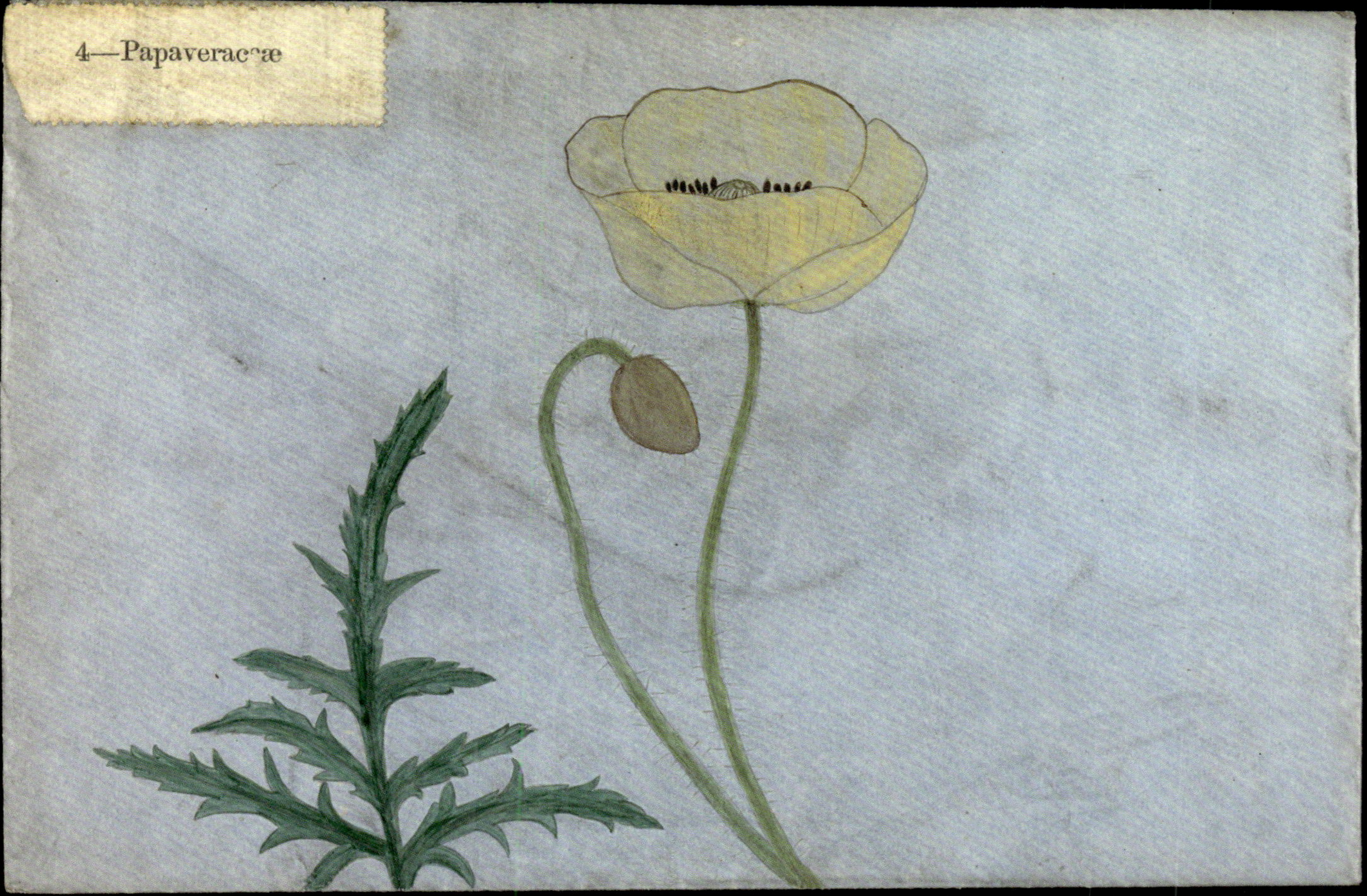
4—Papaveraceæ

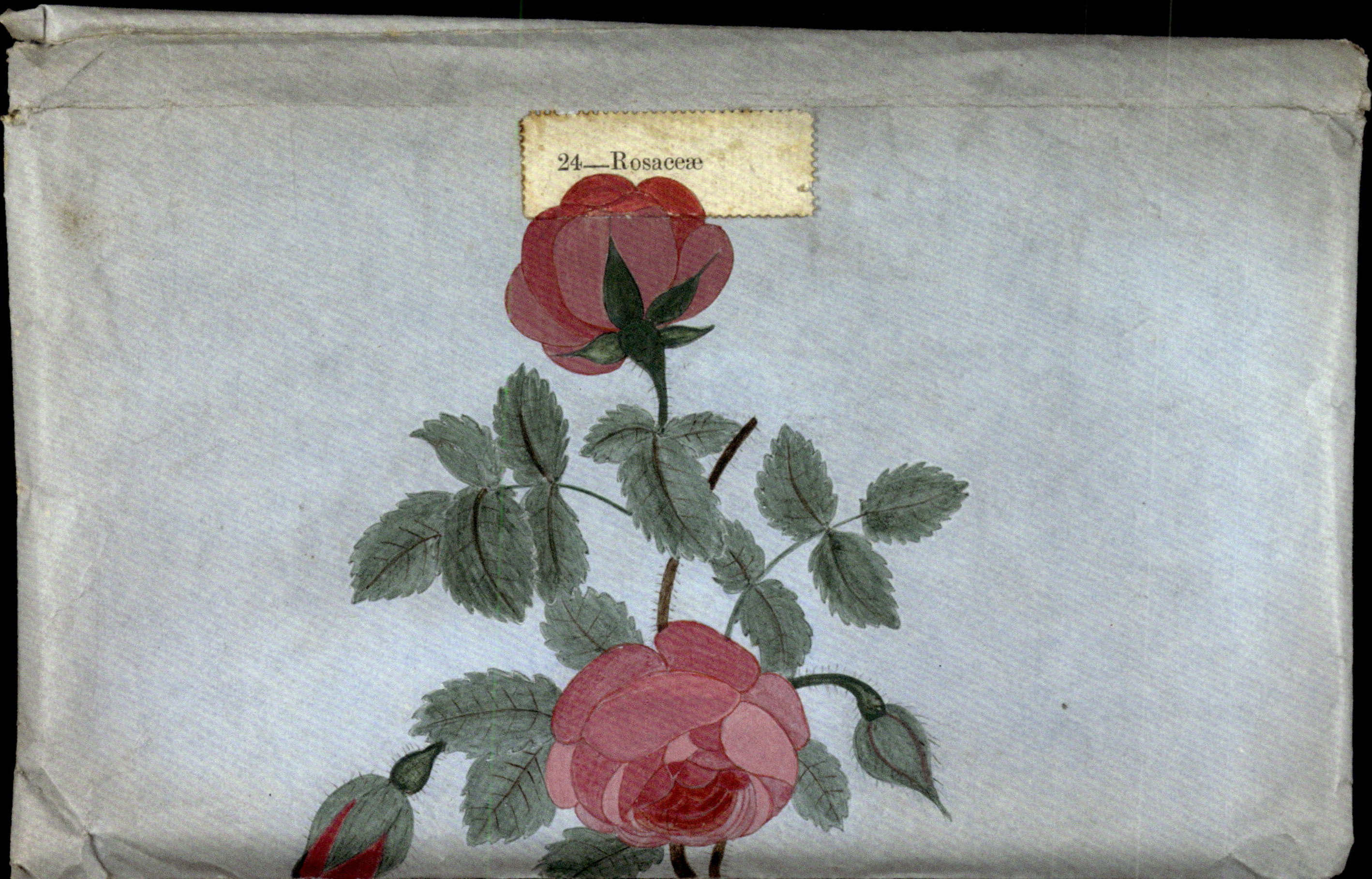
24—Rosaceæ

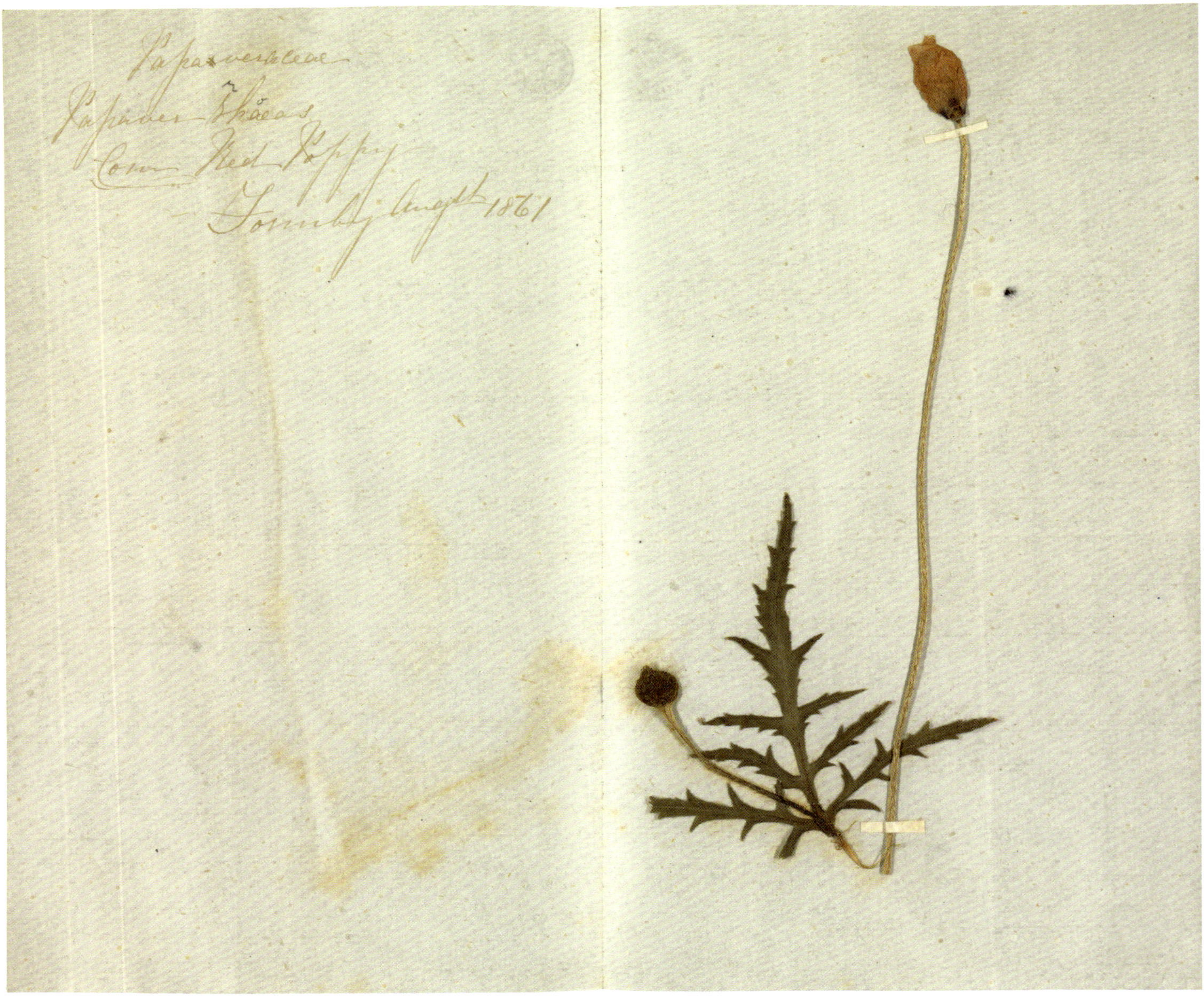

Miss Rowe, *"Papaver rhoeas"* from *Collection of Botanical Specimens*, Liverpool, 1861, natural specimen with pen and ink annotation. Yale Center for British Art, Paul Mellon Fund

OPPOSITE: Mandy Bonnell, "Poppy" from *Wild Flowers Worth Notice*, London, 2012, graphite. Yale Center for British Art, Gift of the artist

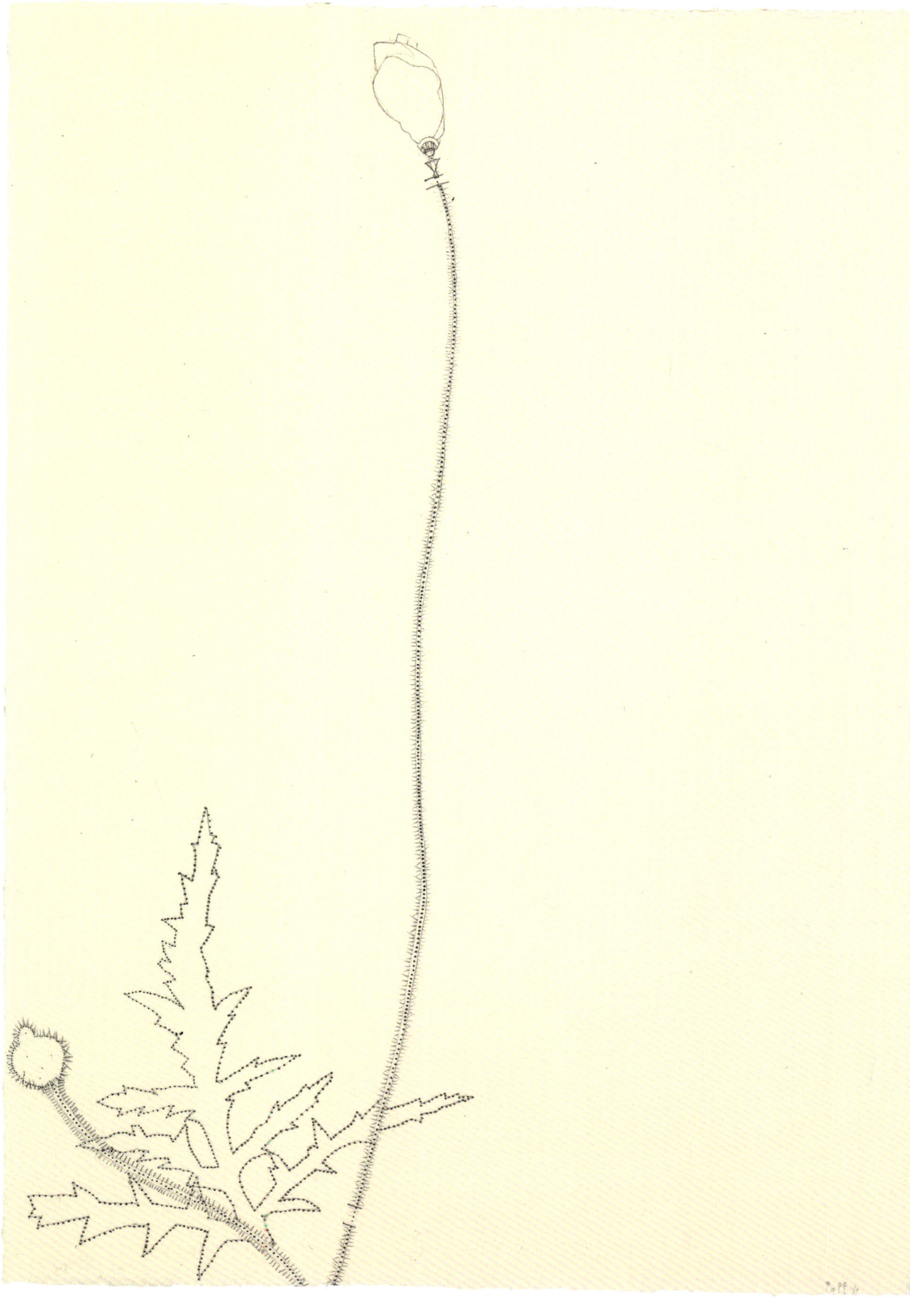

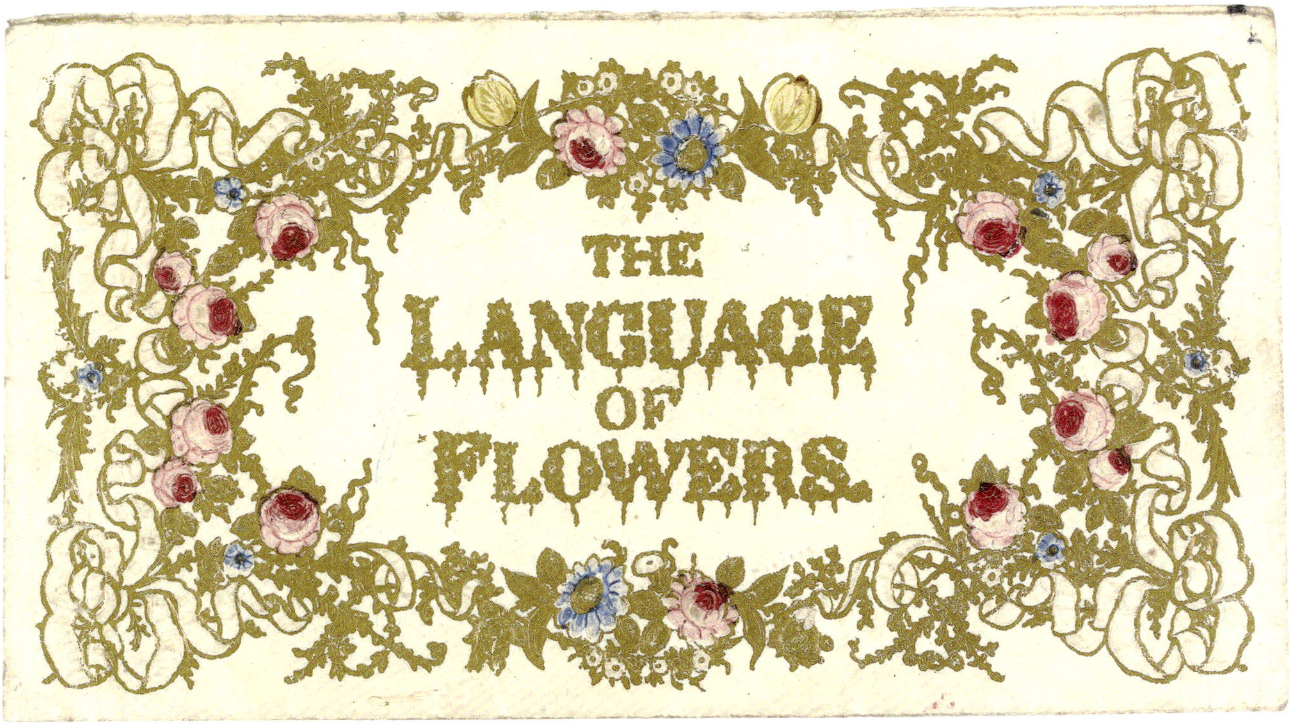

ABOVE AND OPPOSITE: *The Language of Flowers*, England, ca. 1860, chromolithograph, cut and folded paper, and envelope. Yale Center for British Art, Paul Mellon Fund

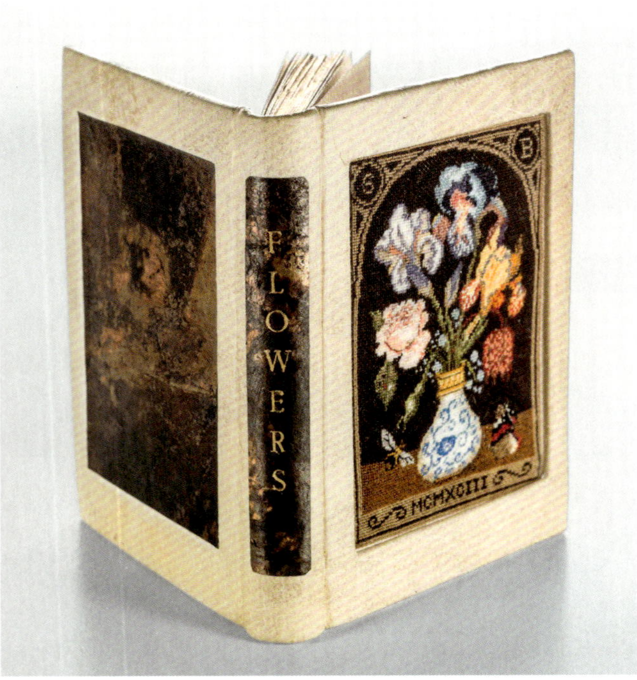

Anna Elizabeth, *Vase of Flowers*, Boston: G. W. Cottrell, 1851, miniature book, with needlepoint cover by Sue Bakker, Peebles, Scotland, 1993, silk thread on sixty-count gauze fabric. Collection of Neale and Margaret Albert

Daisy

Amidst a meadow green and gay, a *Daisy*
reared its head,
And o'er the sward around a glimpse
of cheerfulness was spread;
'Twas like the merry laugh of life,
when innocence is there,
'Twas like the smile of happiness,
dispelling every care.
I love to see thee smile, my love,
and *Cheerfulness* impart,
E'en like the modest daisy, to a
doubting throbbing heart.

The Language of Flowers, England, ca. 1860

OVERLEAF: *Cut-Paper Basket*, England, ca. 1820, hand-cut paper with watercolor, stitched with silk thread. Yale Center for British Art, Paul Mellon Fund

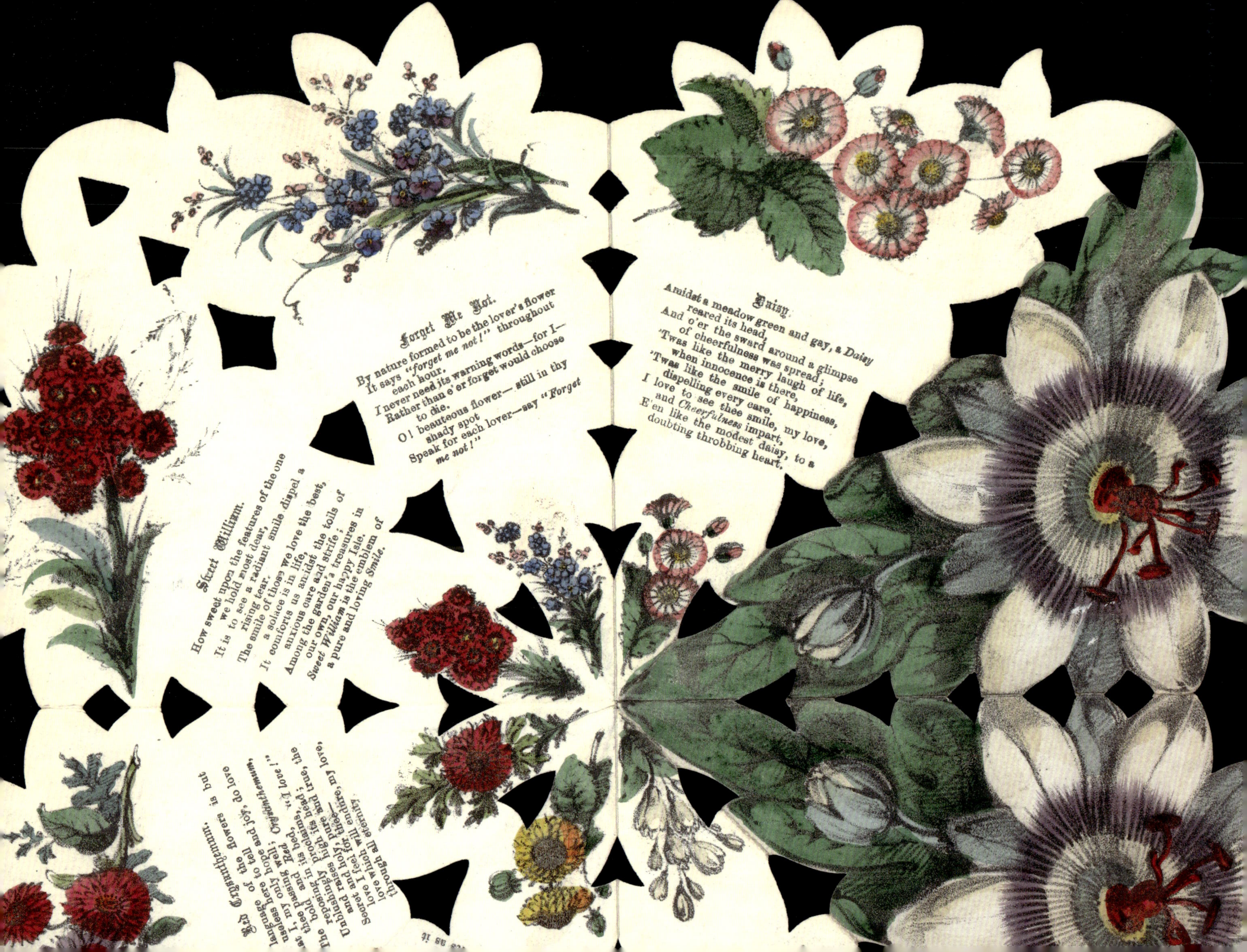
Forget Me Not.
By nature formed to be the lover's flower
It says "forget me not!" throughout
each hour.
I never need its warning words—for I—
Rather than e'er forget would choose
to die.
O! beauteous flower— still in thy
shady spot
Speak for each lover—say "Forget
me not!"
Daisy.
Amidst a meadow green and gay, a Daisy
reared its head,
And o'er the sward around a glimpse
of cheerfulness was spread;
'Twas like the merry laugh of life,
when innocence is there,
'Twas like the smile of happiness,
dispelling every care.
I love to see thee smile, my love,
and Cheerfulness impart,
E'en like the modest daisy, to a
doubting throbbing heart.
Sweet William.
How sweet upon the features of the one
we hold most dear,
It is to see a radiant smile dispel a
rising tear.
The smile of those we love the best,
a solace is in life,
It comforts us amidst the toils of
anxious care and strife;
Among the garden's treasures in
our own, our happy Isle,
Sweet William is the emblem of
a pure and loving Smile.
Red Crysanthemum.
language of the flowers is but
useless here to tell
at I, my only hope and joy, do love
thee passing well;
The bold and Red Crysanthemum,
reposing in its bed,
Unblushingly proclaims, "I love!"
and raises high its head;
Secret and holy, pure and true, the
love I feel for thee—
love which will endure, my love,
through all eternity

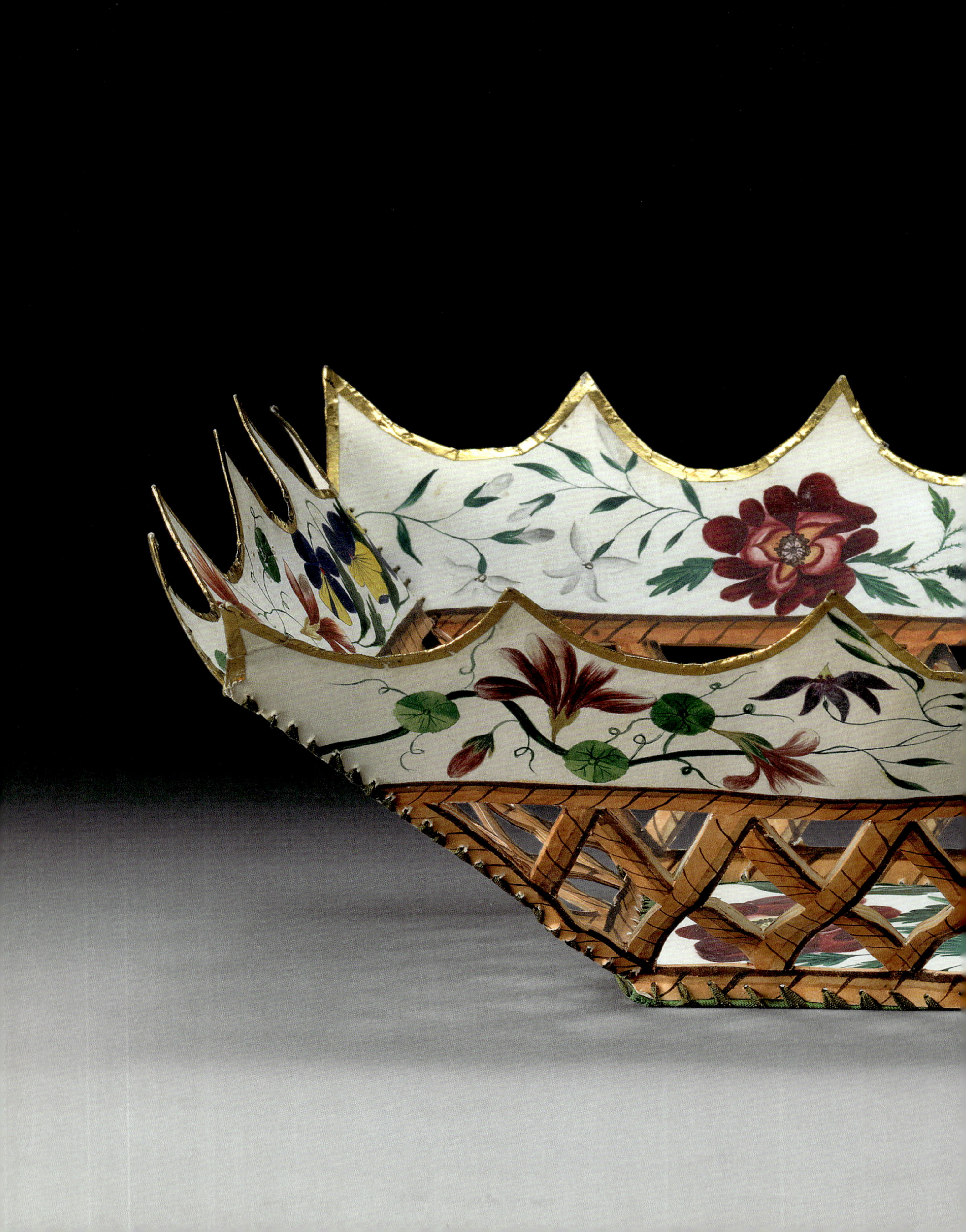

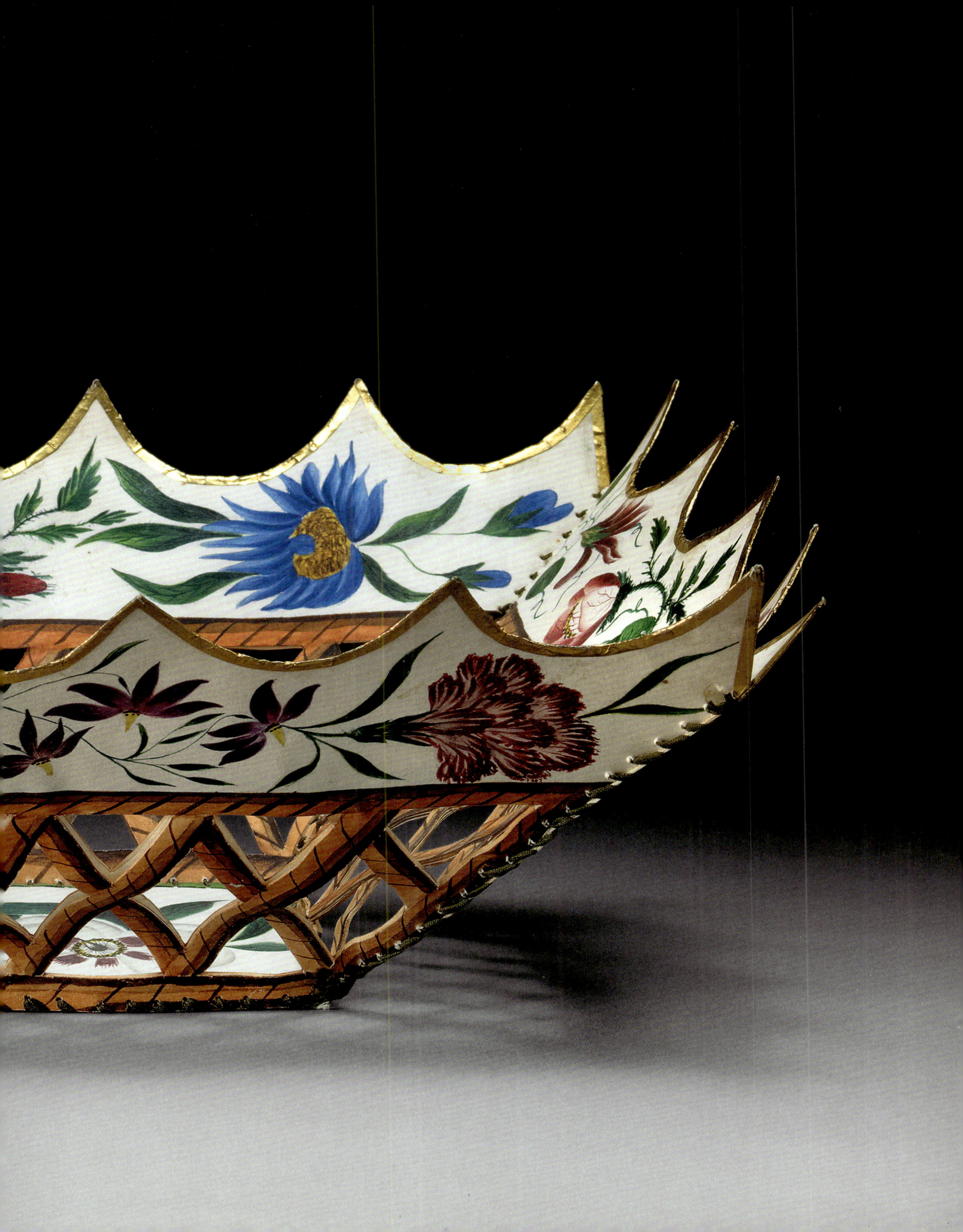

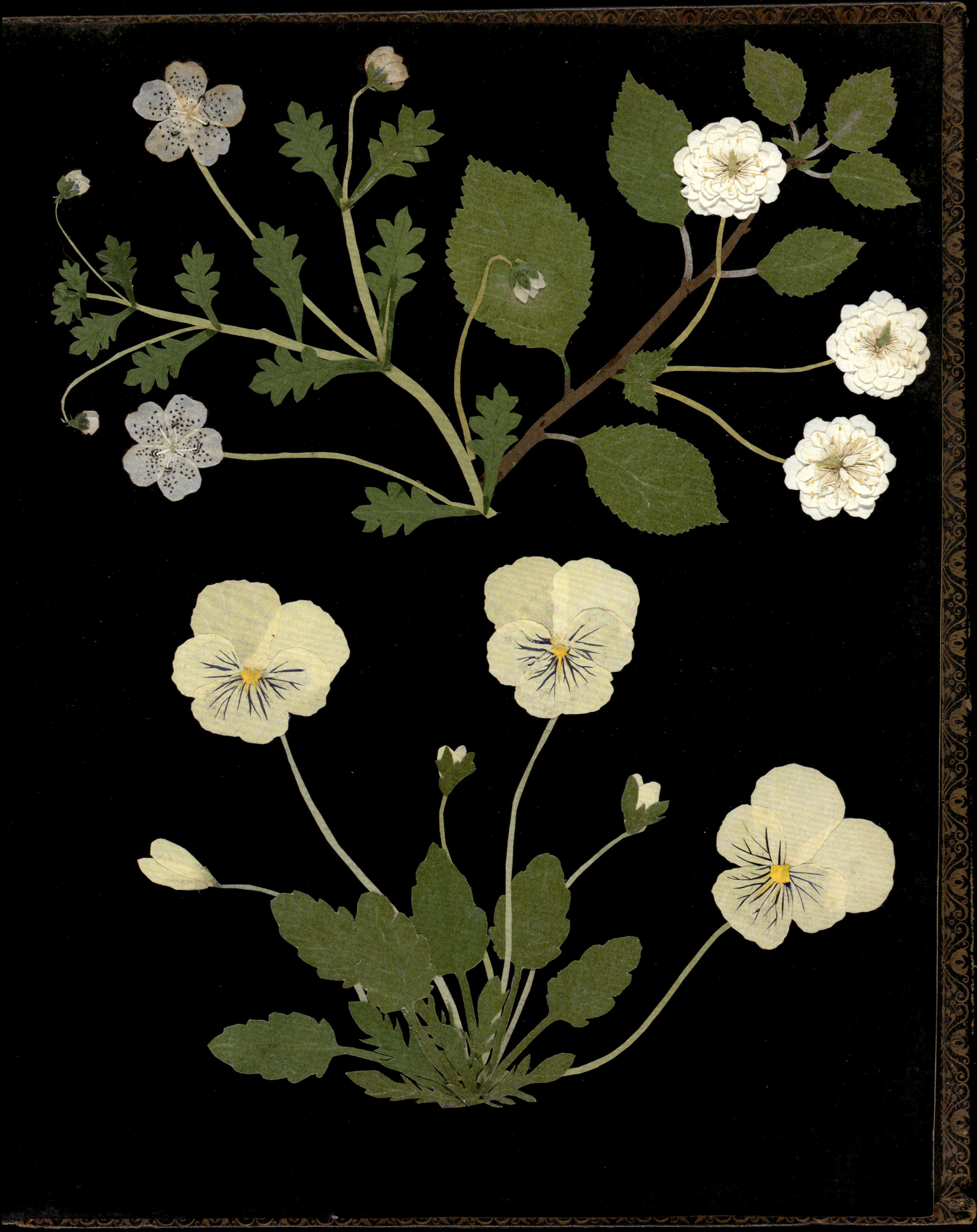

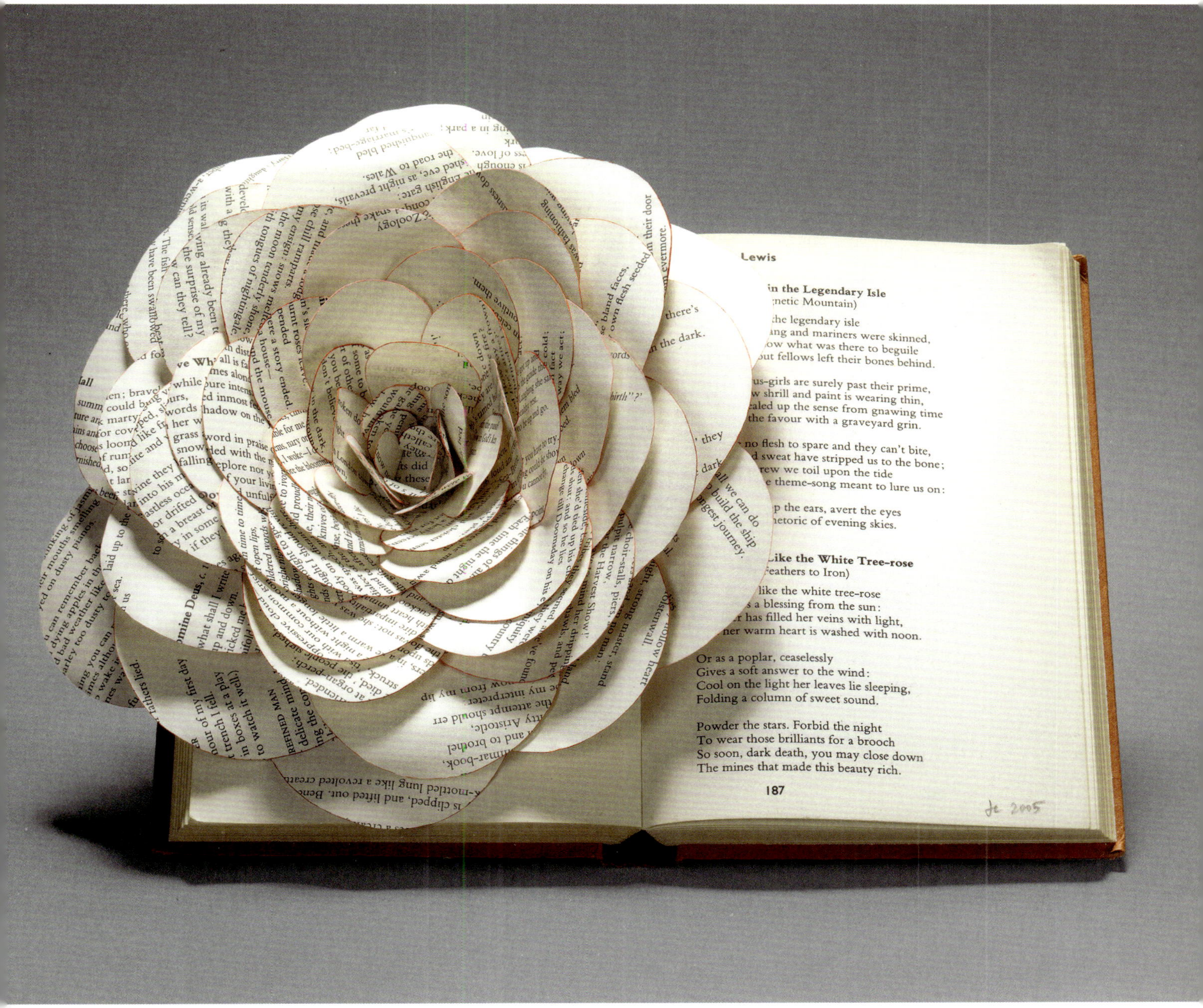

Julie Cockburn, *Flower Book: White Tea-Rose*, 2005, altered book with watercolor. Yale Center for British Art, Friends of British Art Fund

OPPOSITE: Ellen W., *Album of Cut-Paper Flowers*, ca. 1835, hand-cut paper and collage. Yale Center for British Art, Paul Mellon Fund

OVERLEAF: *Helmingham Herbal and Bestiary* (leaves 3v–4r) ca. 1500, watercolor and gouache on parchment. Yale Center for British Art, Paul Mellon Collection

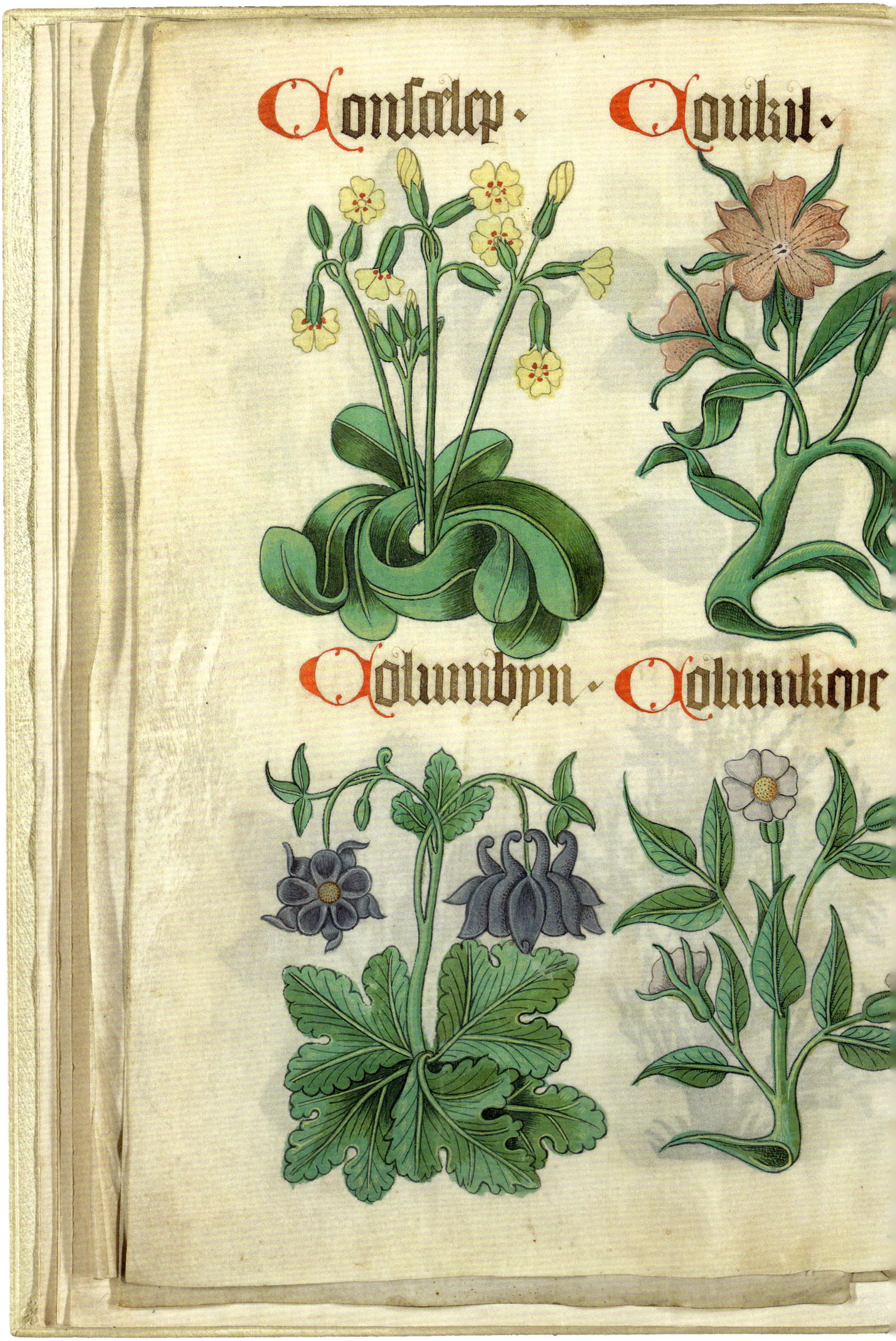

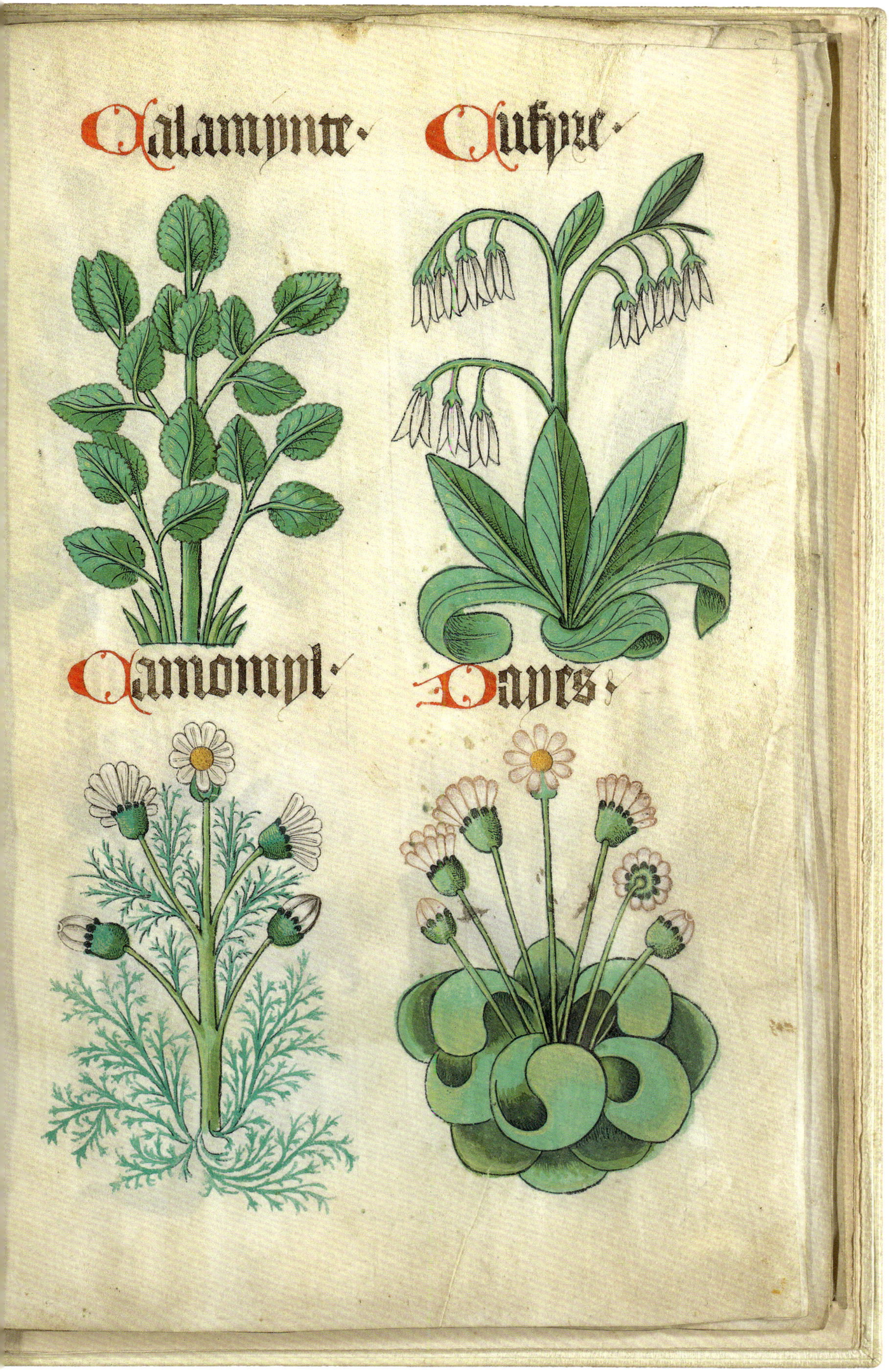
Calamynte·
Camomyl·
Dayes·

Blackberries & Brambles

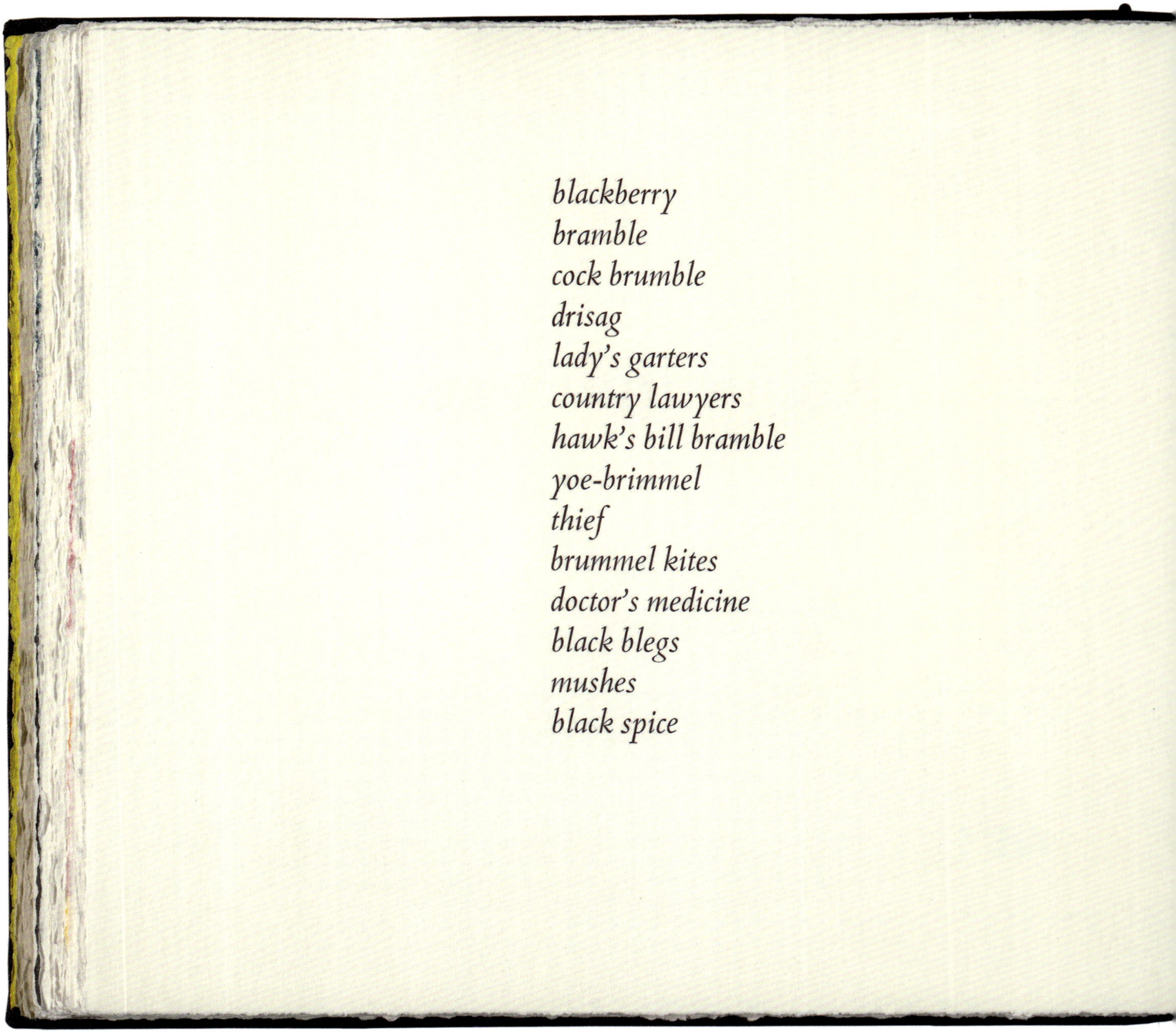

Rosaleen Wain, "Blackberry," etching with colored aquatint in *A Printmaker's Flora: An Anthology of the Names of British Wild Flowers*, Dartington Hall, Devon: Dartington Printmakers, 1996. Yale Center for British Art, Friends of British Art Fund

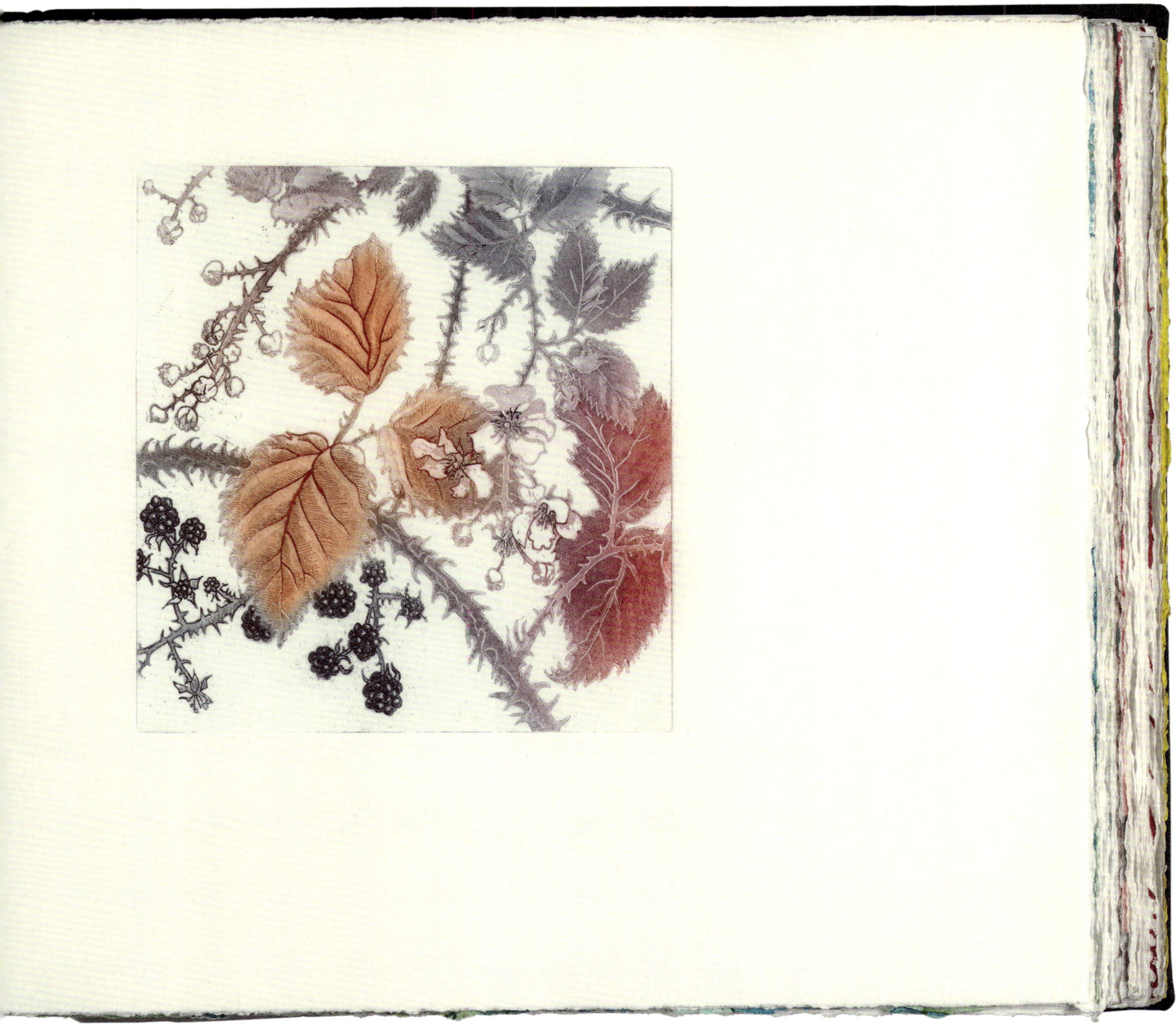

Box·

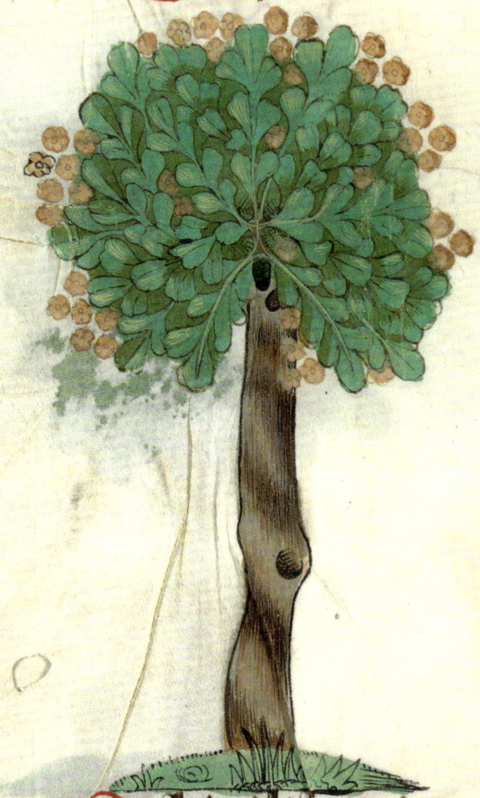

Burch·

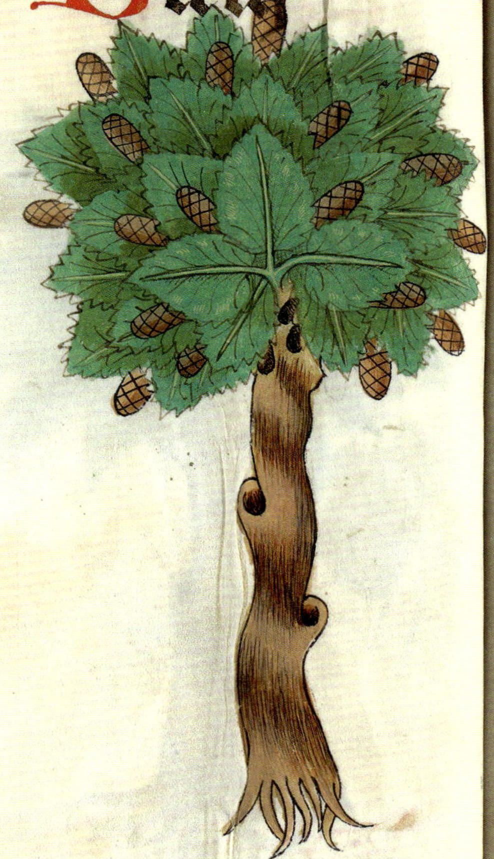

Blackbere·

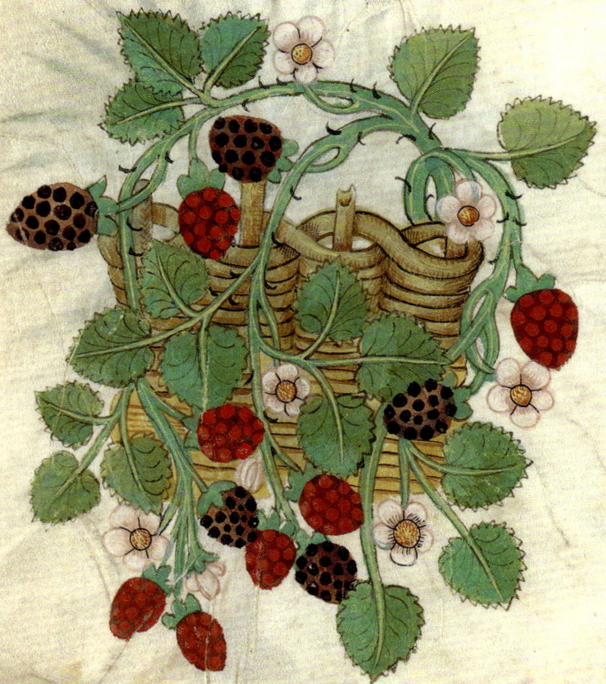

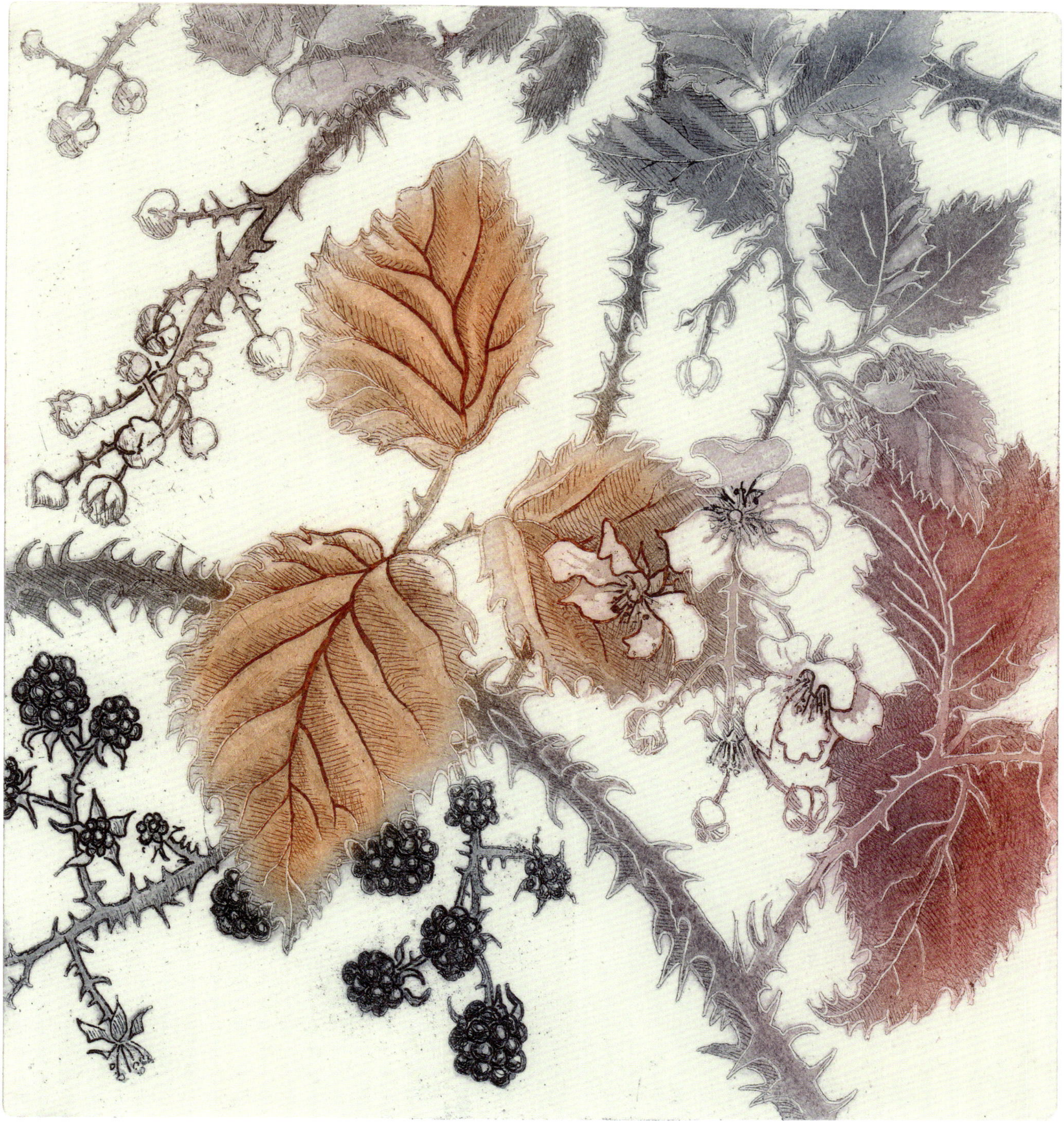

Rosaleen Wain, "Blackberry" (detail), etching with colored aquatint, in *A Printmaker's Flora: An Anthology of the Names of British Wild Flowers*, Dartington Hall, Devon: Dartington Printmakers, 1996. Yale Center for British Art, Friends of British Art Fund

OPPOSITE: *Helmingham Herbal and Bestiary* (leaf 13v), ca. 1500, watercolor and gouache on parchment. Yale Center for British Art, Paul Mellon Collection

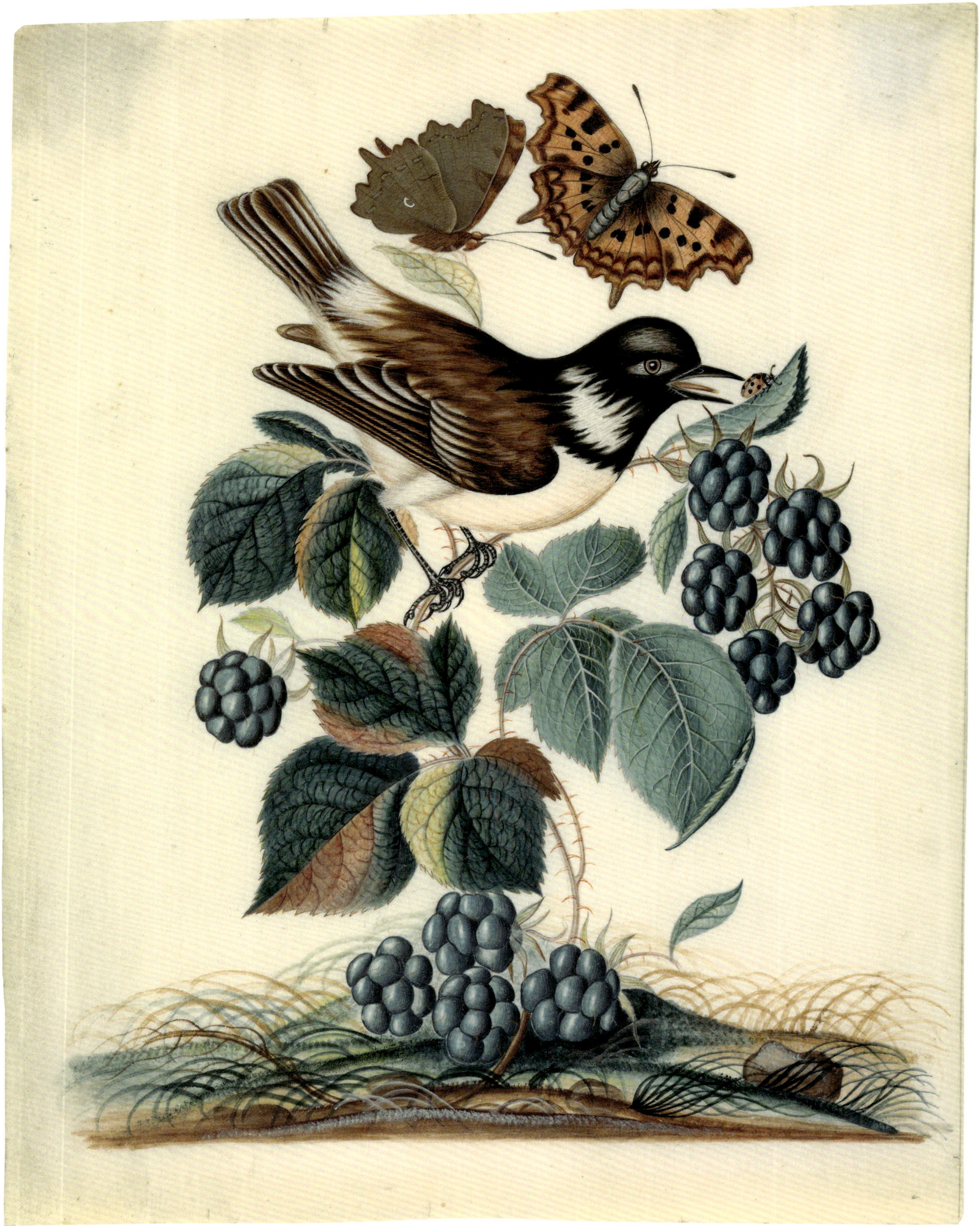

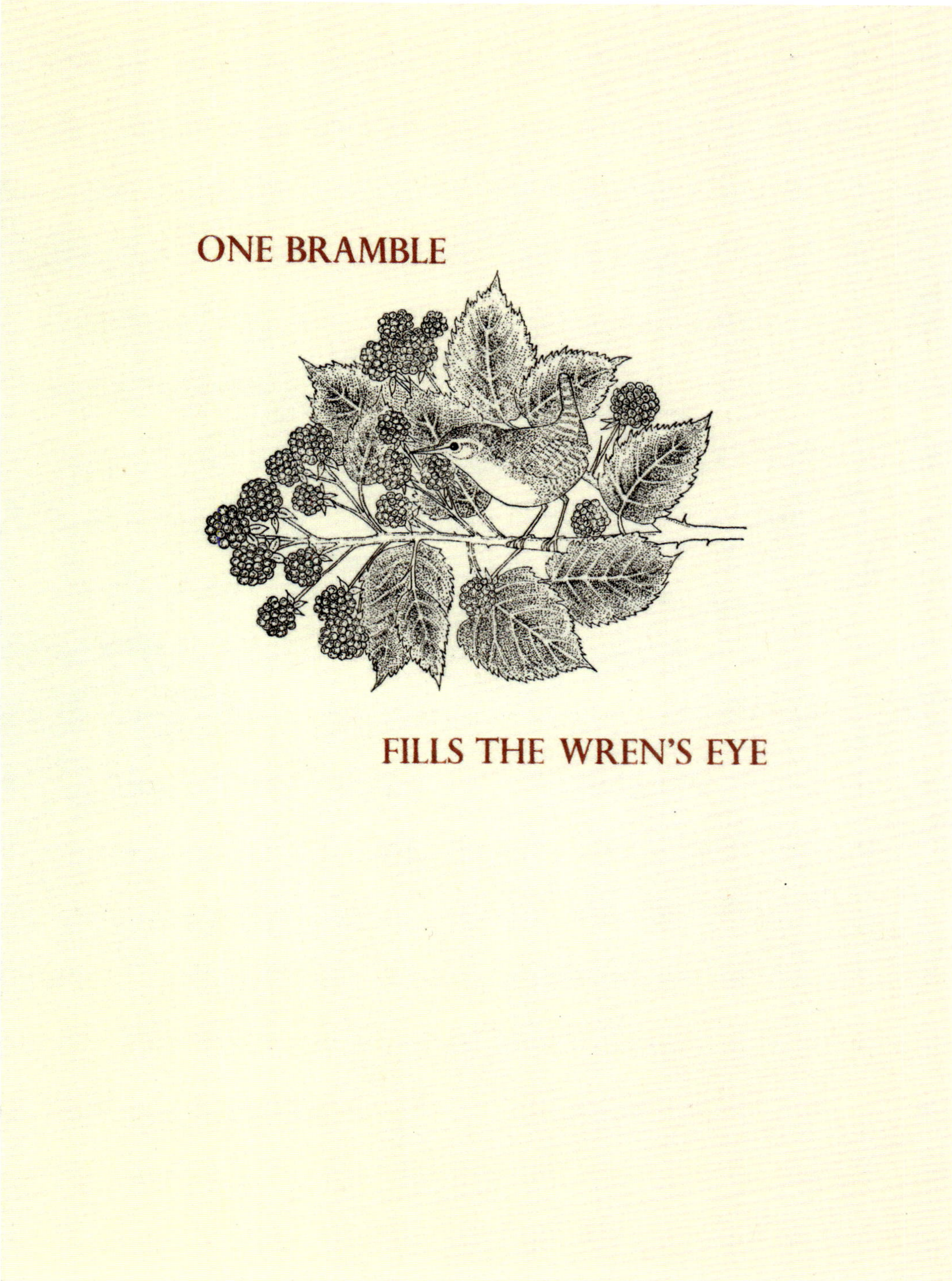

Thomas A. Clark and Laurie Clark, "One bramble fills the wren's eye" from *Proverbs of the Meadow*, Nailsworth, Gloucestershire: Moschatel Press, 1979, line block engraving. Yale Center for British Art, Friends of British Art Fund

OPPOSITE: James Bolton, ?White-cheeked Starling (*Sturnus cineraceus*) with Bramble (*Rubus fruticosus* L.) and Eurasian Comma (*Polygonia c-album)*, both closed and open, and Seven-Spotted Ladybird Beetle (*Coccinella septumpunctata*) on leaf, from the natural history cabinet of Anna Blackburne, ca. 1768, watercolor and gouache over graphite on parchment. Yale Center for British Art, Paul Mellon Fund, in honor of Jane and Richard C. Levin, President of Yale University (1993–2013)

Fruit & Vegetables

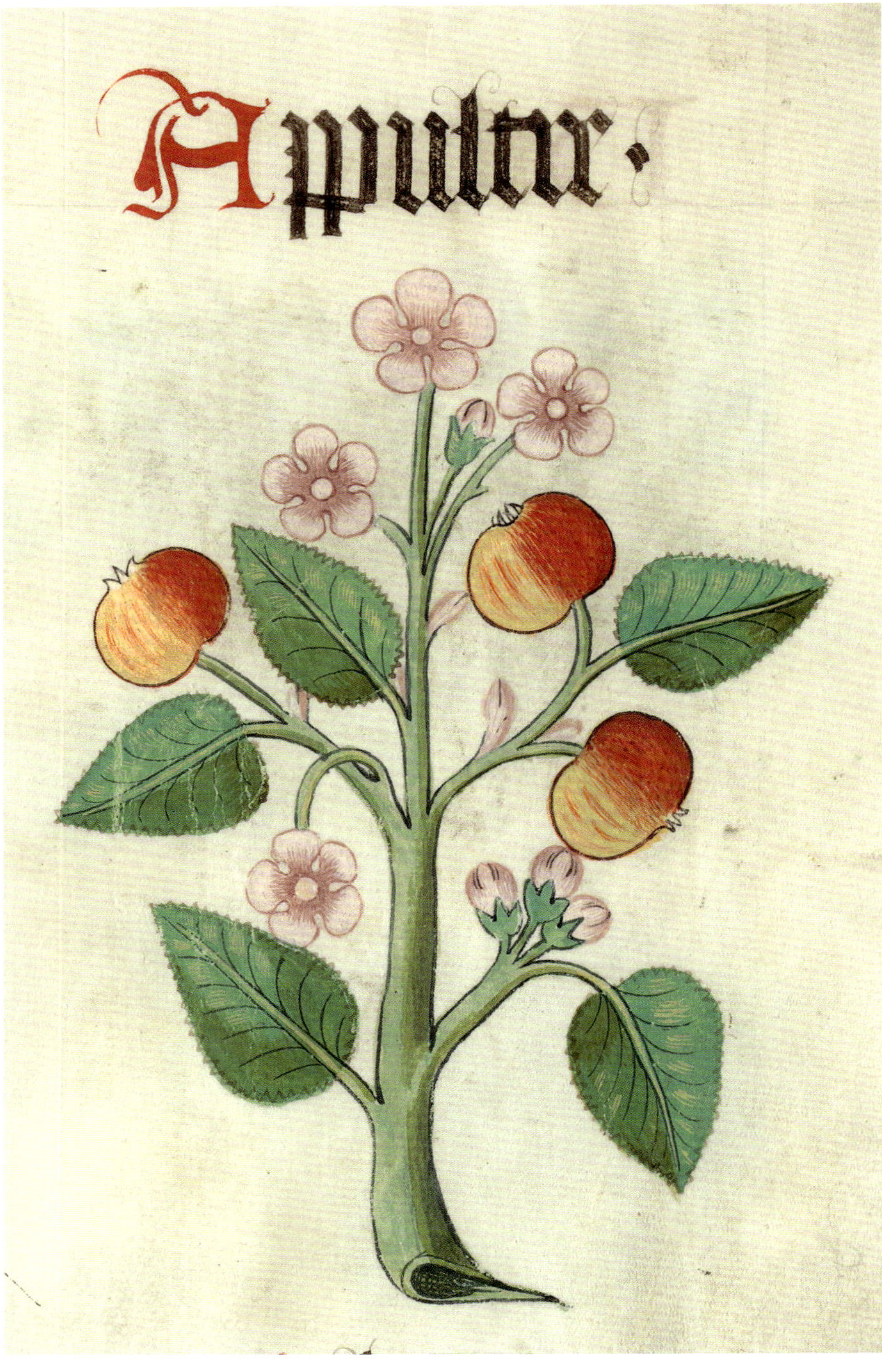

"Appultre" [apple tree] from *Helmingham Herbal and Bestiary* (detail, leaf 11r), ca. 1500, watercolor and gouache on parchment. Yale Center for British Art, Paul Mellon Collection

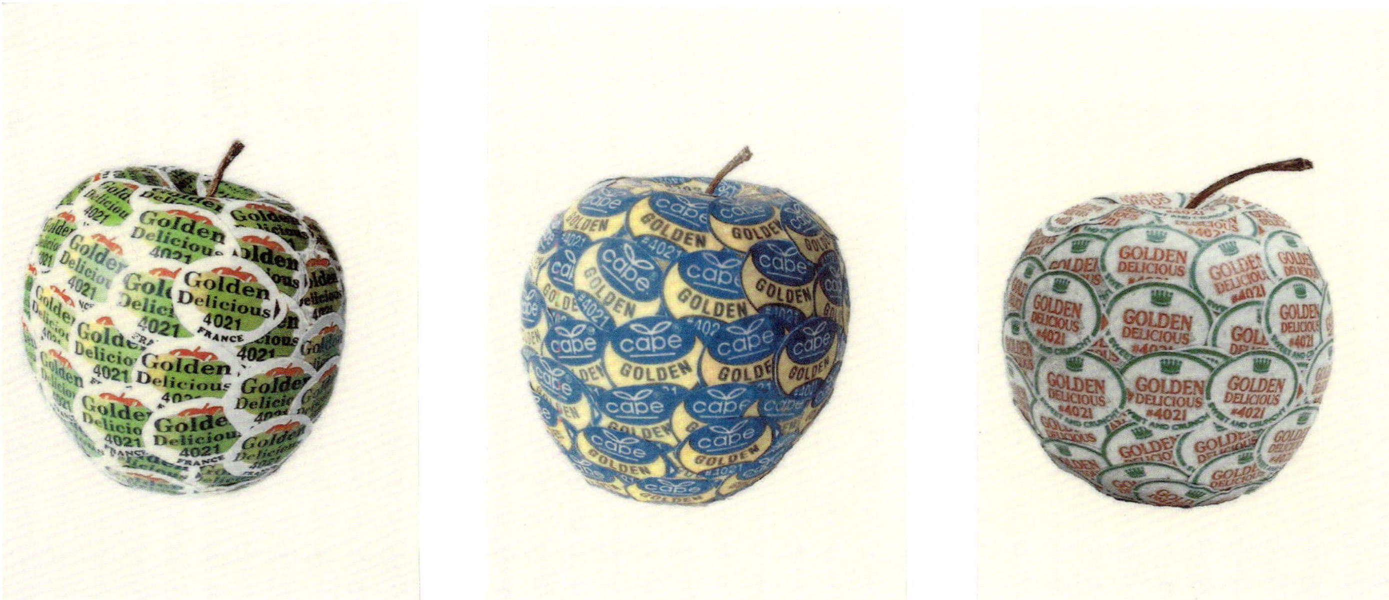

Anne Rook, from *The Book of 4021 & 4020: (Golden Delicious)*, London: MM Visual Catering, 2002, digital print. Yale Center for British Art, Friends of British Art Fund

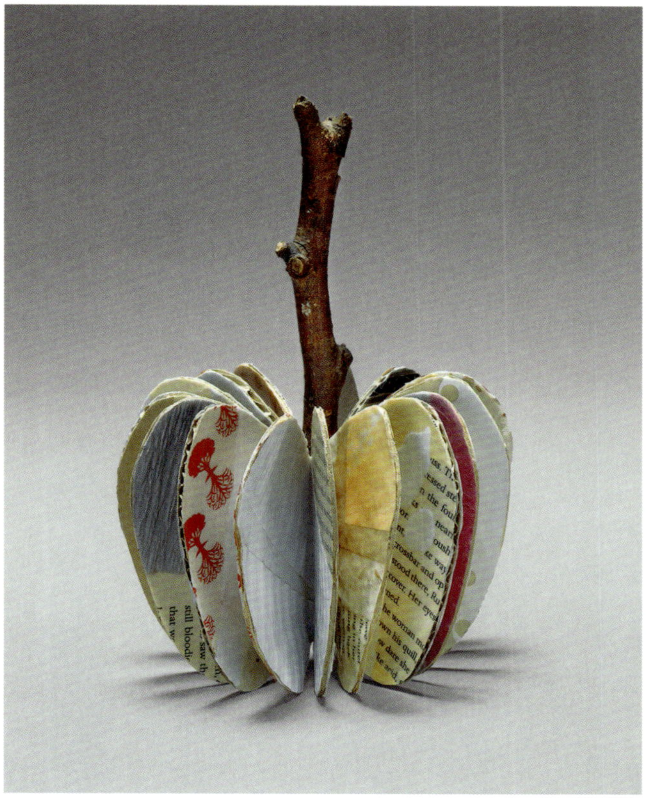

Eiko Yoshida, *Apple*, London, 2013, hand-cut paper and twig. Yale Center for British Art, Friends of British Art Fund

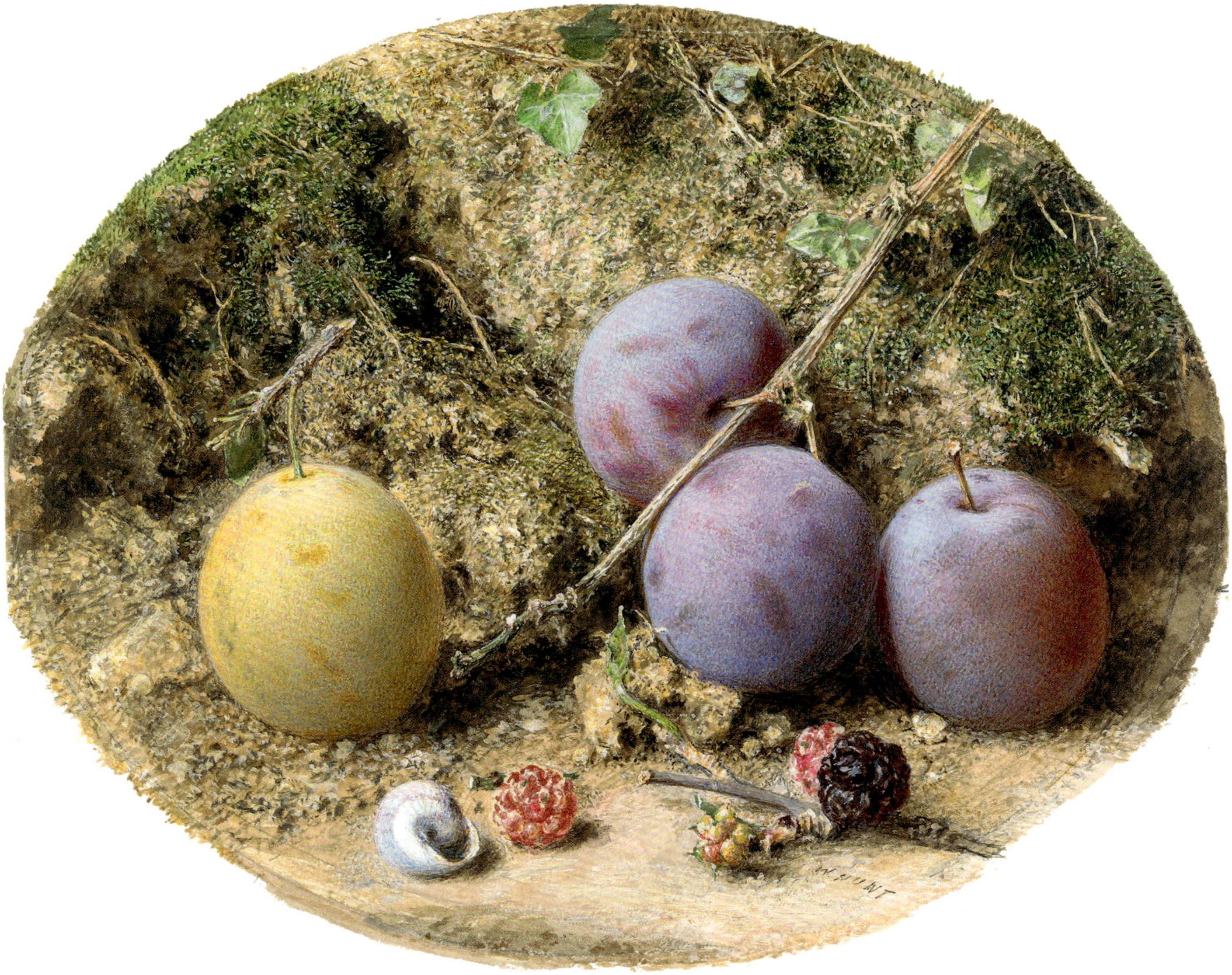

William Henry Hunt, *Plums and Mulberries*, ca. 1860, watercolor, gouache, and gum over graphite. Yale Center for British Art, Paul Mellon Collection

OPPOSITE: James Bolton, European Robin (*Erithacus rubecula*) and eggs, with Wild Strawberry (*Fragaria vesca* L.), from the natural history cabinet of Anna Blackburne, ca. 1768, watercolor and gouache over graphite on parchment. Yale Center for British Art, Paul Mellon Fund, in honor of Jane and Richard C. Levin, President of Yale University (1993–2013)

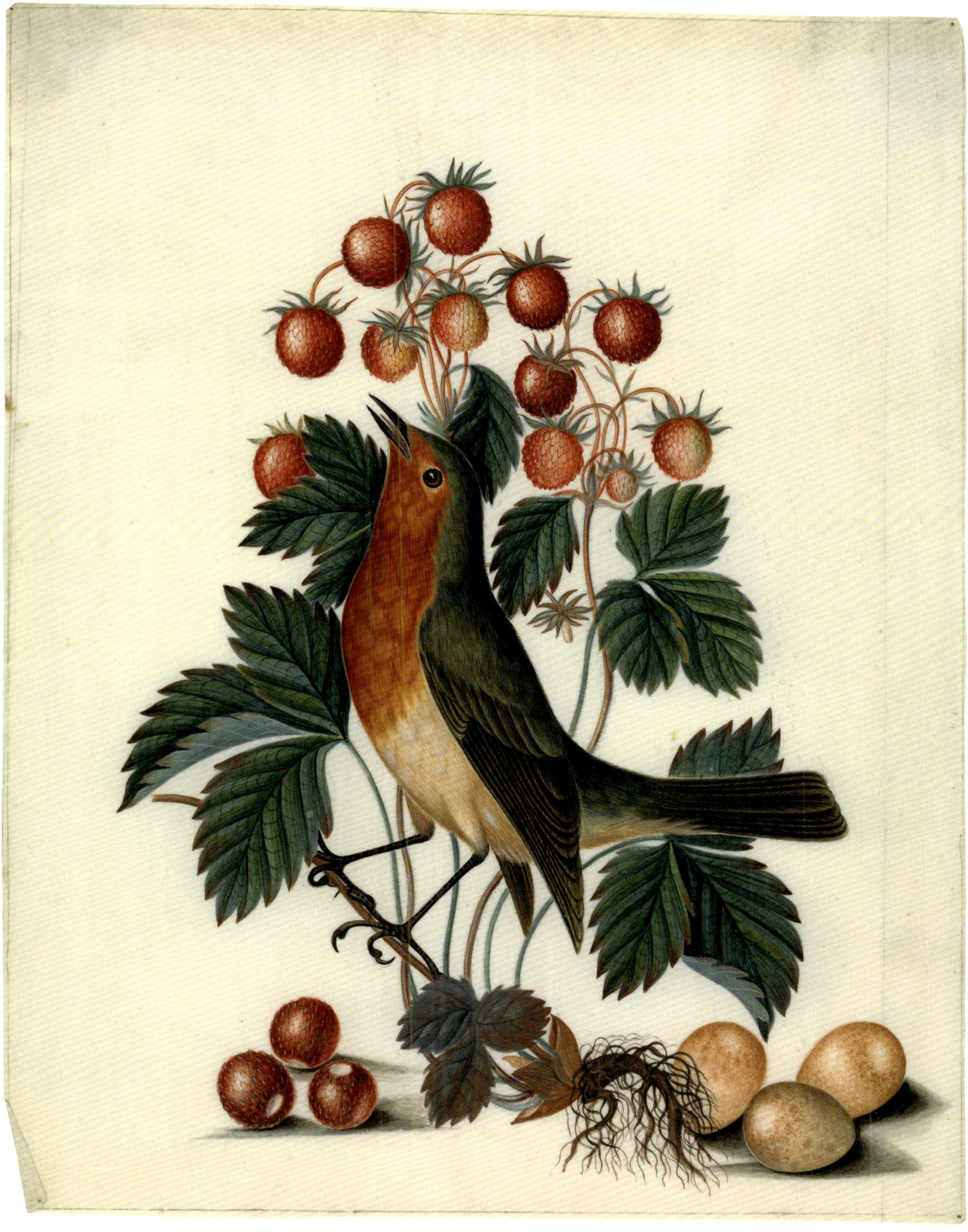

Vegetable Papyrus

My work encompasses papermaking (using vegetables and fruit), book art, and printmaking. I studied printmaking at the University for the Creative Arts and live in the Kent countryside working from my garden studio.

Using techniques similar to those of the papyrus-makers of the ancient Egyptians, I fuse overlapping fruit/vegetable fibers, which laminate to form a unique paper. Each sheet of my papyrus becomes an arrangement of color and tone, retaining the essence of its origins without sacrificing the essential reality of the natural form. It is clear (to me) that the color and patterns created within these tactile works require no further embellishment. The Sussex artist Keith Vaughan said, in reference to his landscapes, "Nature provides living material which is the starting point of art." These unique sheets of papyrus are an exploration of structure, form, and tone based on reality. The fruit and vegetables, the living material, were my starting point.

Within my work I use food in many different guises and find the most exciting thing about this medium is the unexpected results. The colors and textures are randomly created throughout the process. When I begin a piece I am never sure what the outcome will be; the layering of sliced vegetables and fruit forms independent organic imagery. I am attempting to explore the use of food as a metaphorical image of the essence of human nourishment.

Dizzy Pragnell
ARTIST

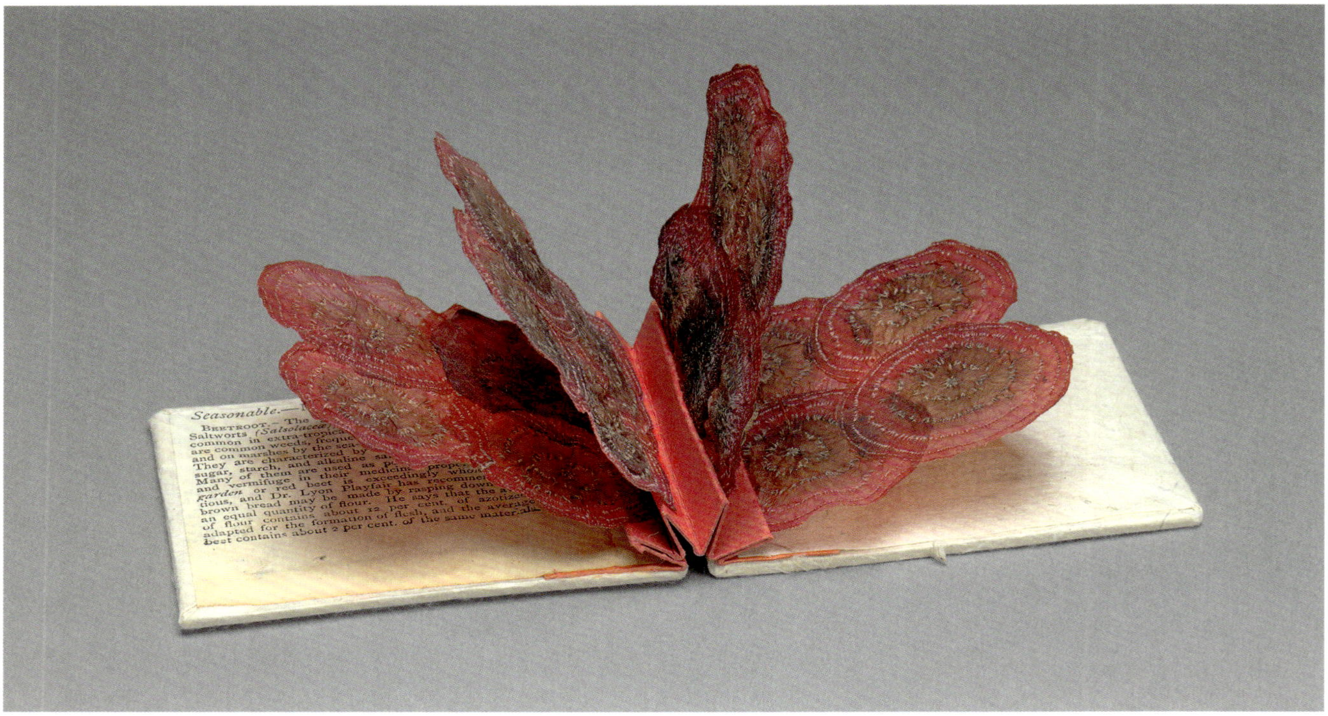

Dizzy Pragnell, *Beetroot*, Canterbury, Kent, 2013, vegetable papyrus with leaf of printed text. Yale Center for British Art, Friends of British Art Fund

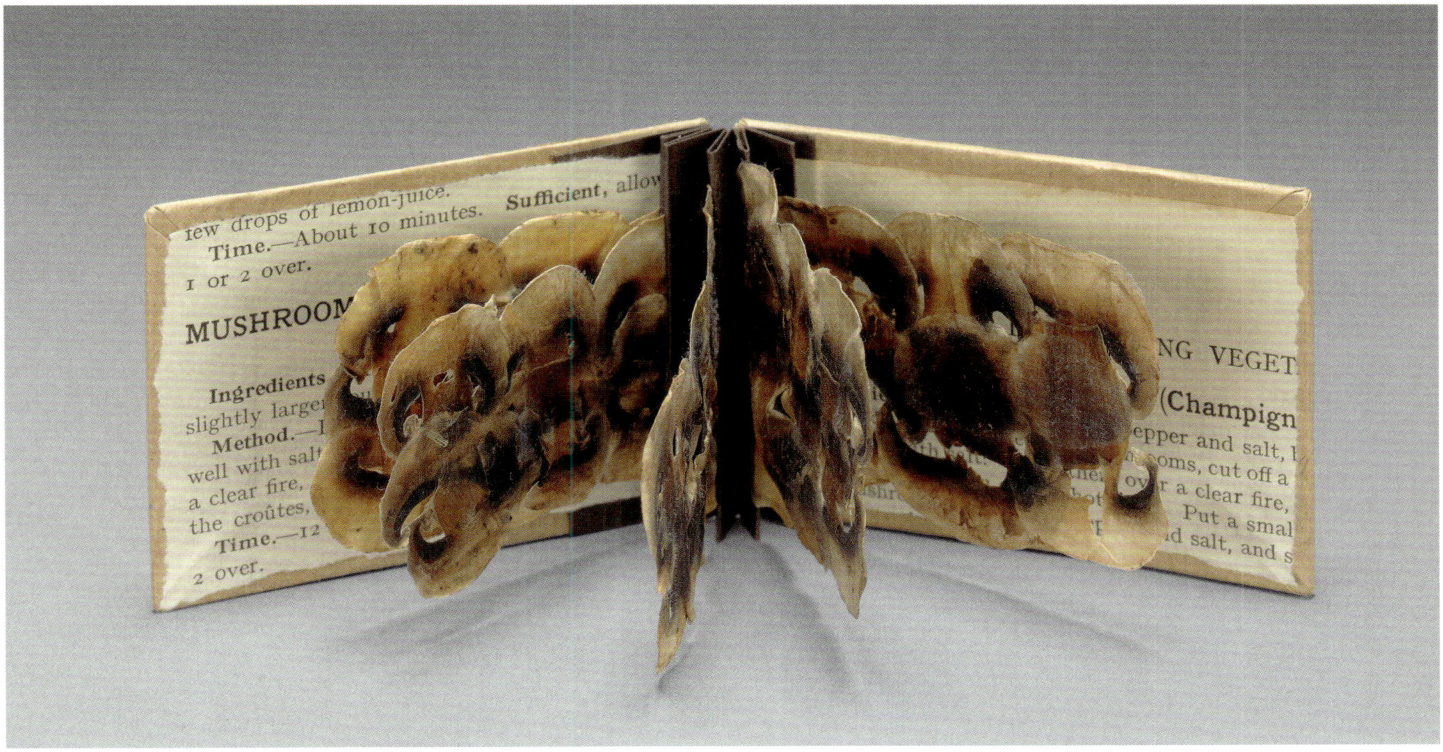

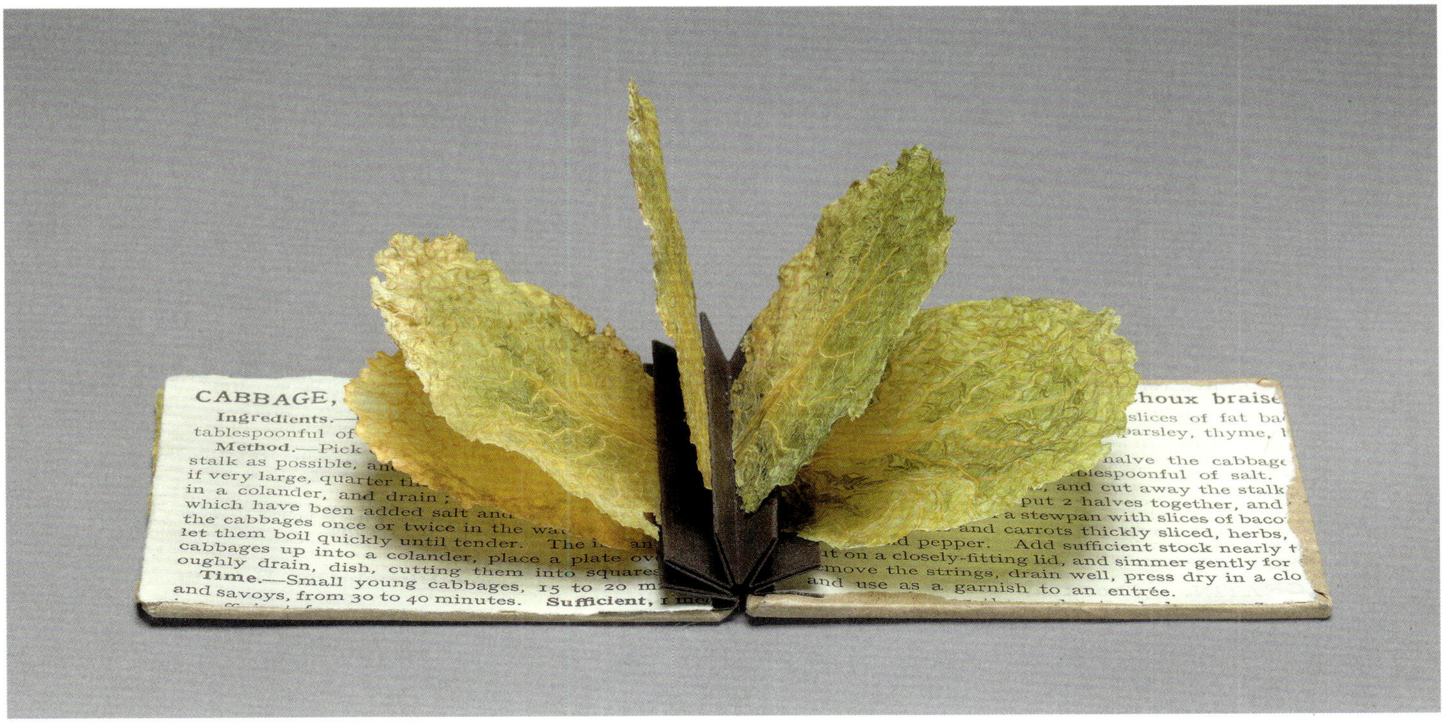

Dizzy Pragnell, "Mushroom" and "Cabbage" from *Provenance: Vegetable Box*, Canterbury, Kent, 2013, vegetable papyrus with leaves of printed text. Yale Center for British Art, Friends of British Art Fund

Grasses

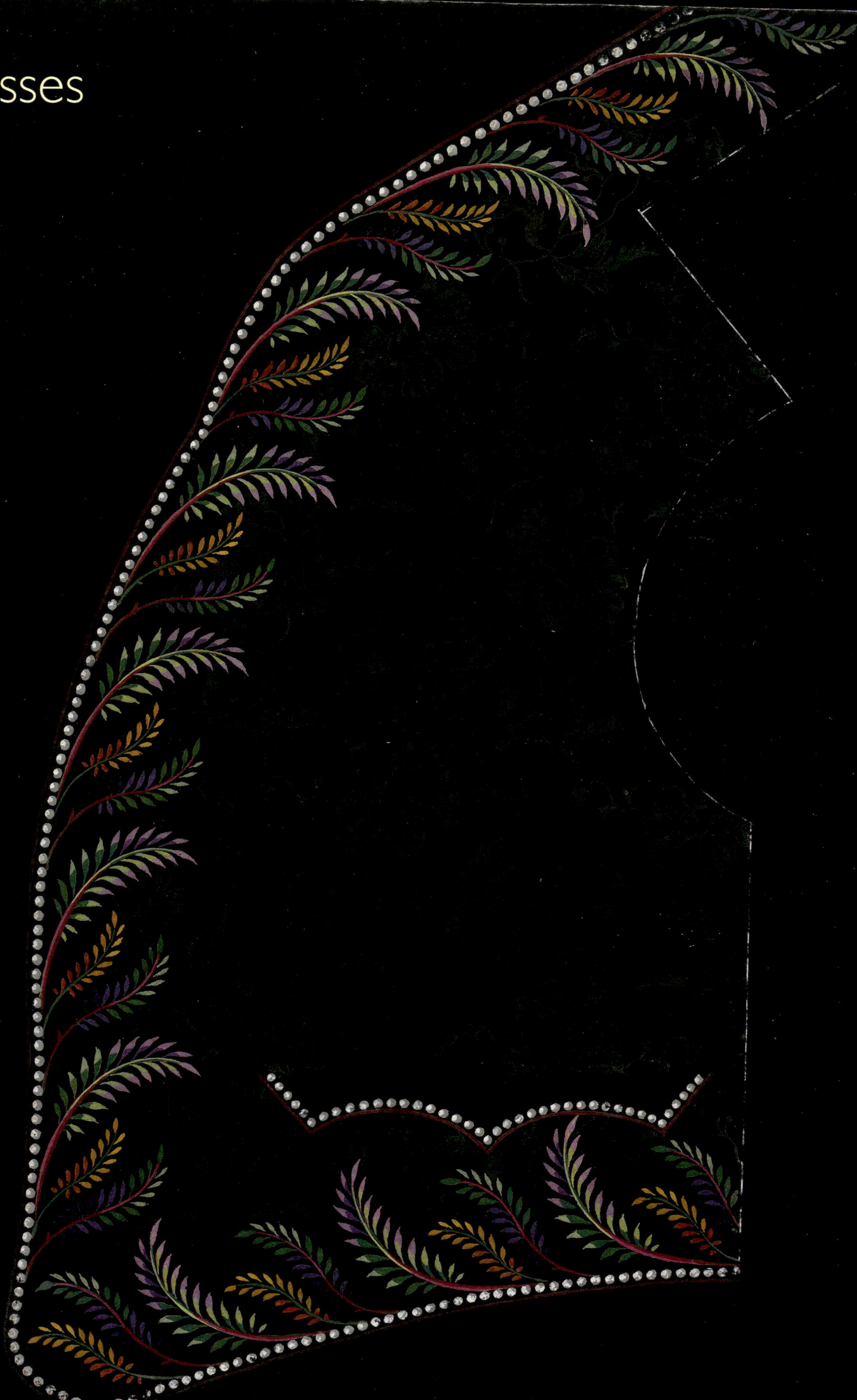

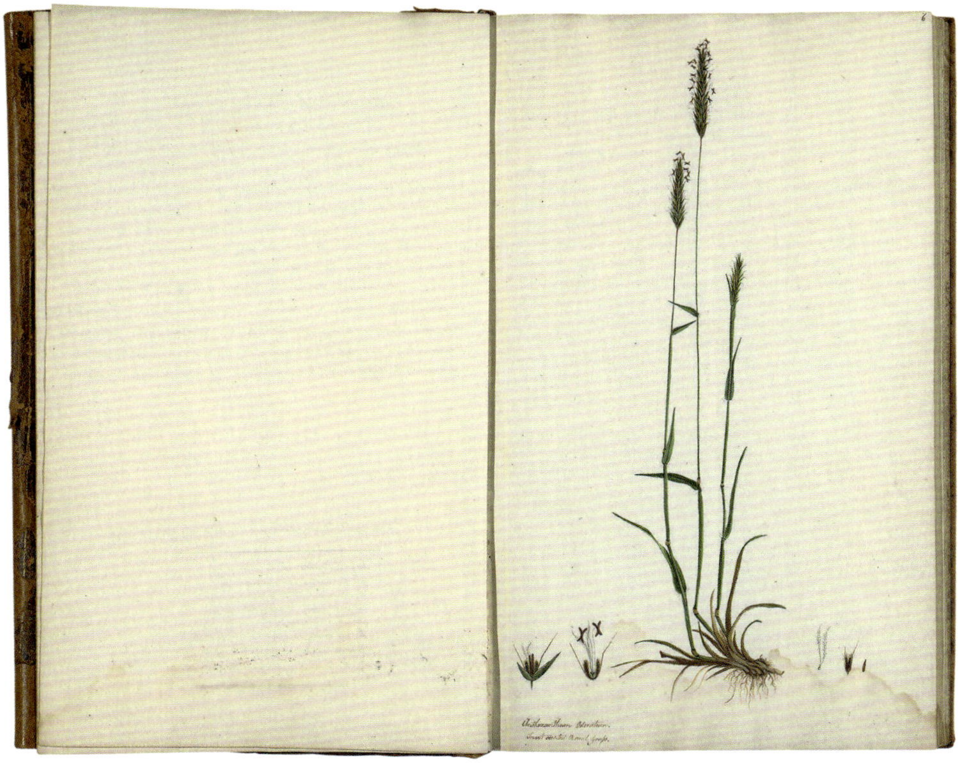

Harriet Wickham, *A Collection of English Grasses*, ca. 1792–1808, pen and ink with watercolor. Yale Center for British Art, Paul Mellon Fund

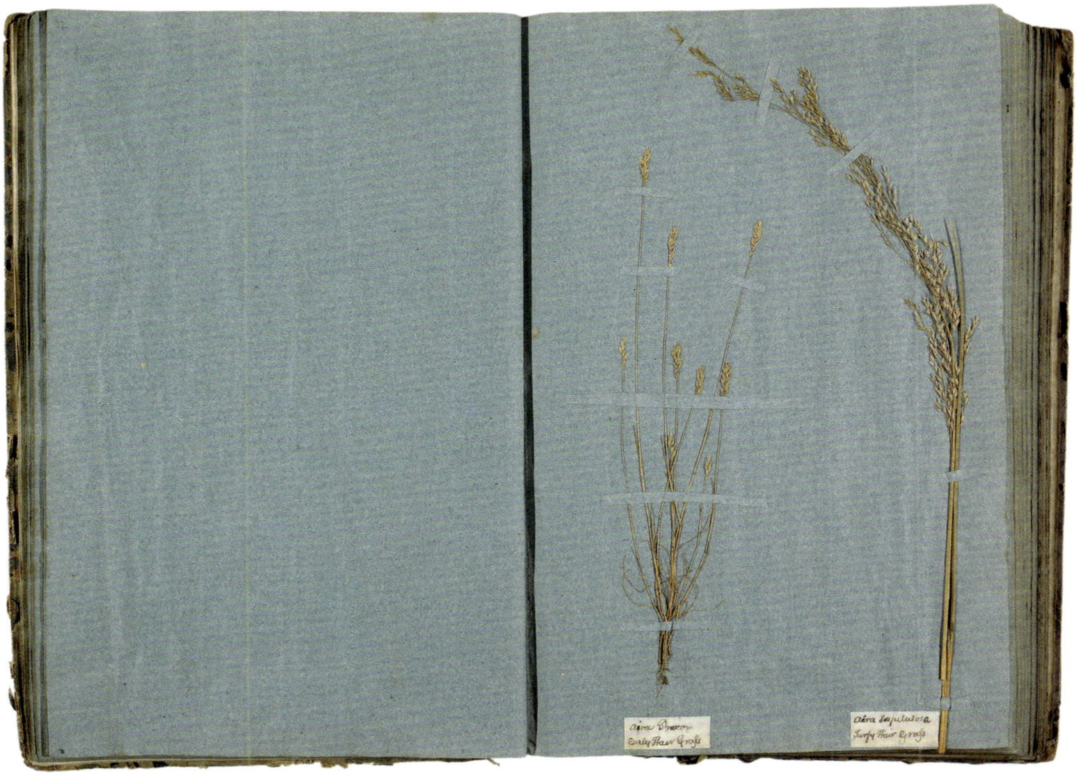

Album of British Grasses, England, ca. 1800, natural specimens with pen and ink labels. Yale Center for British Art, Paul Mellon Fund

OPPOSITE: *Design for Embroidered Waistcoat*, Britain or France, ca. 1845, watercolor on black coated paper. Yale Center for British Art, Paul Mellon Collection

Trees

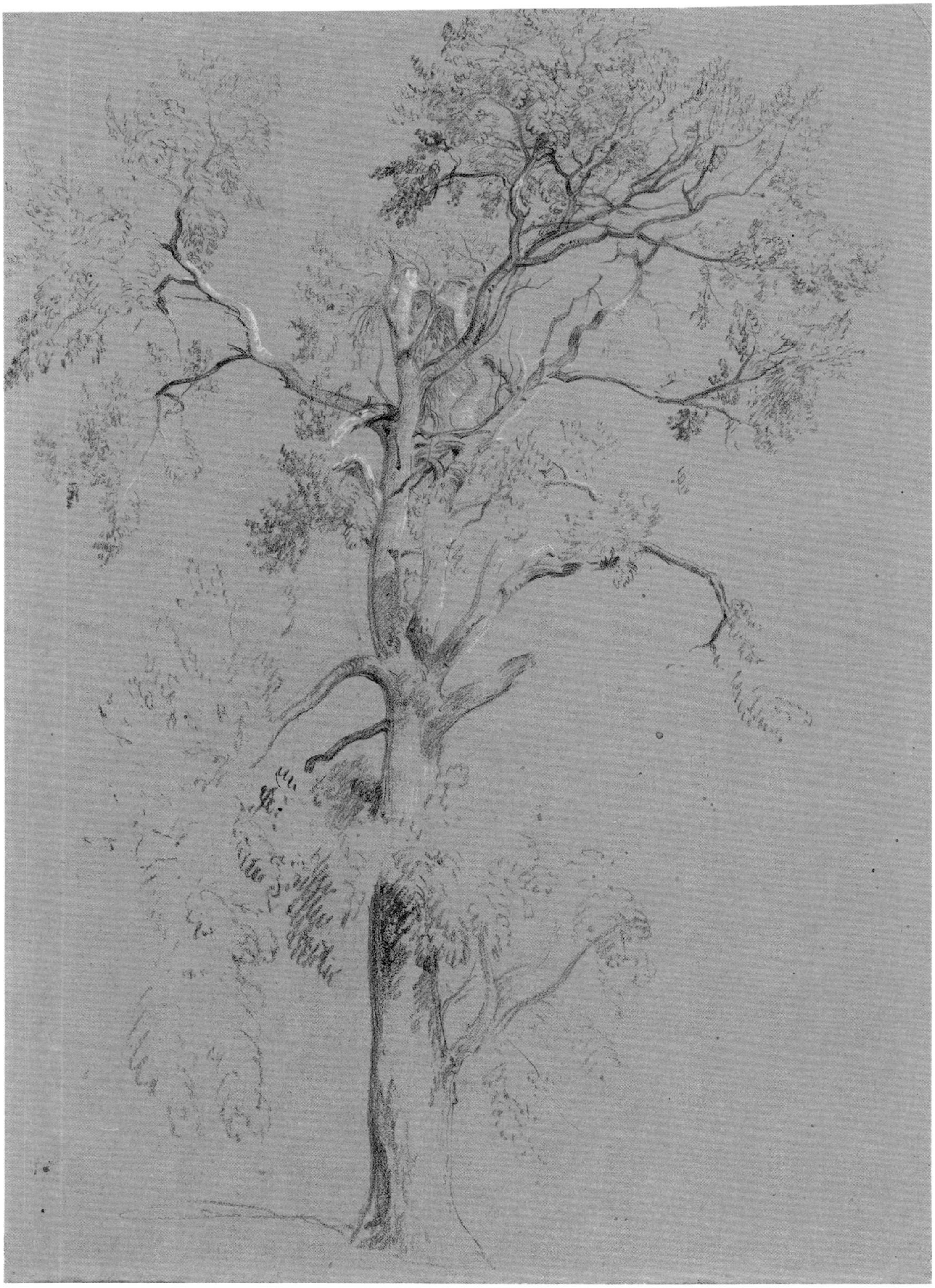

Peter De Wint, *Ash Tree in Ashstead Park*, ca. 1847, black chalk, heightened with white. Yale Center for British Art, Paul Mellon Collection

OPPOSITE: Leaves and Lord's Prayer, in *Commonplace Book, with Drawings of Insects*, ca. 1830, nature prints, pen and ink and letterpress. Yale Center for British Art, Paul Mellon Fund

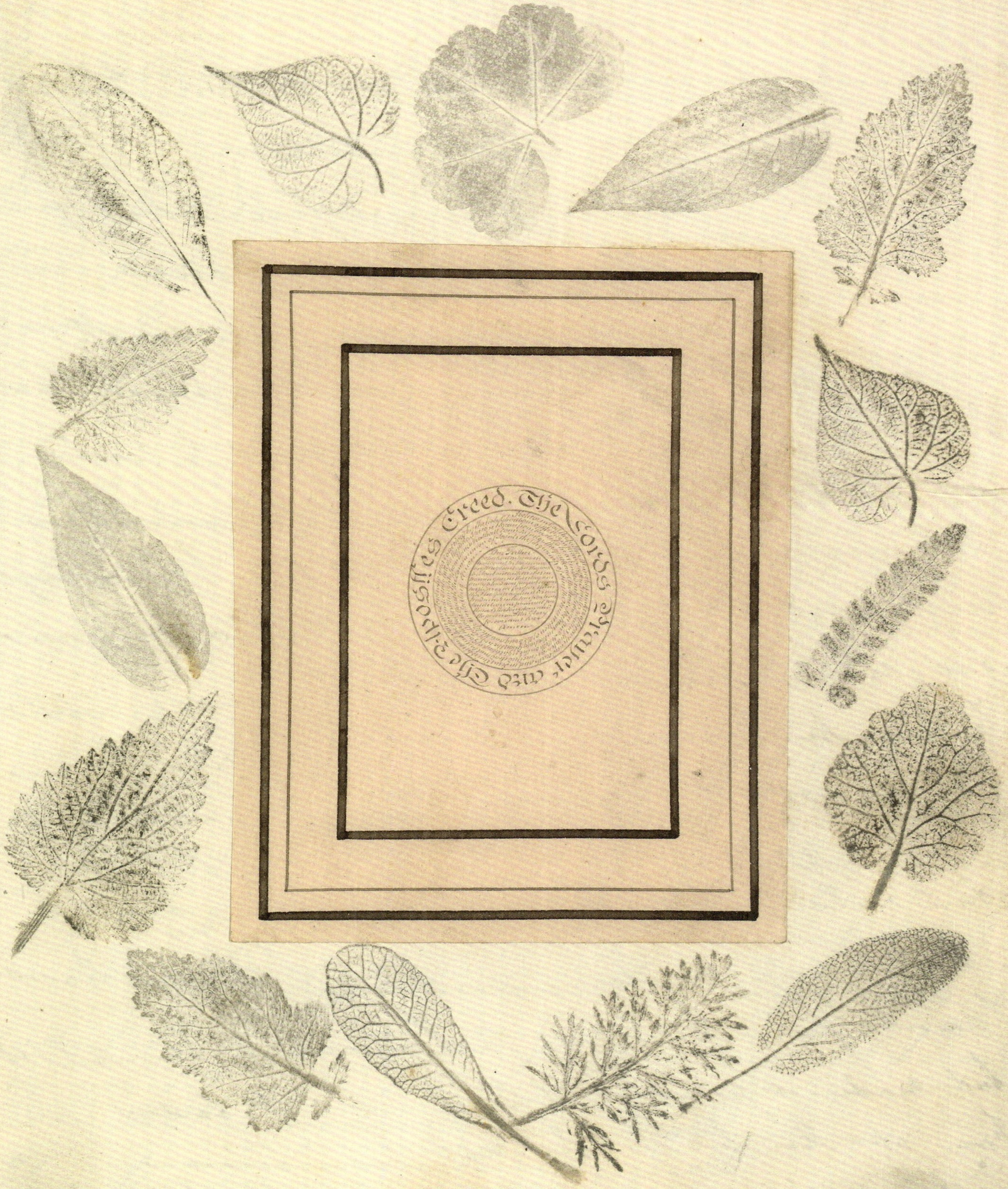
The Lords Prayer and The Apostles Creed.
Our Father

Rex Vicat Cole and the Teaching of Drawing

The study of trees was the central artistic preoccupation of the Edwardian landscape painter Rex Vicat Cole (1870–1940), and a mission to promulgate this lifelong fascination dominated his work as a teacher and writer. His book *British Trees,* which appeared in two substantial volumes in 1907, features closely observed studies, mainly in pencil, of the forms of individual species, with details of leaves and fruit. While Cole aspired to botanical accuracy, his motivations were, finally, aesthetic: the book was "intended specially as a book of reference for art students and designers."[1] Cole produced a lavishly illustrated technical treatise, *The Artistic Anatomy of Trees*, in 1915, leading art students through the complexities of representing tree-forms. Sheets of the original drawings for the book survive, providing a testimony to Cole's technical expertise.

Although his summers were spent sketching in Yorkshire or Sussex, Cole taught in the Women's Department of King's College, London, where he converted a suite of attic rooms that he "filled with small trees, branches, casts, and still-life objects."[2] Unusually, Cole taught landscape painting as a separate subject, making use of his still-life collections, but also leading weekly outings to Kew and Chelsea for outdoor sketching. For the study of trees, Cole adapted classic techniques for teaching the drawing and painting of the human body. Accuracy was prized. Yet underlying the anatomical empiricism was a mystic approach to nature: "[No one who] has felt the magic of the woods, and all but seen a fairy where some leaf twinkled, or startled the revels of elves and pixies from the mossy mound at the roots of a hollow beech," wrote Cole in *British Trees*, "can wonder at the tales of mystery that have grown up everywhere about trees."[3]

Tim Barringer
ART HISTORIAN

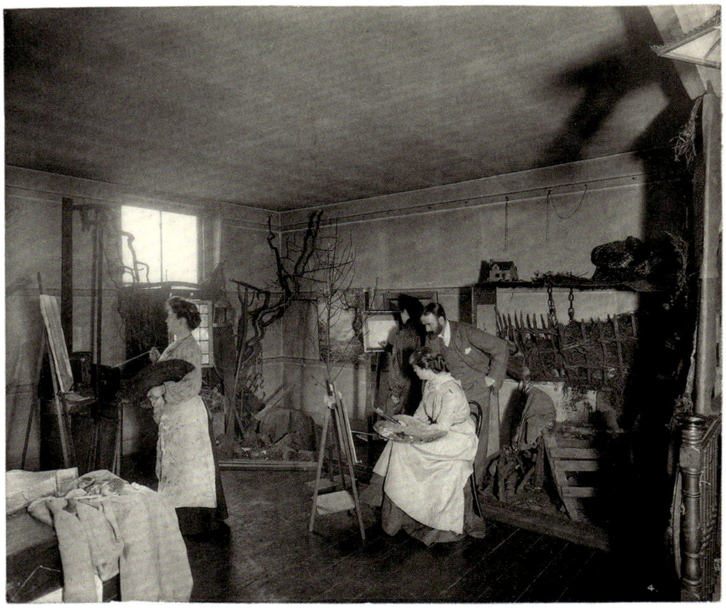

Rex Vicat Cole in his teaching room at Kings College for Women, Kensington. Private collection

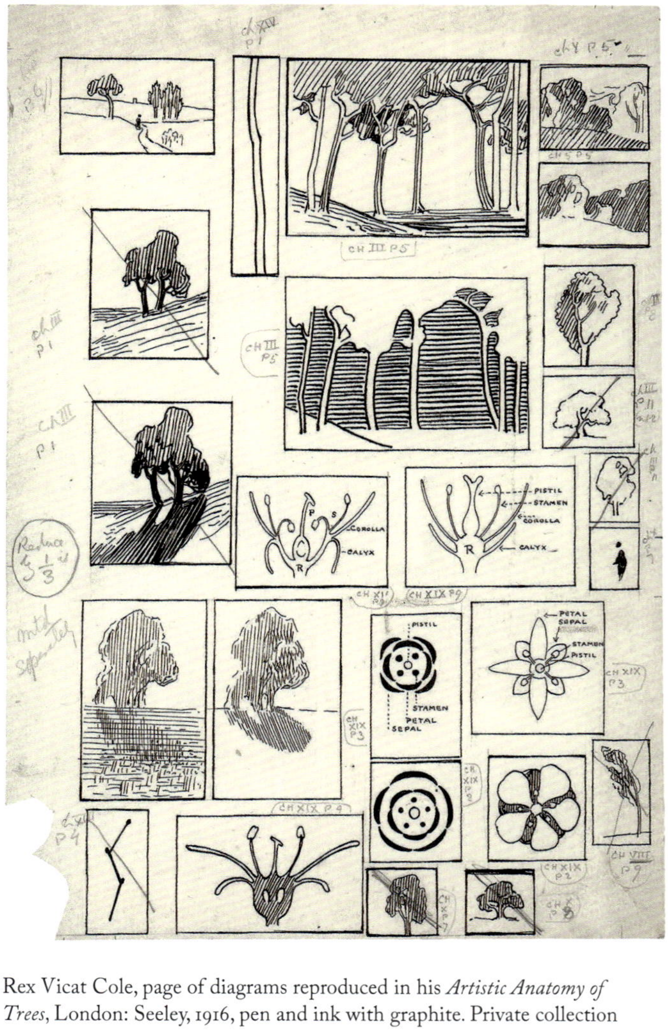

Rex Vicat Cole, page of diagrams reproduced in his *Artistic Anatomy of Trees*, London: Seeley, 1916, pen and ink with graphite. Private collection

1. Tim Barringer, *The Cole Family: Painters of the English Landscape* (Portsmouth: Portsmouth City Museums, 1988), 122. See also Tim Barringer, *Before and After Modernism: Byam Shaw, Rex Vicat Cole, Yinka Shonibare* (London: Central St. Martin's, 2010), 50–56.
2. Rex Vicat Cole, *The Art and Life of Byam Shaw* (London: Seeley Service, 1932), 134.
3. Rex Vicat Cole, *British Trees* (London: Hutchinson & Co., 1907), 2: 555.

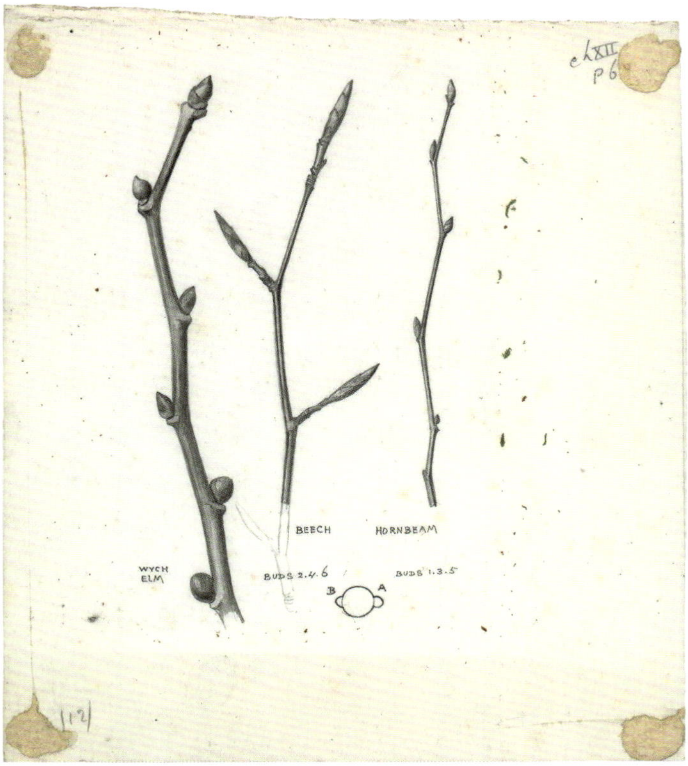

Rex Vicat Cole, drawing reproduced in his *Artistic Anatomy of Trees*, London: Seeley, 1916, pen and ink with graphite. Private collection

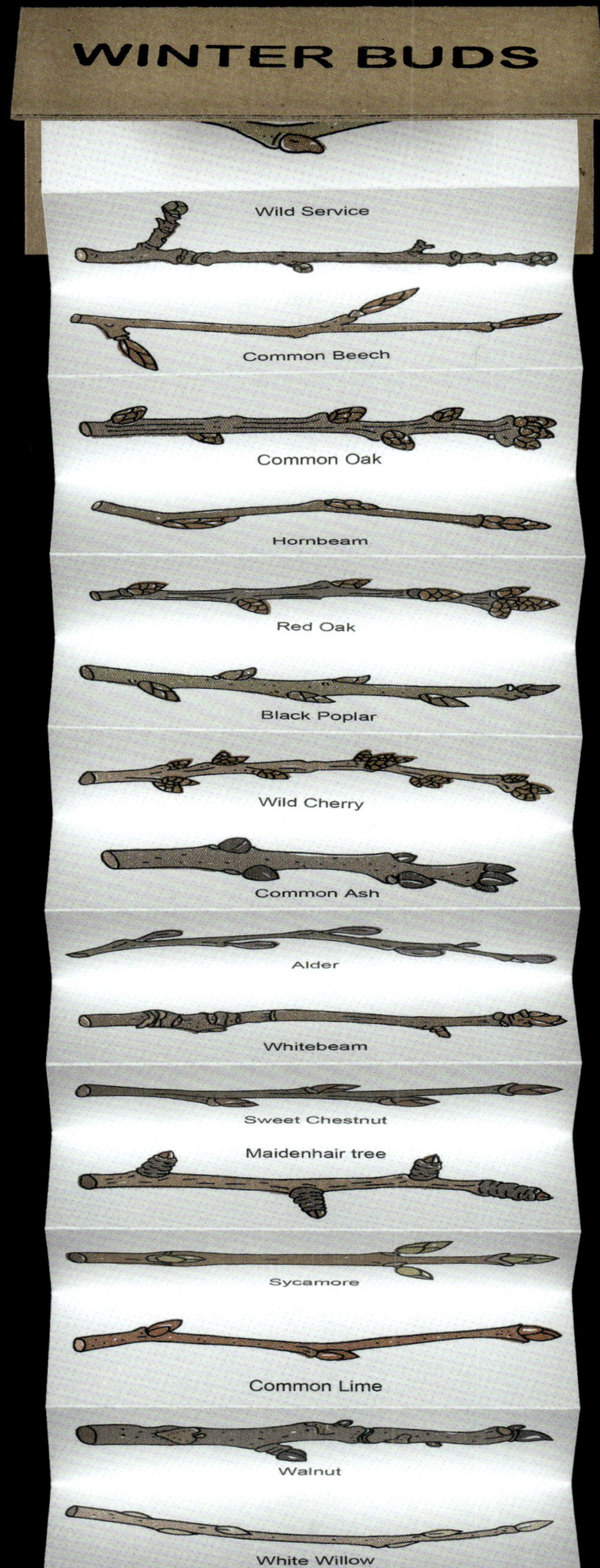

RIGHT: John Dilnot, *Winter Buds*, Brighton, 2008, screen print. Yale Center for British Art, Friends of British Art Fund

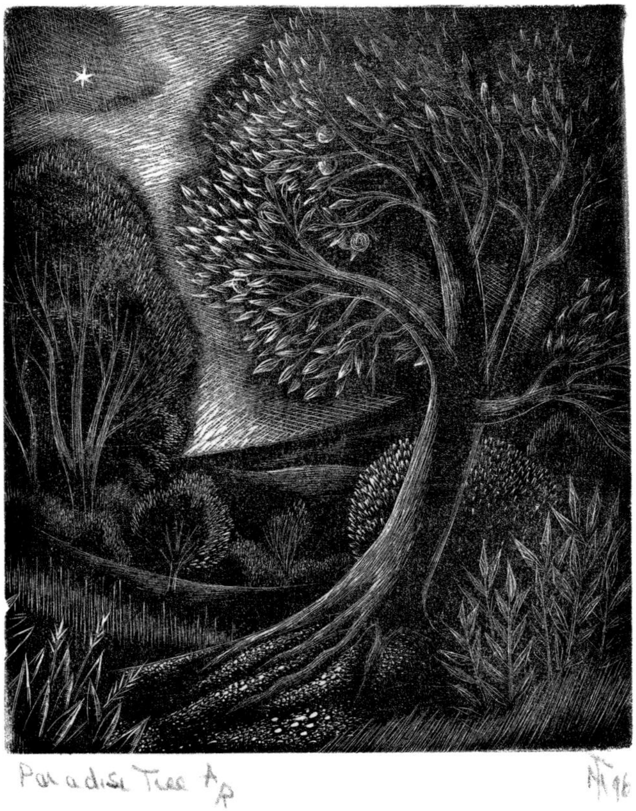

Sister Margaret Tournour, *Paradise Tree*, 1996, wood engraving. Collection of David Burnett

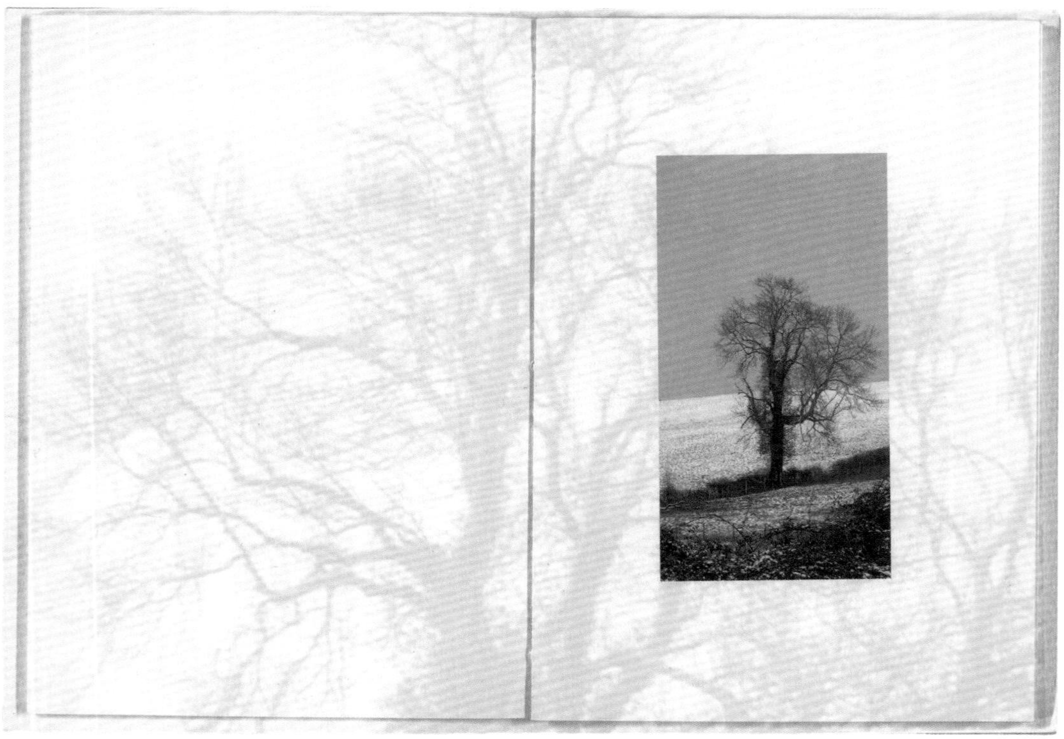

Pauline Lamont-Fisher, *Daylight*, Wormelow, England, 2006, digital print. Yale Center for British Art, Friends of British Art Fund

OPPOSITE: John Martin, *Characters of Trees*, London: Ackermann, 1817, soft ground etching. Yale Center for British Art, Paul Mellon Collection

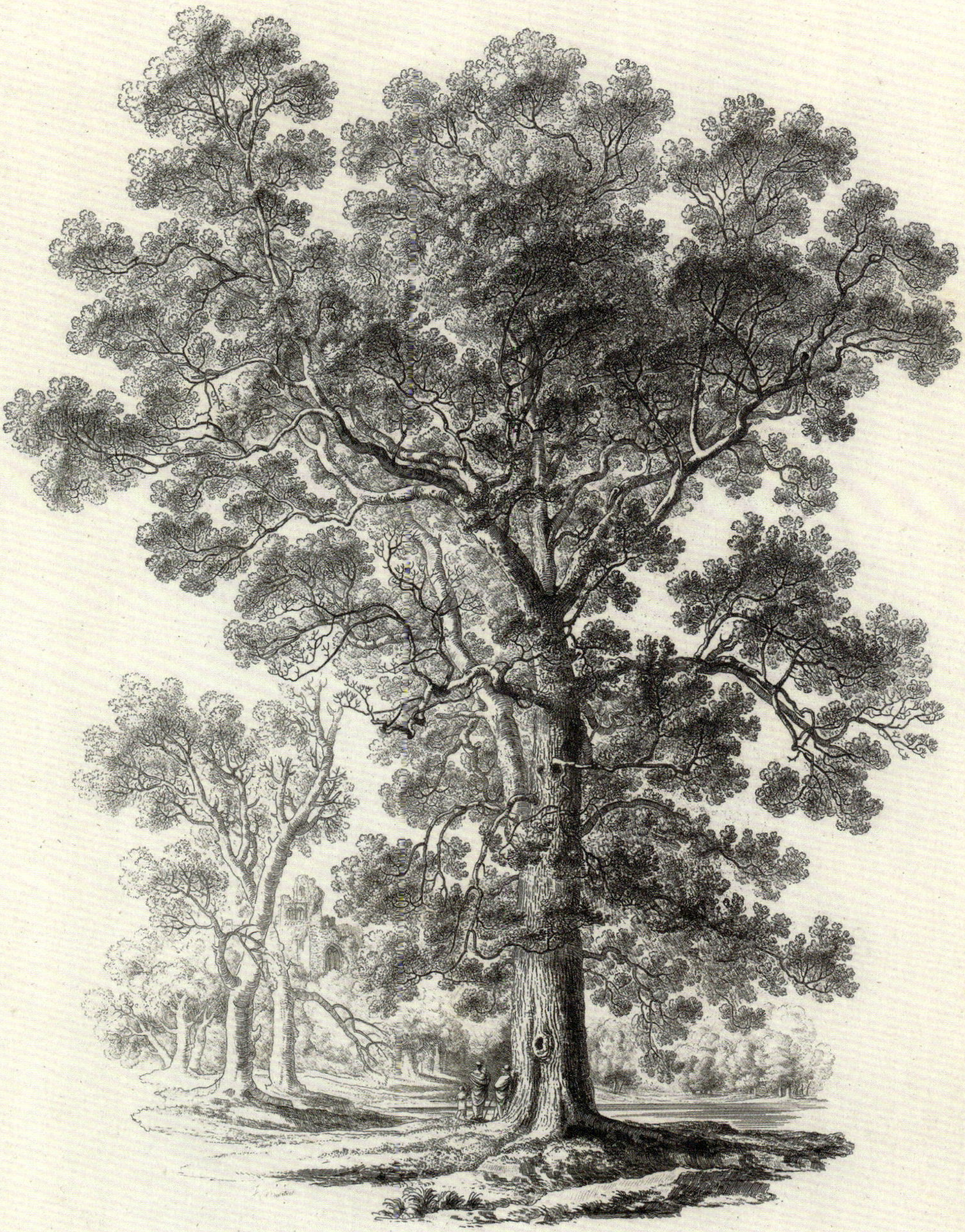

Elm

London, Published May 15th 1817, at R. ACKERMANN's Repository of Arts, 101, Strand.

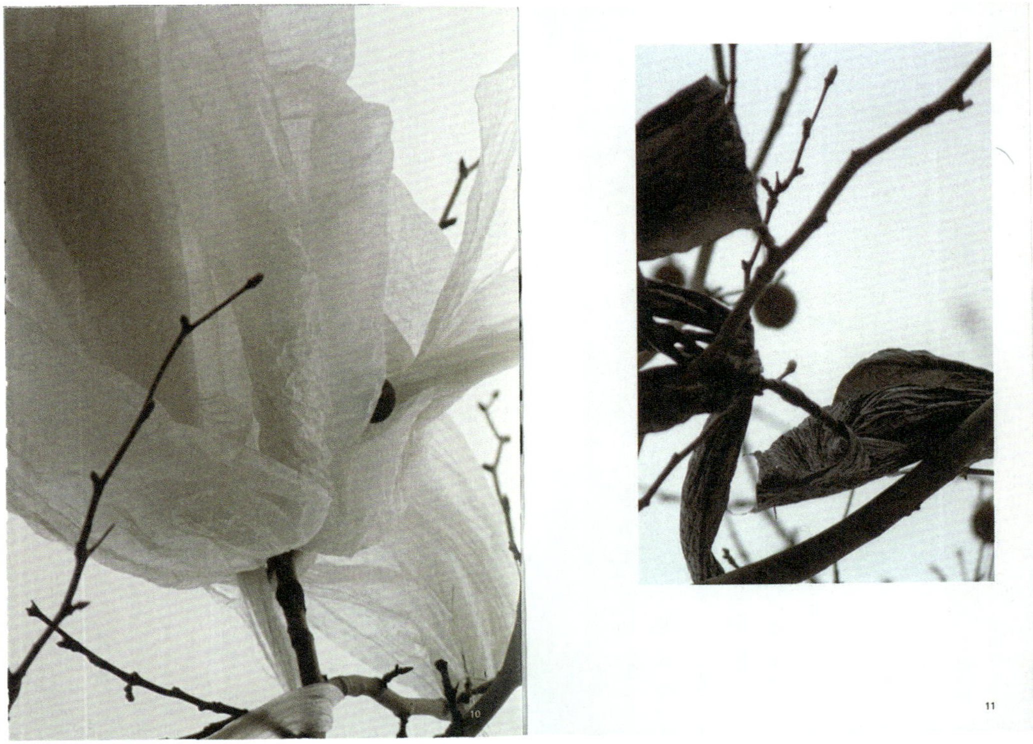

Steve Perfect, *Wind Flowers*, London, 2008, digital print. Yale Center for British Art, Friends of British Art Fund

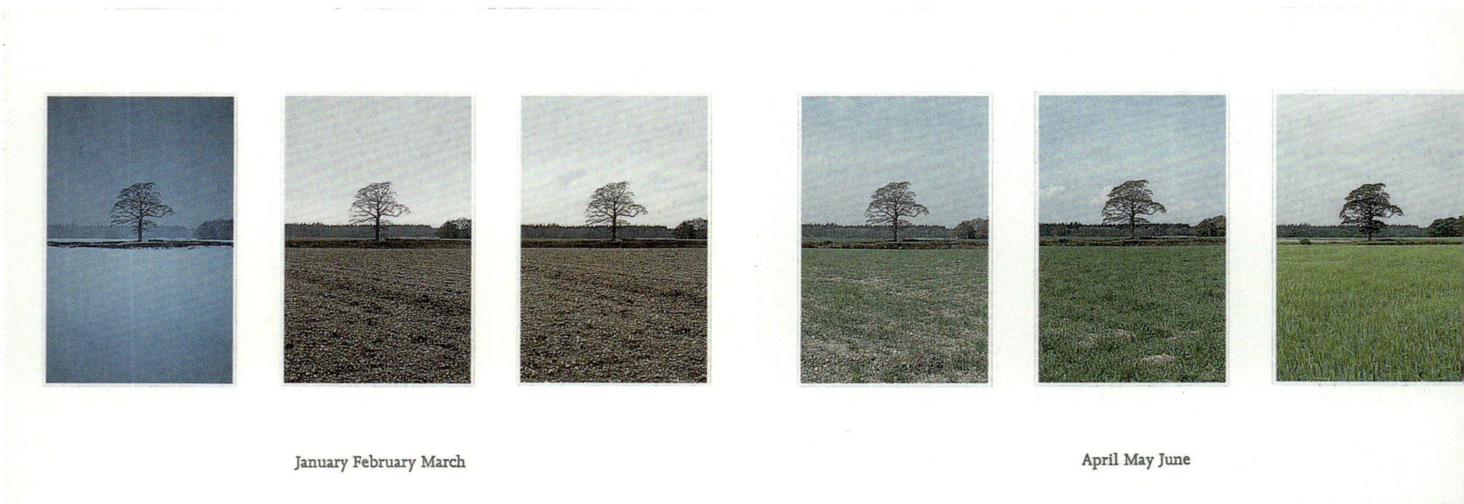

Robert Maude, *English Oak Tree and Wheat Field*, London, 1988, digital print. Yale Center for British Art, Friends of British Art Fund

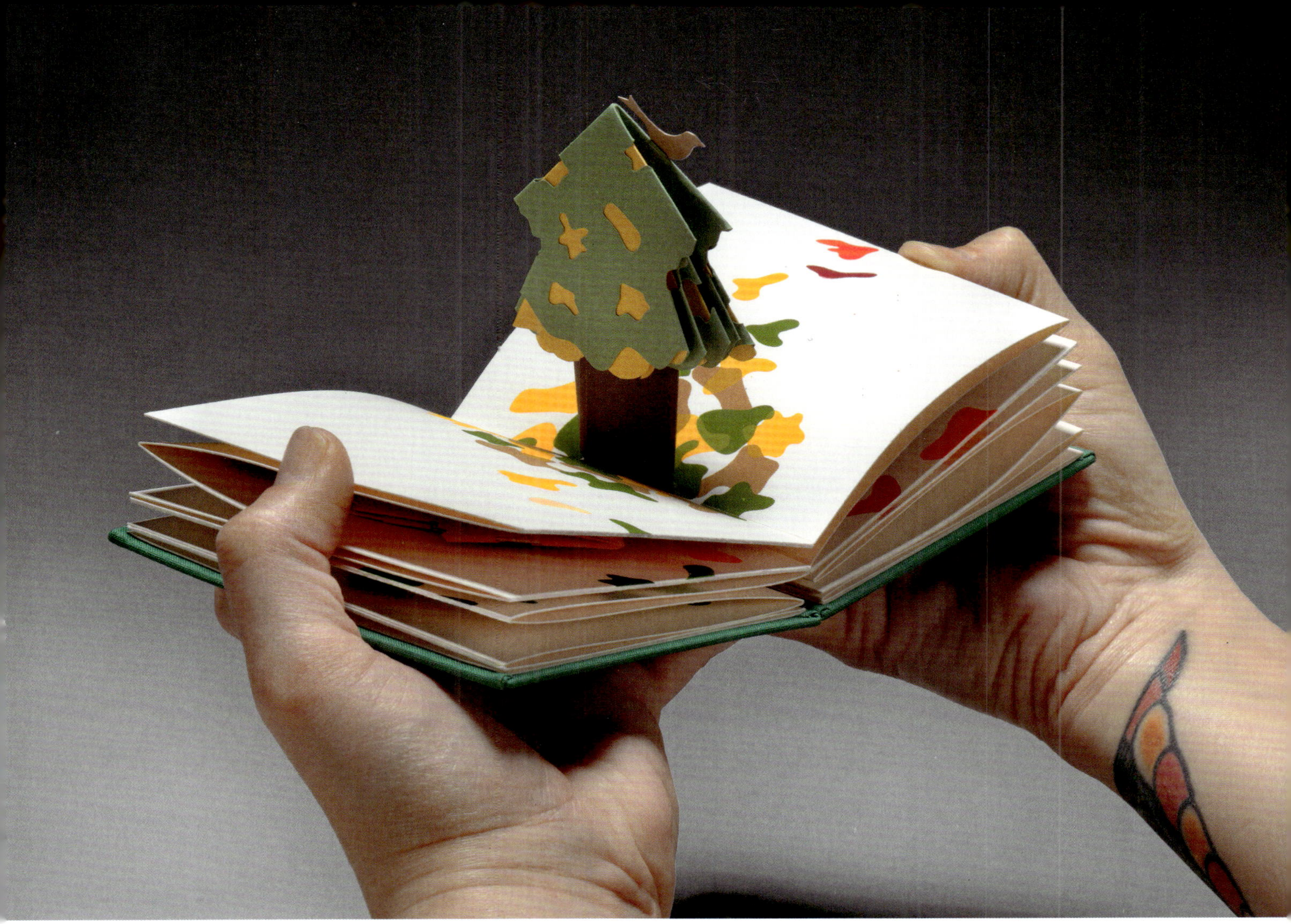

Chisato Tamabayashi, *Branches*, London, 2007, screen print and cut and folded paper. Yale Center for British Art, Friends of British Art Fund

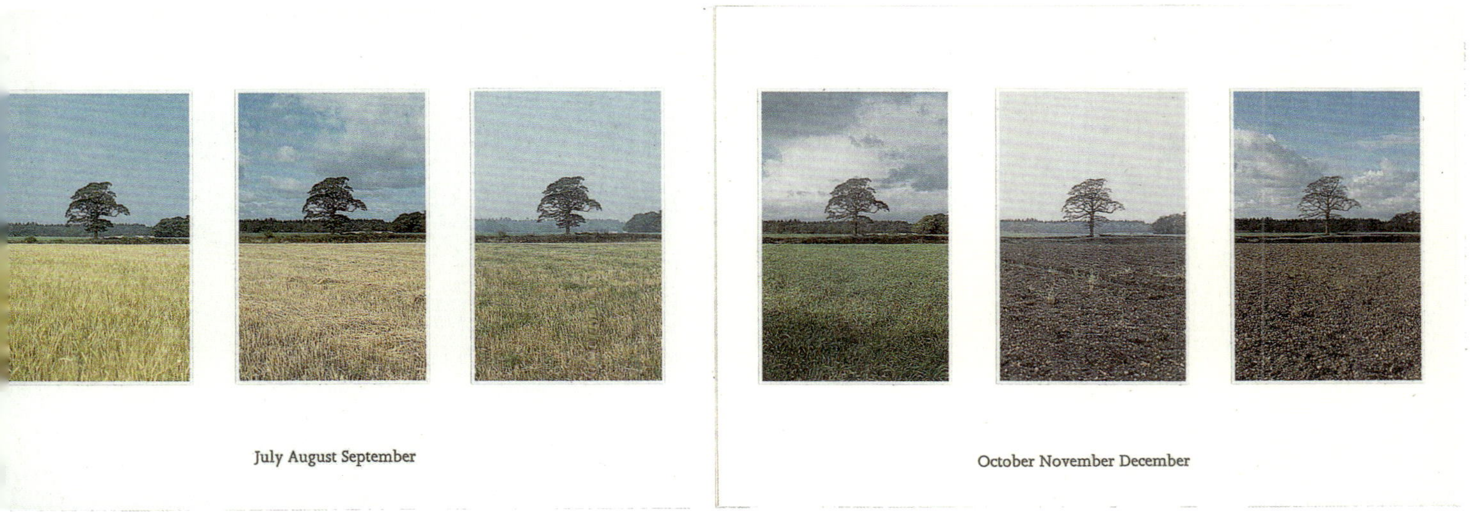

Log-Books

I began these log-books in 1995—an ongoing project for a series of bookworks intended to explore the possibilities of sawing a log of wood into book sections or pages, which can be reassembled into their original log form. The designs vary in size and structure, but the larger they get, the more impressive a presence they take on. In fact, both of the big log-books come from the same trunk—a 14-foot cedar of Lebanon tree (6' and 8' pared down to 5½' and 7½')—which have now both been moved to our own garden and have turned a beautiful silver grey. *Codex II* is the taller one, shown here at West Dean College, West Sussex, in December 2007. I am directing its assembly with Joe Thompson (back to camera), who teaches and works at the Weald and Downland Open Air Museum near Chichester. He helps me find the trunks and the crews to transport and cut according to my designs; he is indispensable!

Ron King
ARTIST

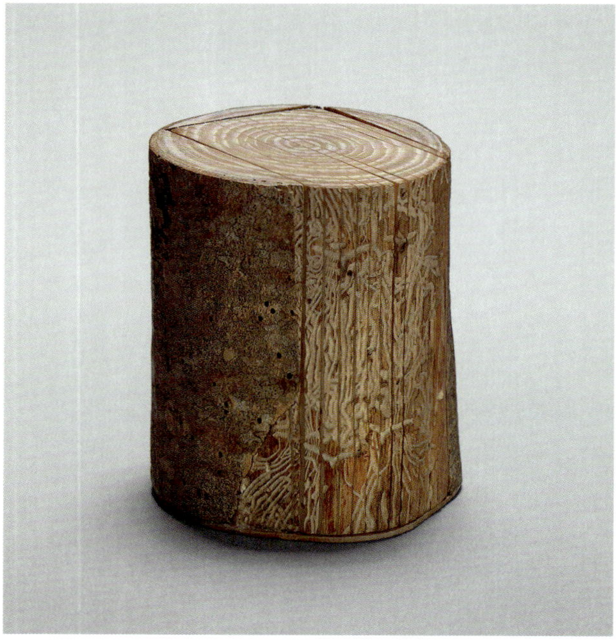

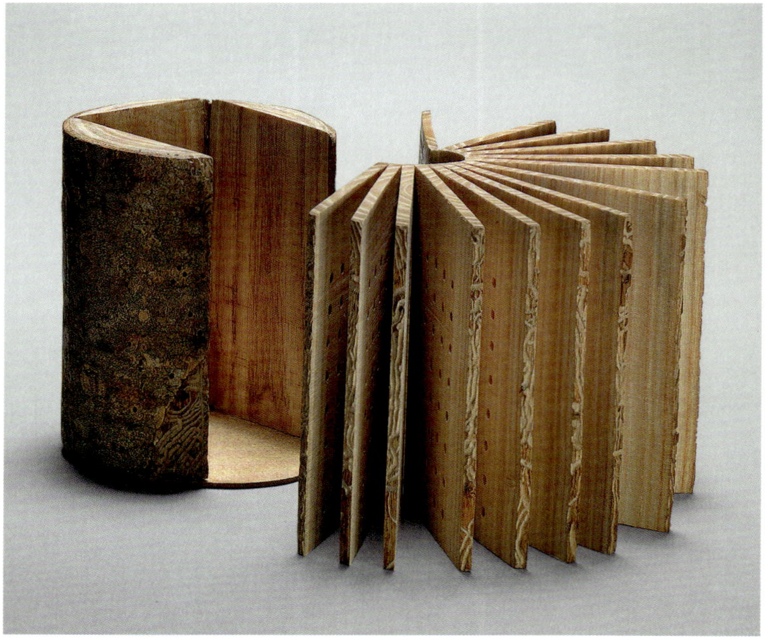

Ron King, *Termite Grid* (closed and open), London: Circle Press, 1998, cut and drilled log, bound with linen. Yale Center for British Art, Friends of British Art Fund

OPPOSITE ABOVE: Ron King, *Codex II*, 2007, cut log book, as displayed at West Dean College, West Sussex. Collection of the artist.

OPPOSITE BELOW: Installation of *Codex II* by Ron King (left) and Joe Thompson, West Dean College, West Sussex, 2007

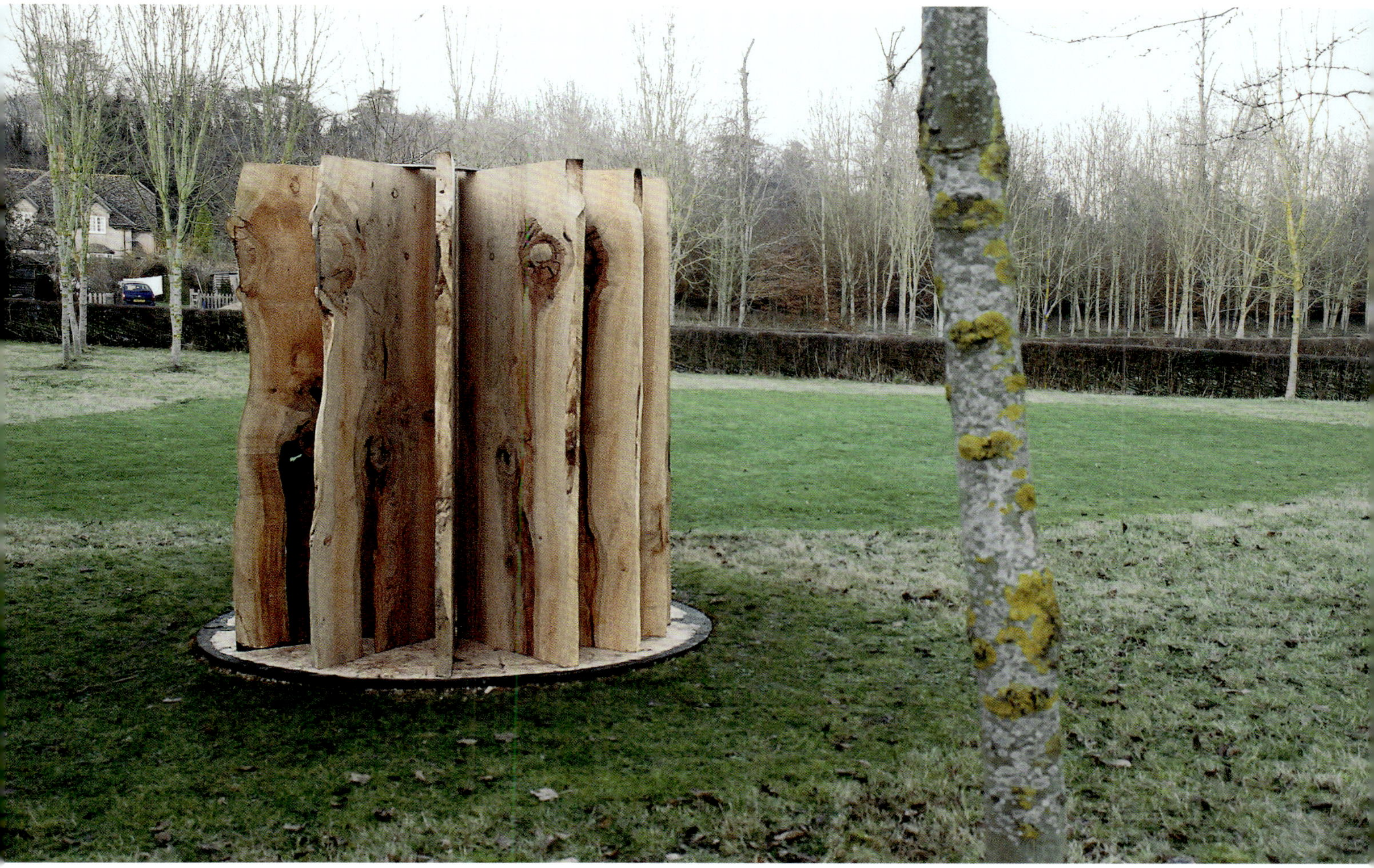

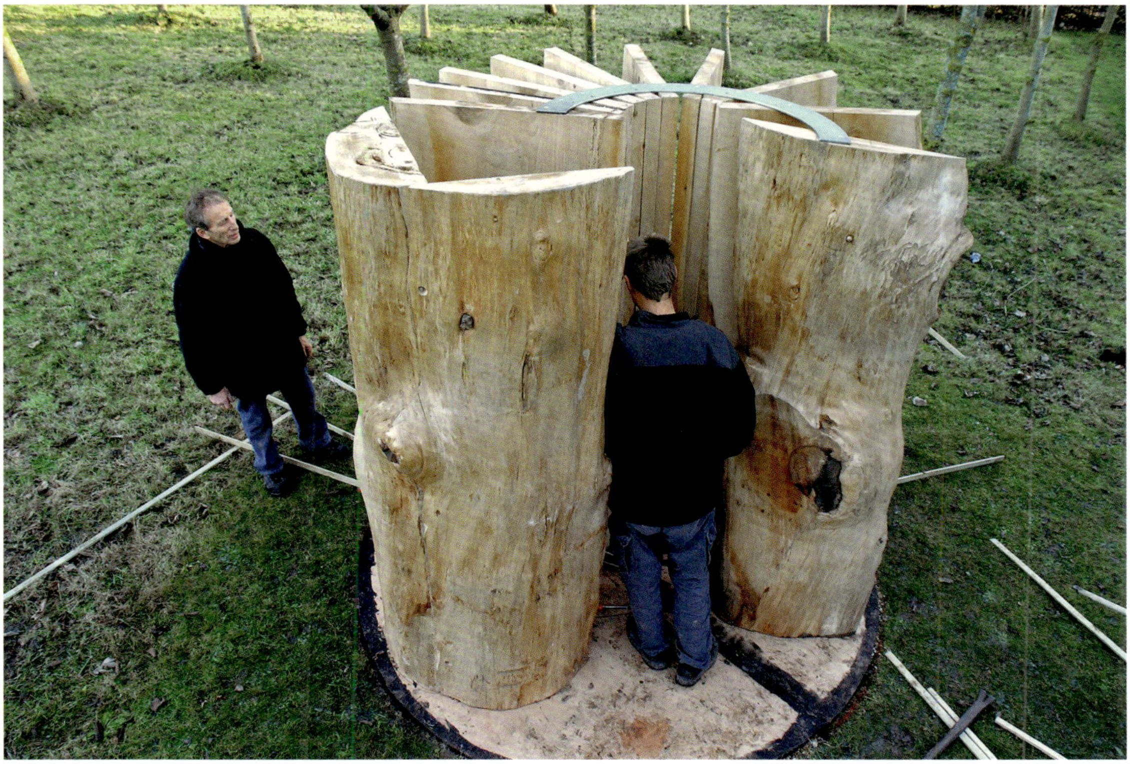

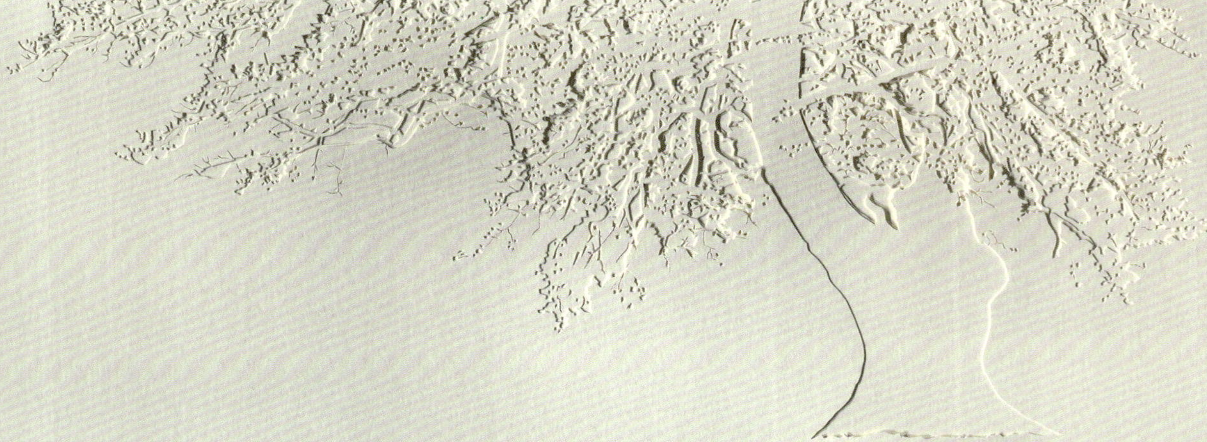

Cut-Paper Trees

My work involves the manipulation and cutting of paper. I make scalpel drawings in what appear on the surface to be blank book structures, in which the turn of a page starts to reveal a narrative or image. The processes involved, choice of materials, binding structure, cutting away, and the spaces left behind are integral to the work as a whole.

The origins of this method and interest in surface may have started when I made some work triggered by seeing the stone that surrounded a doorway in Gower Street, London, that had been repaired and filled in with blocks so that the stonework took on a surface texture akin to a Gustav Klimt painting. I photographed the doorway and made a triptych of prints. I then cut away parts of the image to echo the way that accretions of time had changed the doorway. Similarly, seeing a book in a display cabinet at University College London that had been eaten away by bookworms led to initial ideas for a piece of work, *city* (2001), which became a pivotal piece in terms of method and ideas.

Recurring themes of place, scale, surface, and natural forms are the origins of my work. This was intensified by a month-long artist's residency at West Dean College, West Sussex, in 2003, where, immersed in the surrounding grounds and landscape, I began to generate and amass ideas and photographs of the flora, trees, lichen, skyline, and landscape, which continue to influence my work to date.

Clare Bryan
ARTIST

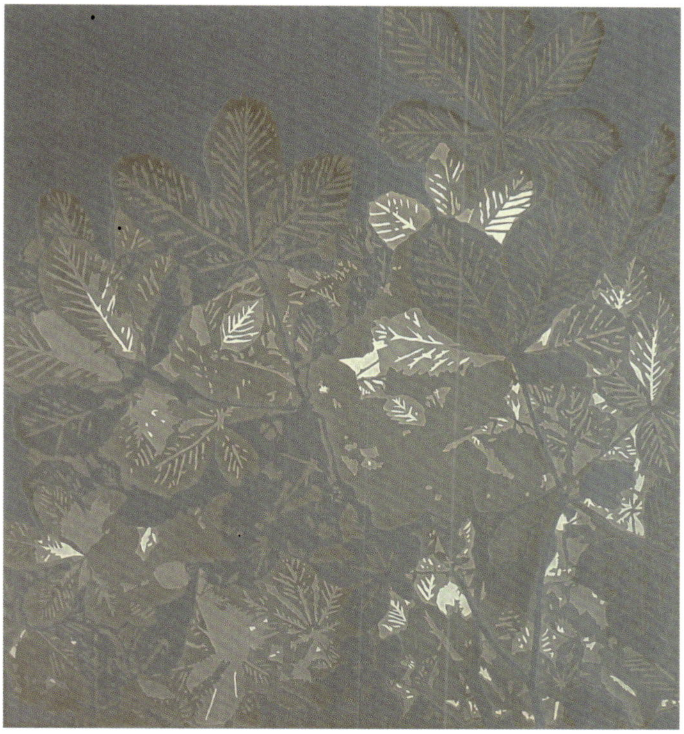

Clare Bryan, "Horse Chestnut" (detail) from *Canopy: Sycamore, Beech, Horse Chestnut*, 2006, hand cut paper. Yale Center for British Art, Friends of British Art Fund

OPPOSITE: Clare Bryan, *Beech* (detail), England, 2004, hand-cut paper. Yale Center for British Art, Friends of British Art Fund

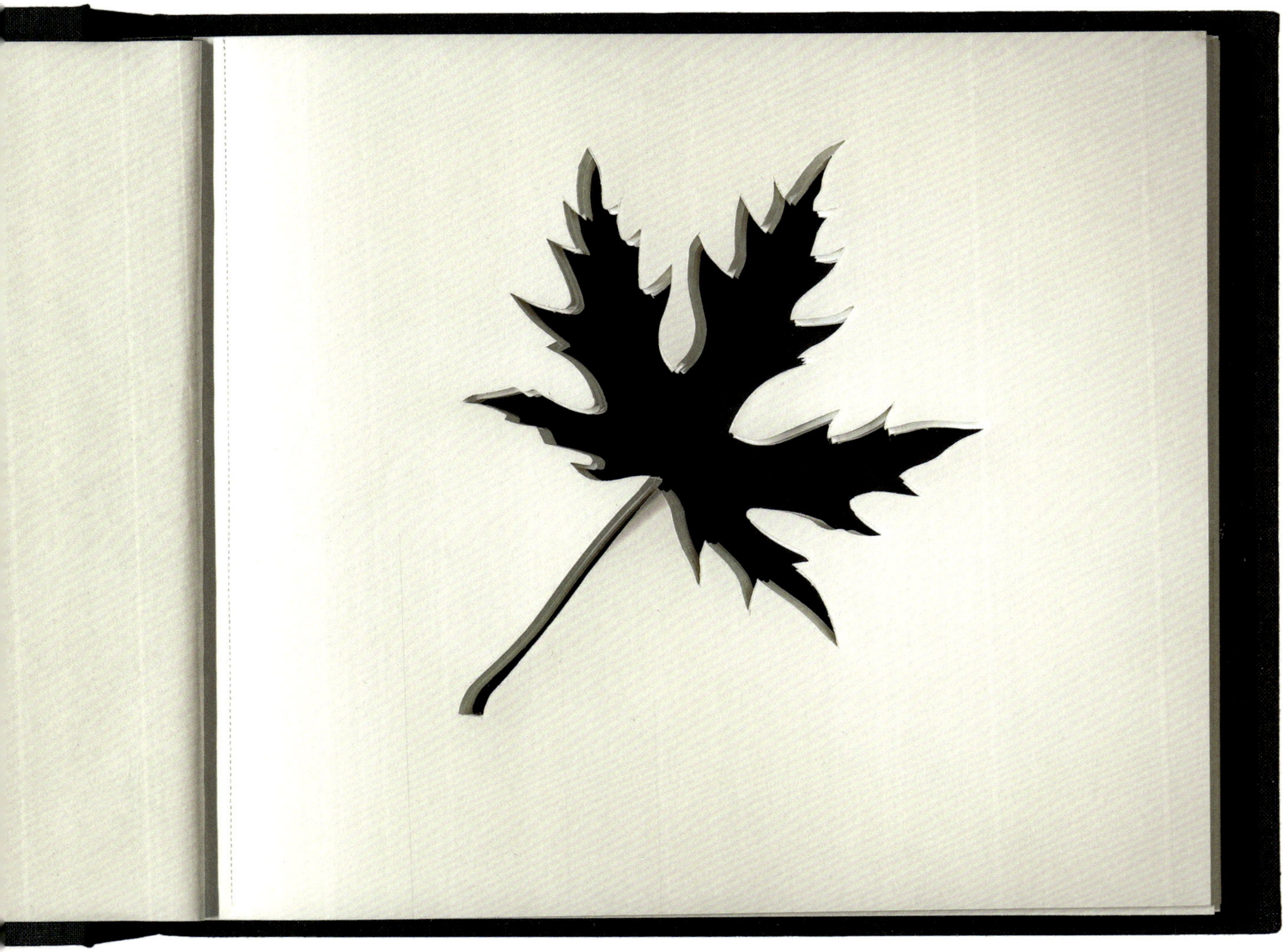

Anne Lydiat, *Without* (detail), Sheffield, England, 1999, cut paper. Yale Center for British Art, Gift of the artist

OPPOSITE: Andrew Norris, *POETree*, Zagreb, Croatia, 2005, natural specimens and rubber stamping. Yale Center for British Art, Friends of British Art Fund

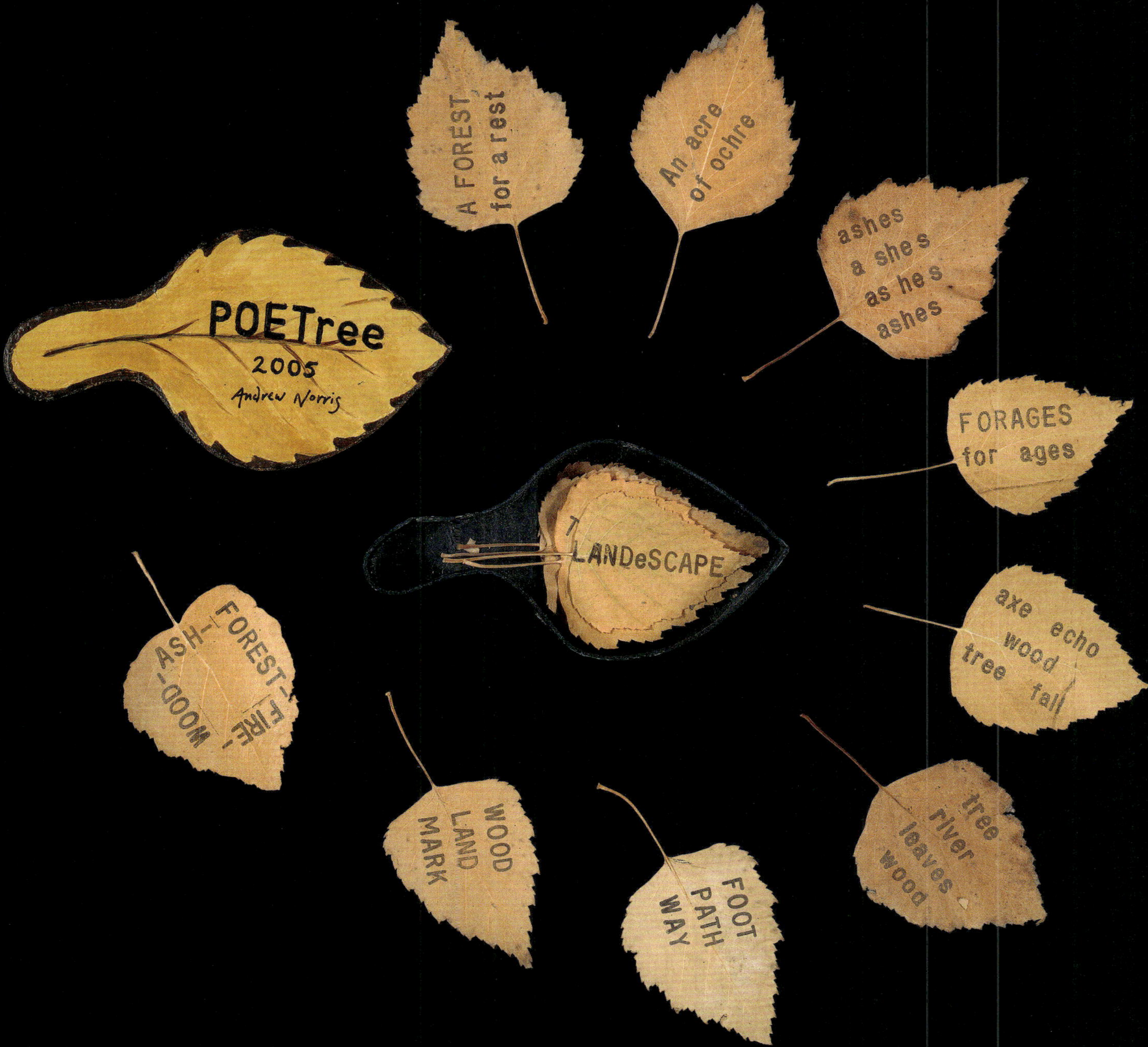
POETree
2005
Andrew Norris
A FOREST
for a rest
An acre
of ochre
ashes
a she s
as he s
ashes
FORAGES
for ages
LANDeSCAPE
axe echo
wood
tree fall
tree
river
leaves
wood
FOOT
PATH
WAY
WOOD
LAND
MARK
FOREST-
FIRE-
ASH-
WOOD-

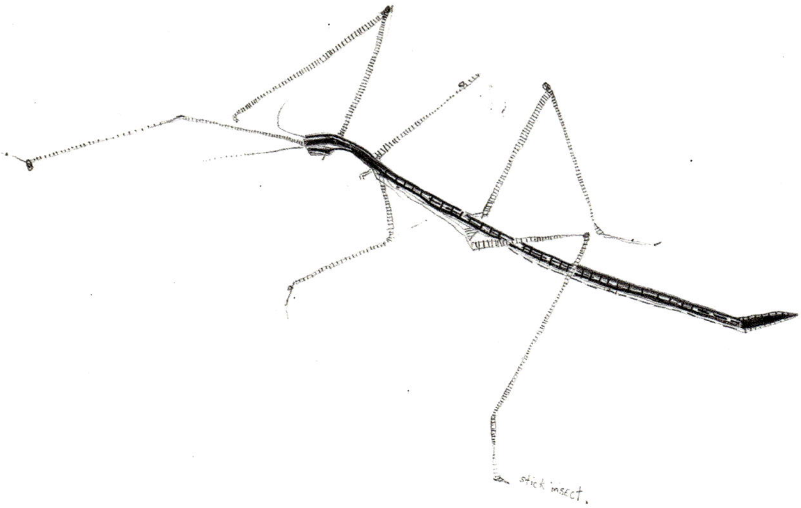

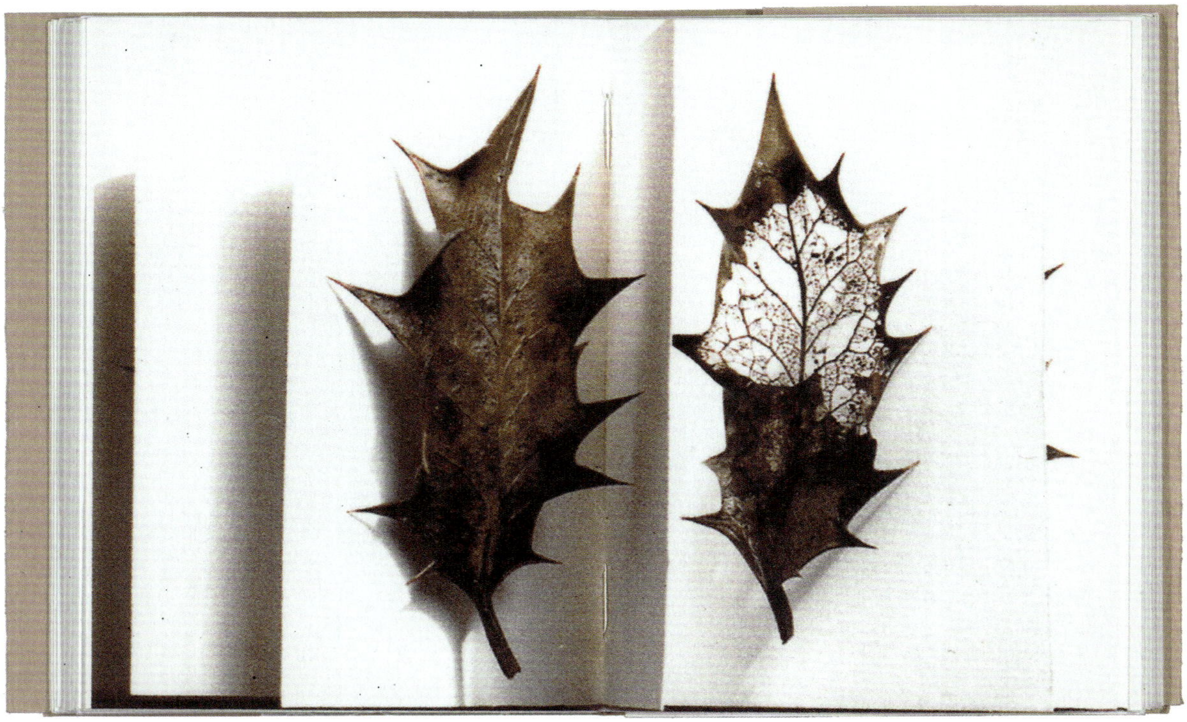

Helen Douglas, *Field notes*, Yarrow, Scotland: Weproductions, 2006, digital print. Yale Center for British Art, Friends of British Art Fund

ABOVE: Mandy Bonnell, *Stick Insect*, 2006, graphite. Yale Center for British Art, Gift of the artist

OPPOSITE: Mary Gibbs Shapter, *Notes on Various Trees*, England, 1879–1880, pen and ink, graphite, and natural specimens. Yale Center for British Art, Paul Mellon Fund

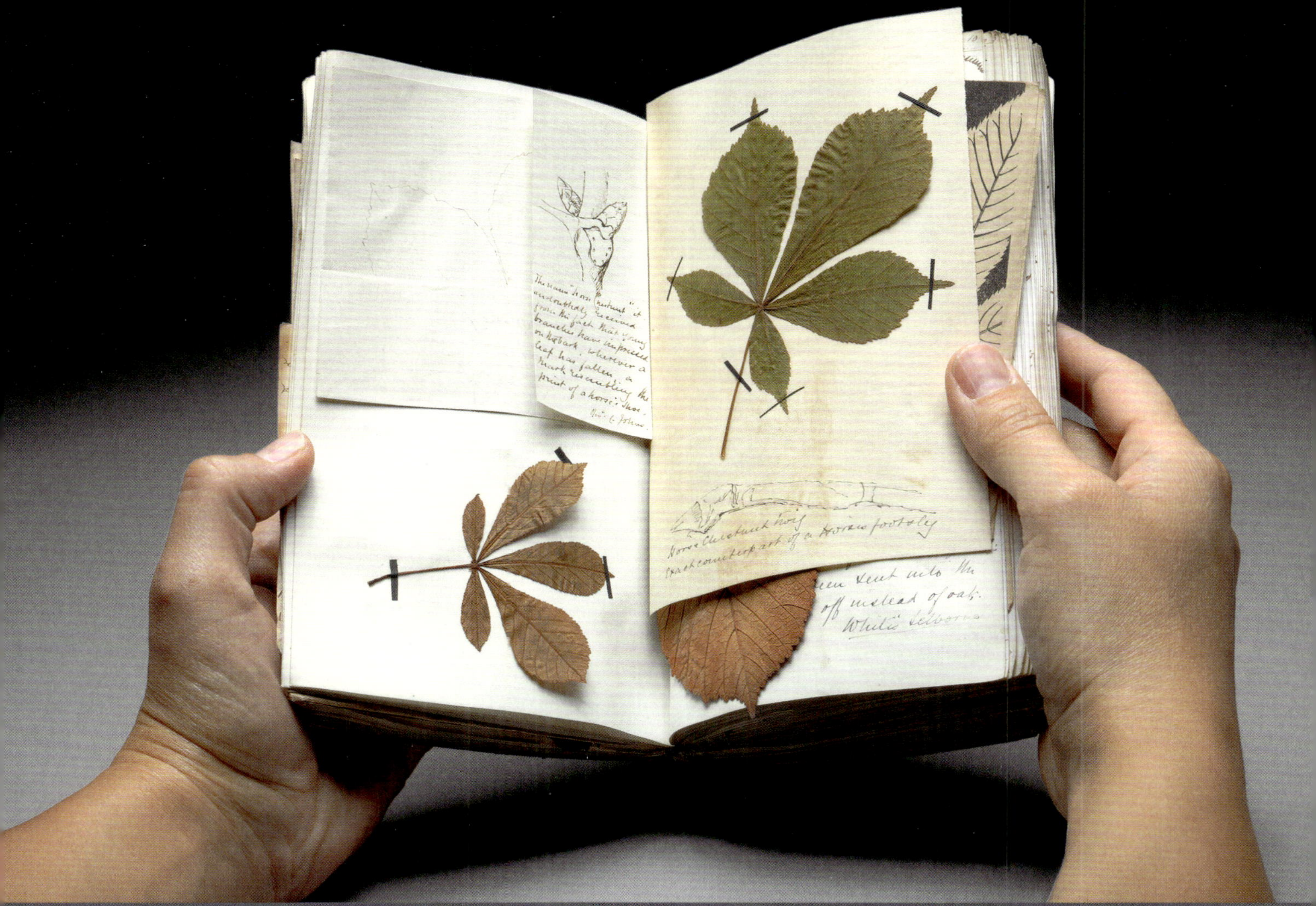
The name "Horse chestnut" is undoubtedly received from the fact that young branches have impressed on the bark, wherever a leaf has fallen, a mark resembling the print of a horse's shoe.
Rev. C. Johns
Horse Chestnut twig
Exact counterpart of a horse's foot & leg
off instead of oak.
White's Selborne

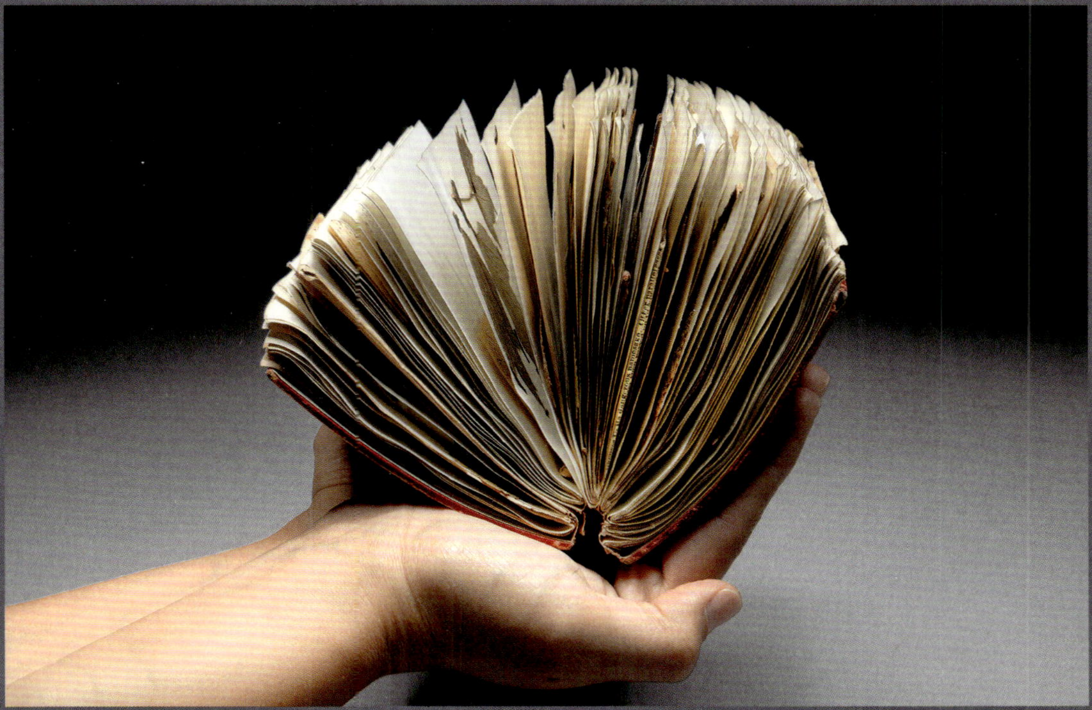

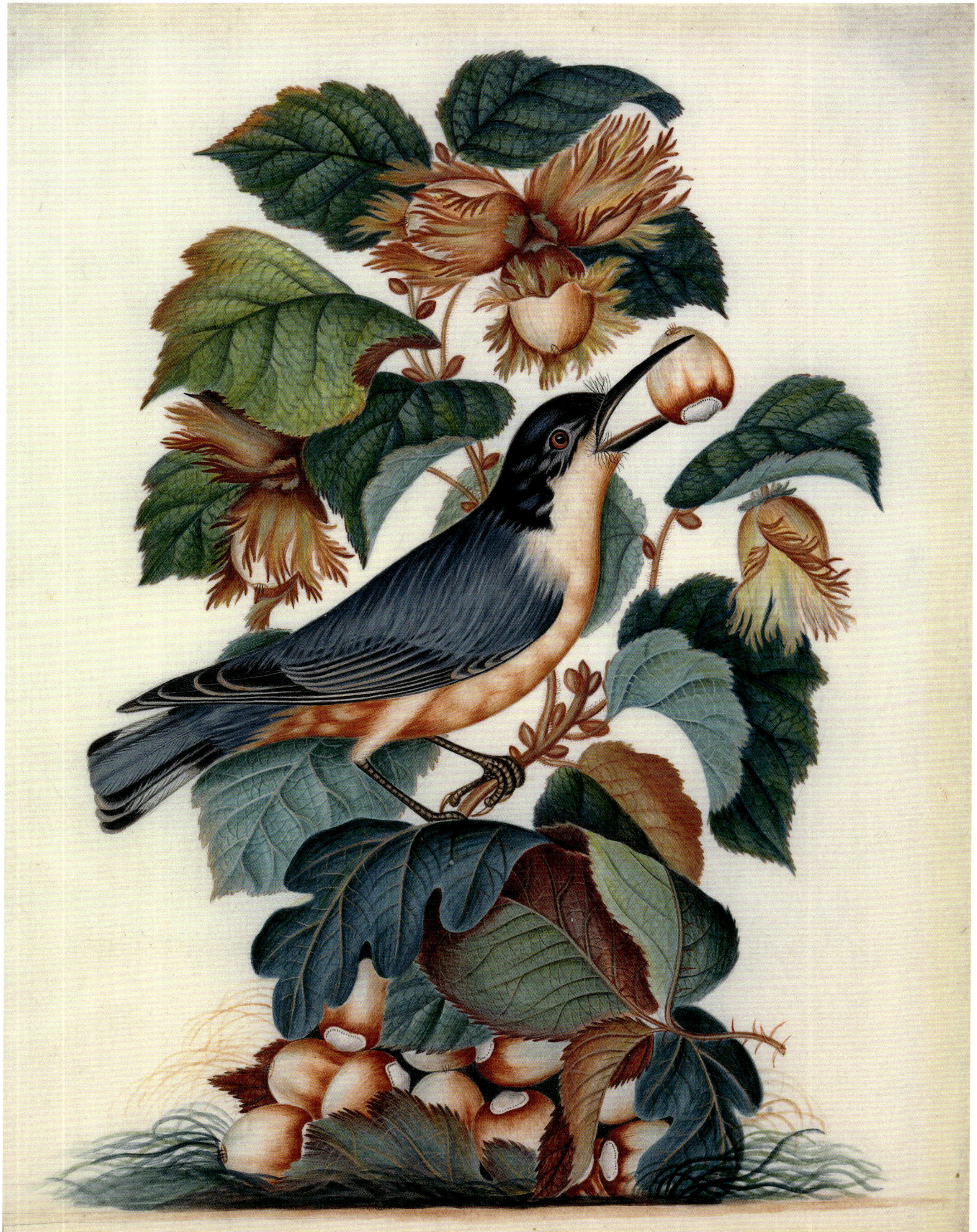

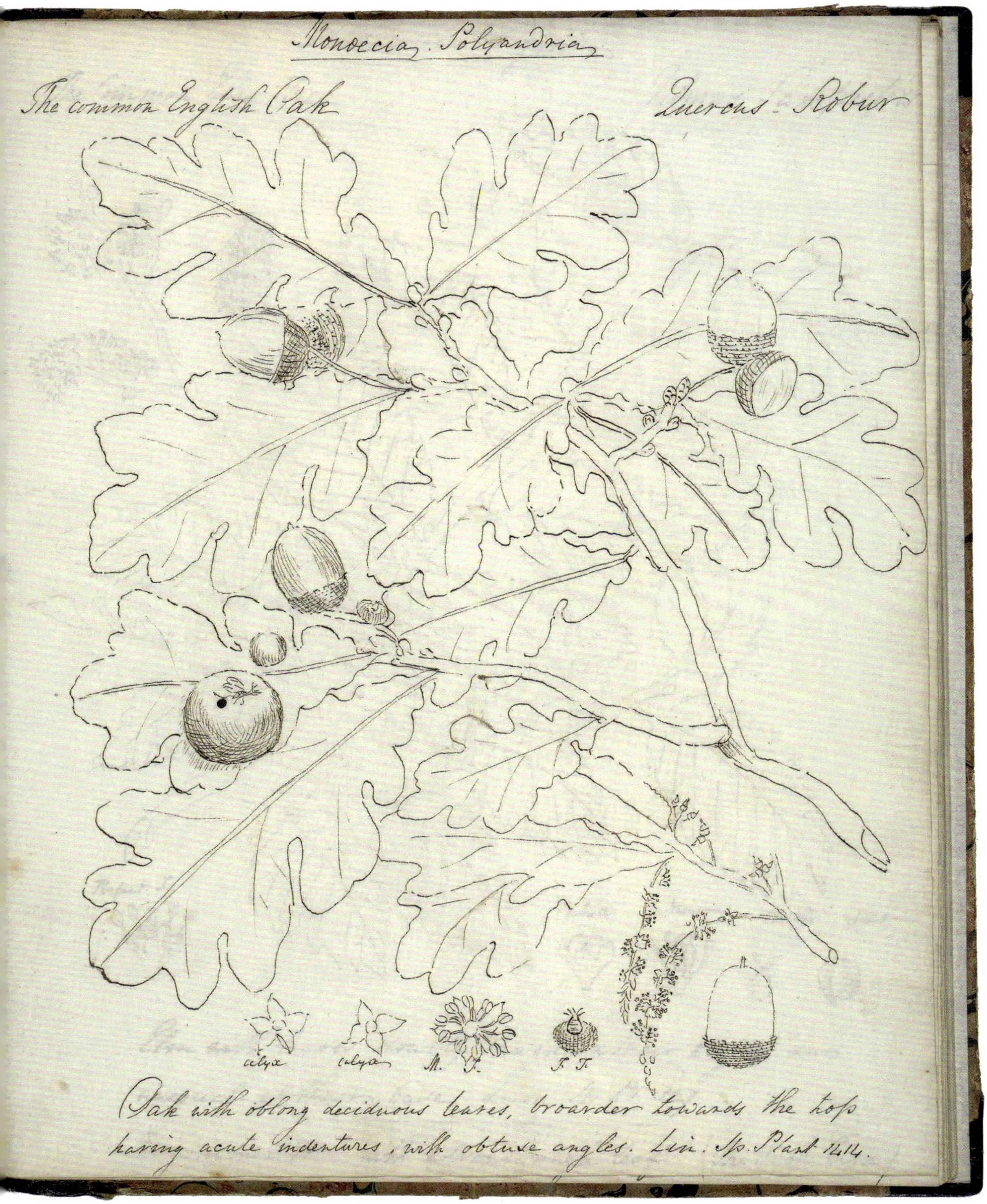

Character of Forest Trees, ca. 1817, pen and ink. Yale Center for British Art, Paul Mellon Fund

OPPOSITE: James Bolton, Beautiful Nuthatch (*Sitta Formosa*) with Hazel (*Corylus avellana* L.), Oak (*Quercus* L.), and Bramble (cf. *Rubus*), from the natural history cabinet of Anna Blackburne, ca. 1768, watercolor and gouache over graphite on parchment. Yale Center for British Art, Paul Mellon Fund, in honor of Jane and Richard C. Levin, President of Yale University (1993–2013)

OVERLEAF: Mandy Bonnell, *Bethany Seeds*, with poem by Gabriel Gbadamosi, London: EMH Arts, 2012, wood engraving. Yale Center for British Art, Friends of British Art Fund

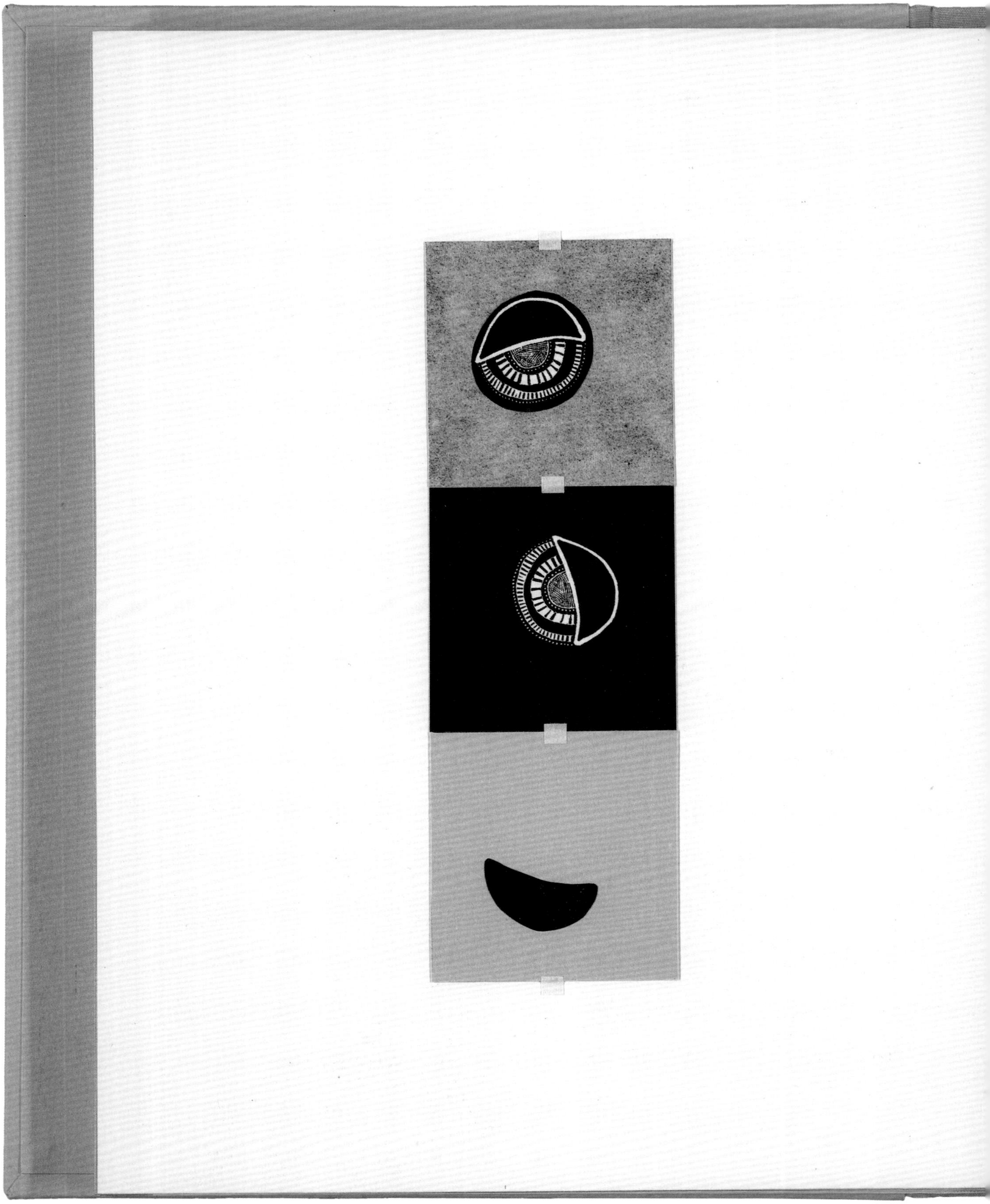

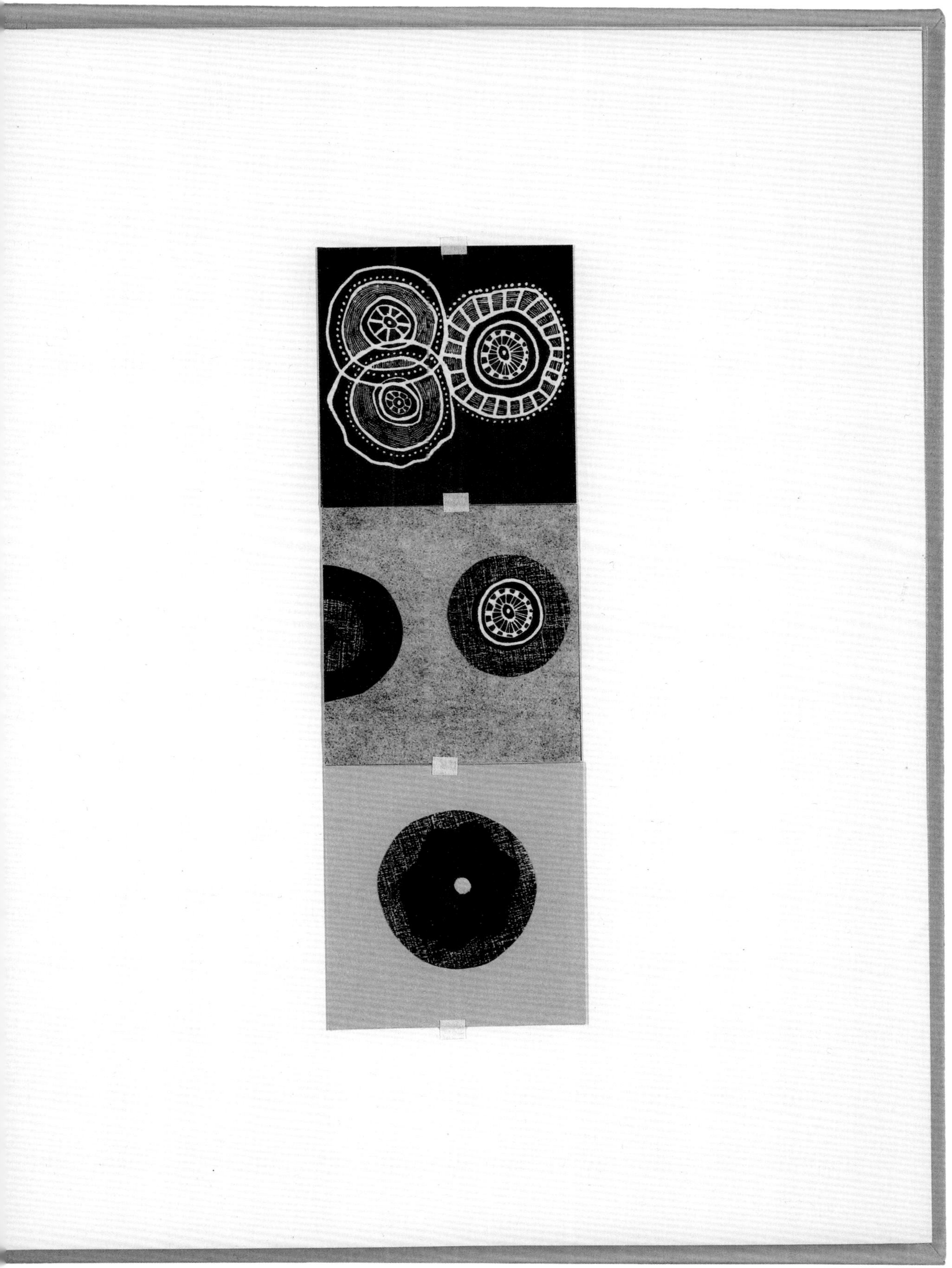

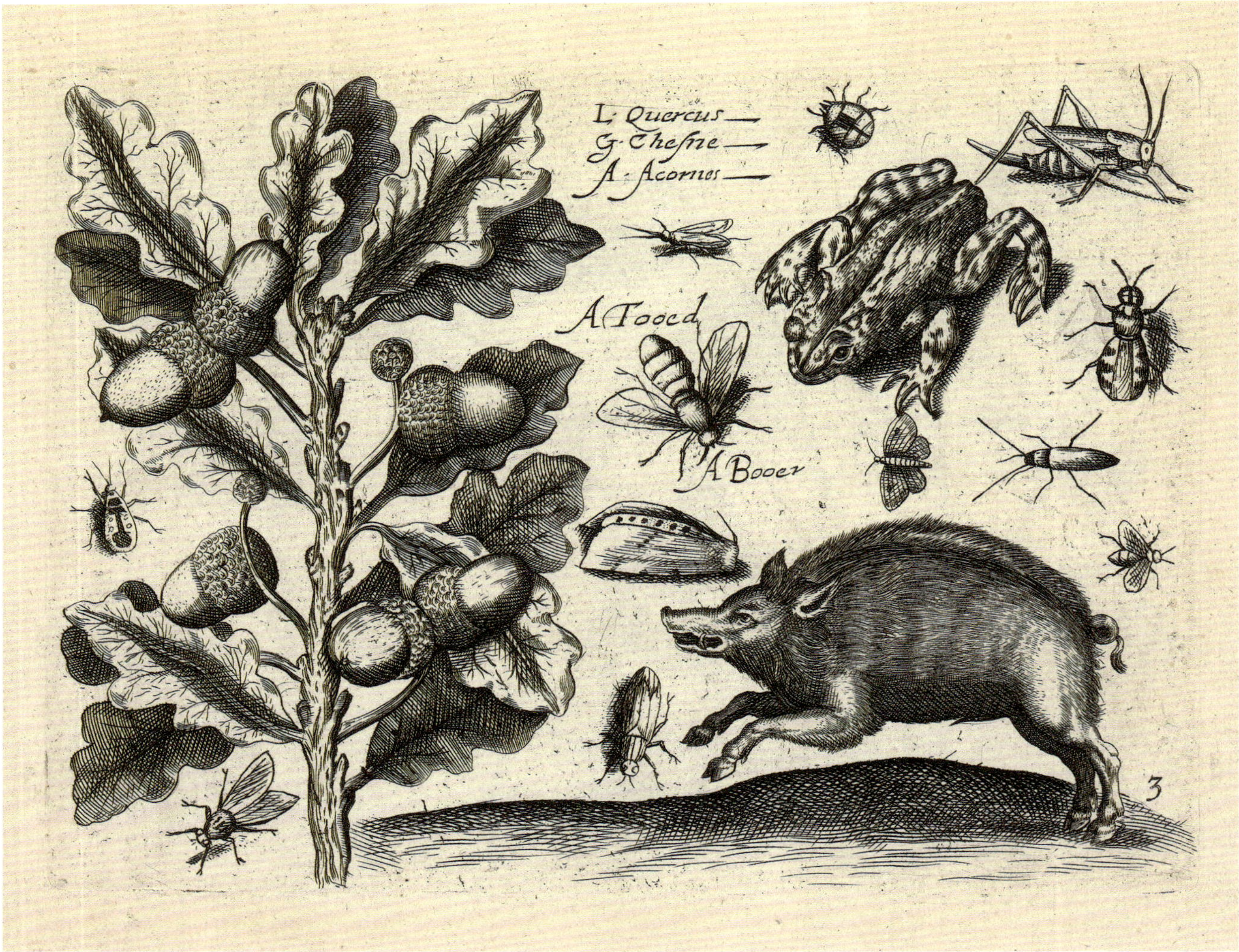

John Dunstall, *The Therd Booke of Flowers Fruits Beastes Birds and Flies Exactly Drawne*, London: Peter Stent, 1661, etching. Yale Center for British Art, Paul Mellon Collection

Lo! on each SEED within its slender rind
  Life's golden threads in endless circles wind;
  Maze within maze the lucid webs are roll'd,
  And, as they burst, the living flame unfold.
The pulpy acorn, ere it swells, contains
  The Oak's vast branches in its milky veins;
  Each ravel'd bud, fine film, and fibre-line
  Traced with nice pencil on the small design.

Erasmus Darwin

"The Economy of Vegetation," from *The Botanic Garden; A Poem, in Two Parts*, London: Printed for J. Johnson, 1791, canto IV: 381–388

OPPOSITE: Herbert Ernest Bates, *Through the Woods: The English Woodland, April to April*, with wood engravings by Agnes Miller Parker, New York: Macmillan, 1936. Yale Center for British Art, Gift of Janet and James Sale, Yale BA 1960

OVERLEAF: *"Lichen islandius"* (Eryngo-leaved Liverwort) in *Album of Drawings of English Moths, Butterflies, Flowers, and Mollusks*, 1805–1822, pen and ink with watercolors. Yale Center for British Art, Paul Mellon Fund

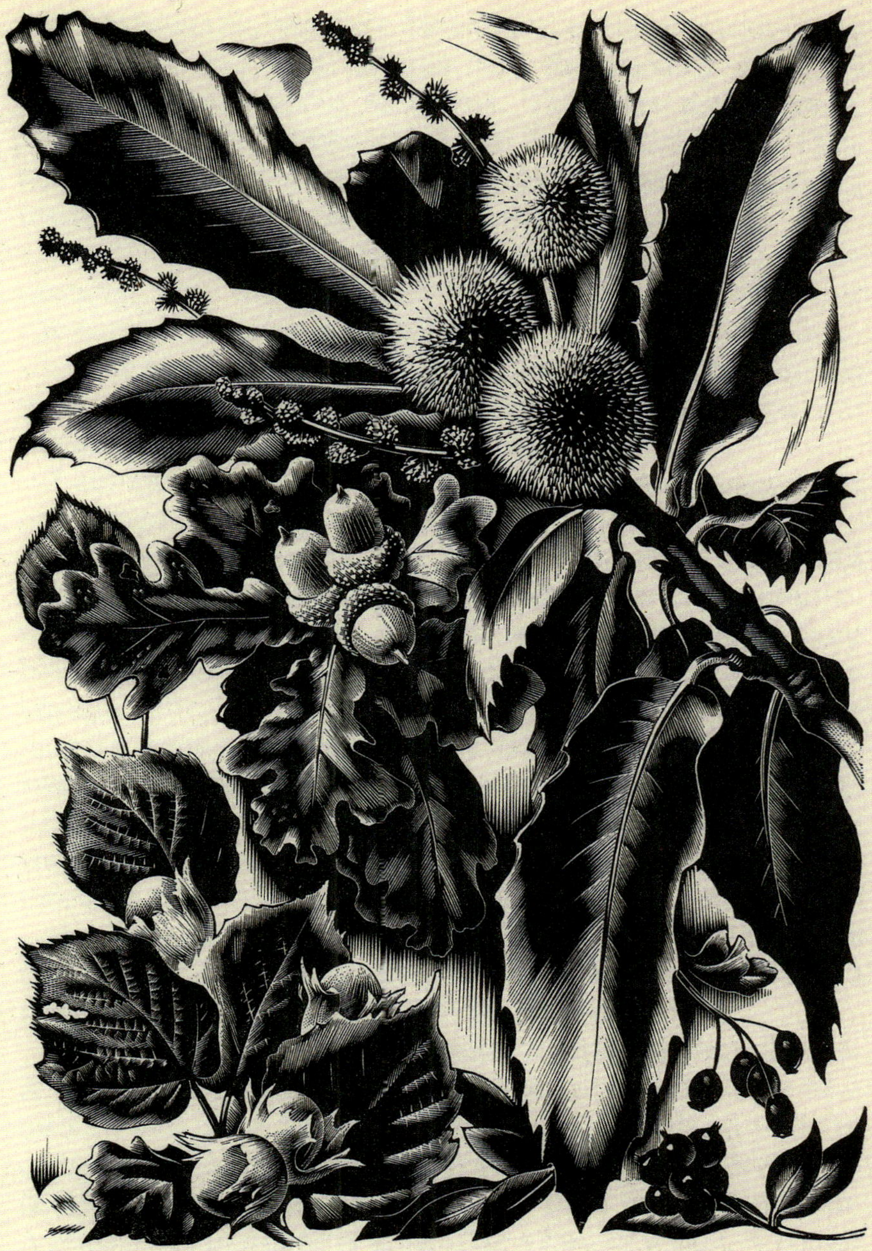

Lichen

meal moth

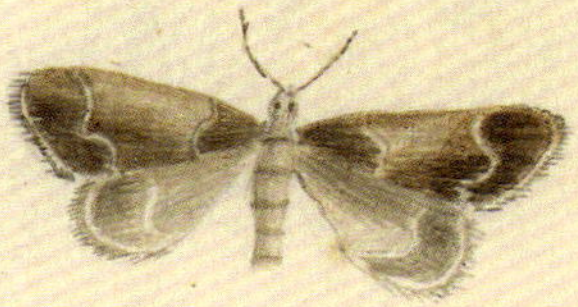

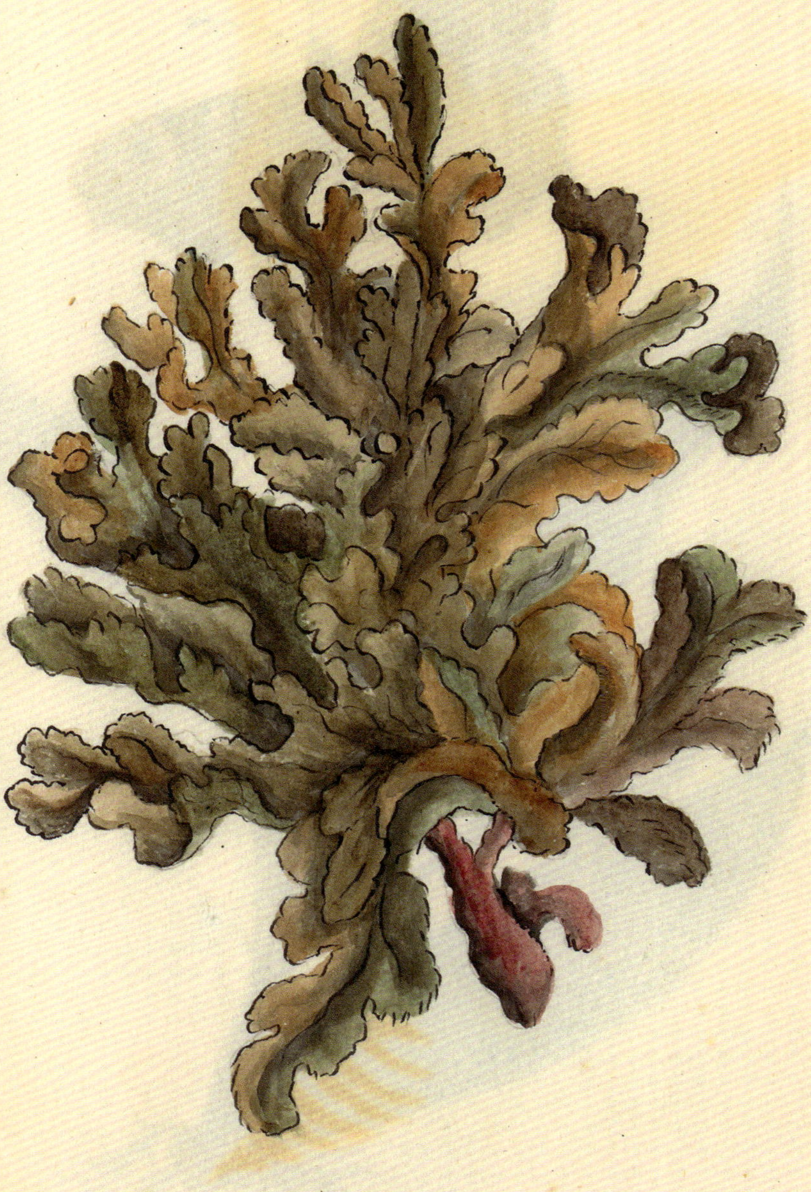

Lichen islandicus
Eryngo-Leaved Lichen

enlichenlichenlichenlichenliche
lichenlichenlichenlichenlichenlic

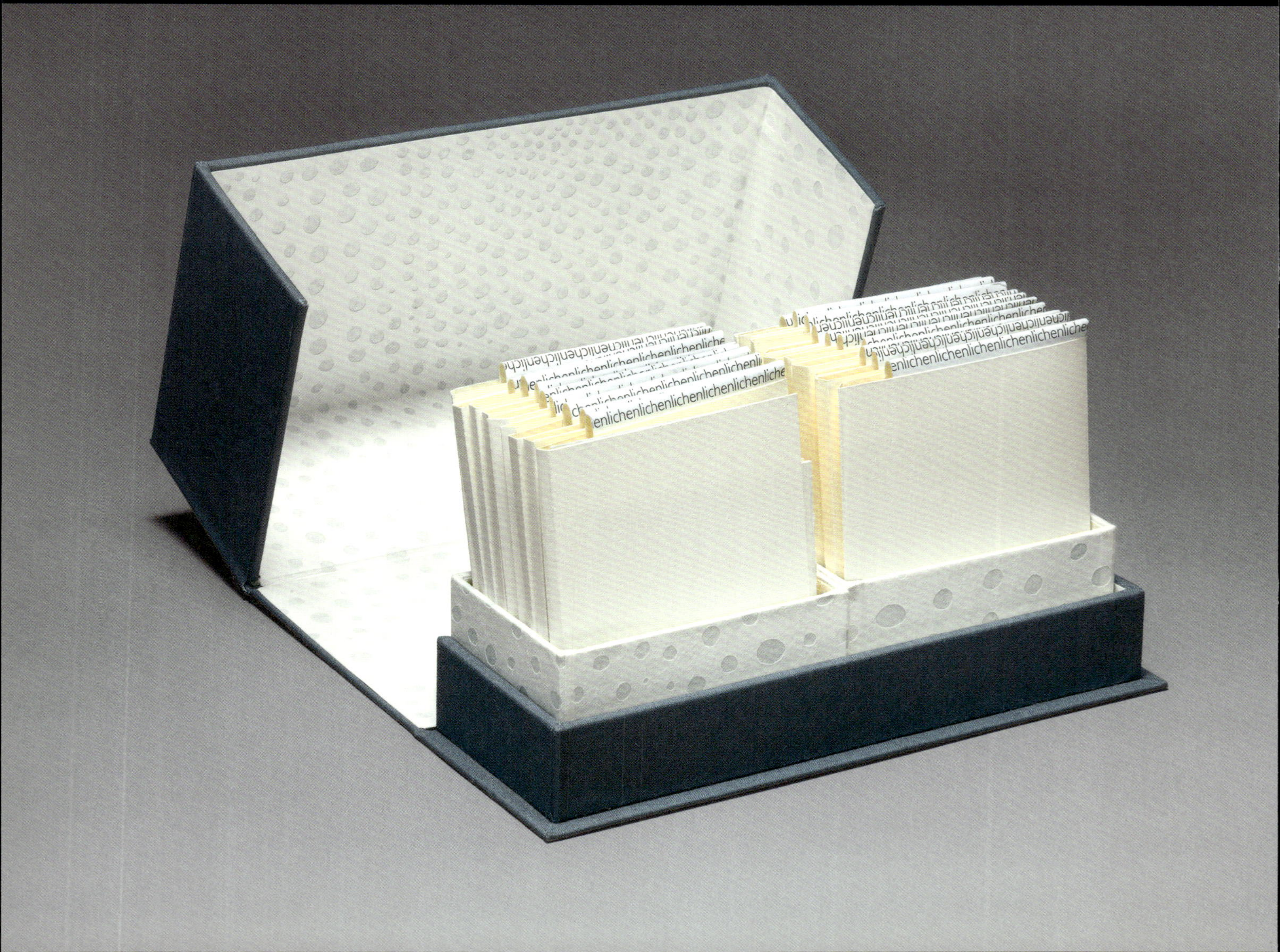

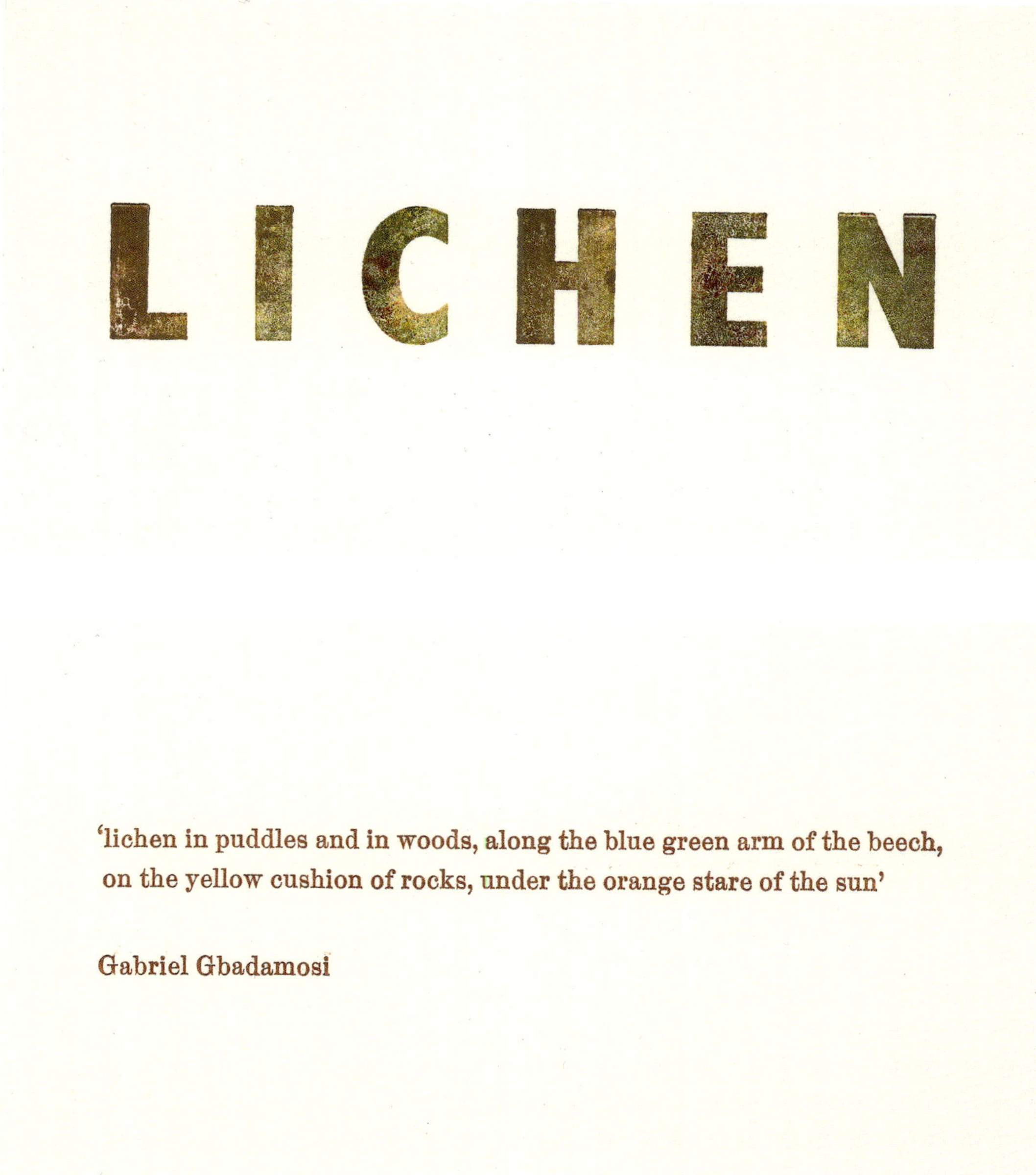

OPPOSITE AND ABOVE: Mandy Bonnell, *Lichen*, with poem by Gabriel Gbadamosi, London, 2012, graphite and hand-cut paper in twenty folded packets, in cloth-covered box with hand-cut paper interior. Yale Center for British Art, Friends of British Art Fund

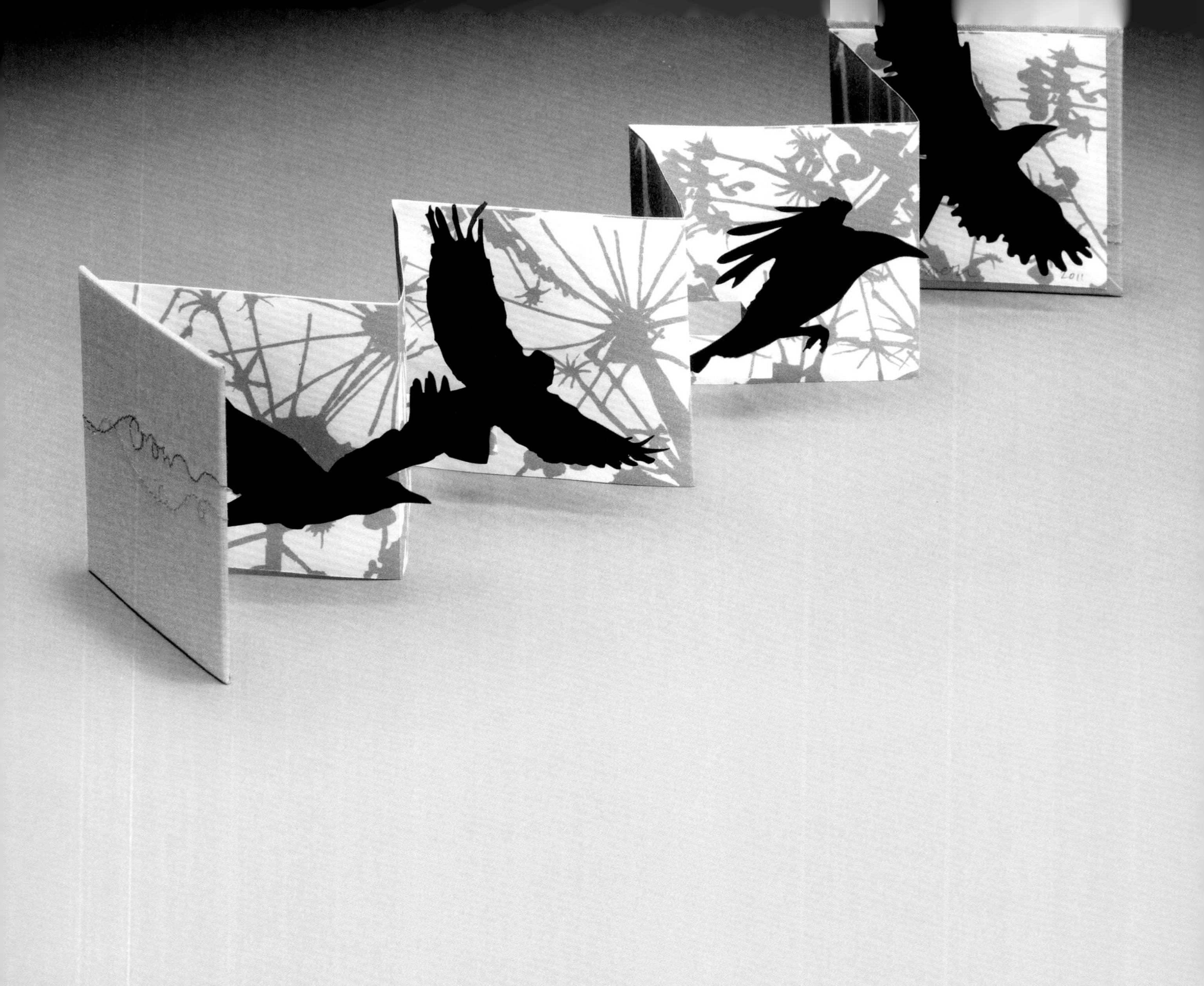

Sarah Morpeth, *Crow*, Elsdon, Northumberland, 2011, cut and folded paper, thread, and digital prints. Yale Center for British Art, Friends of British Art Fund

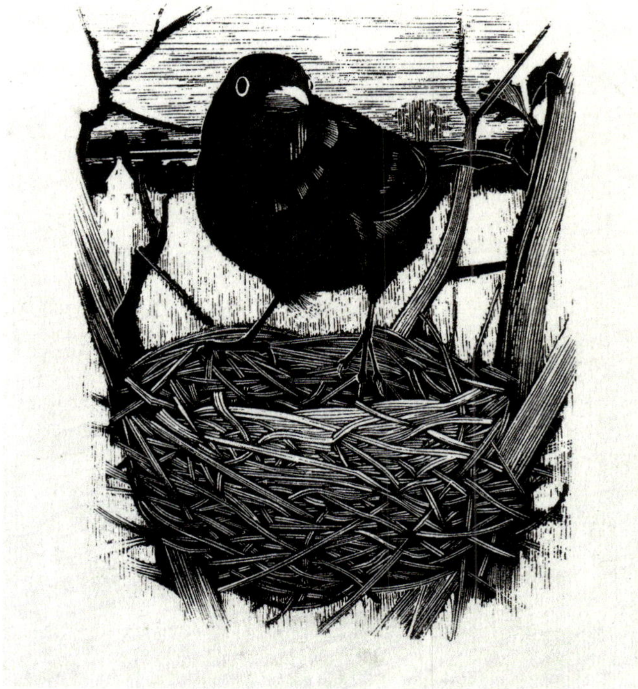

Donald Myall, *Blackbird*, 1988, wood engraving. Collection of David Burnett

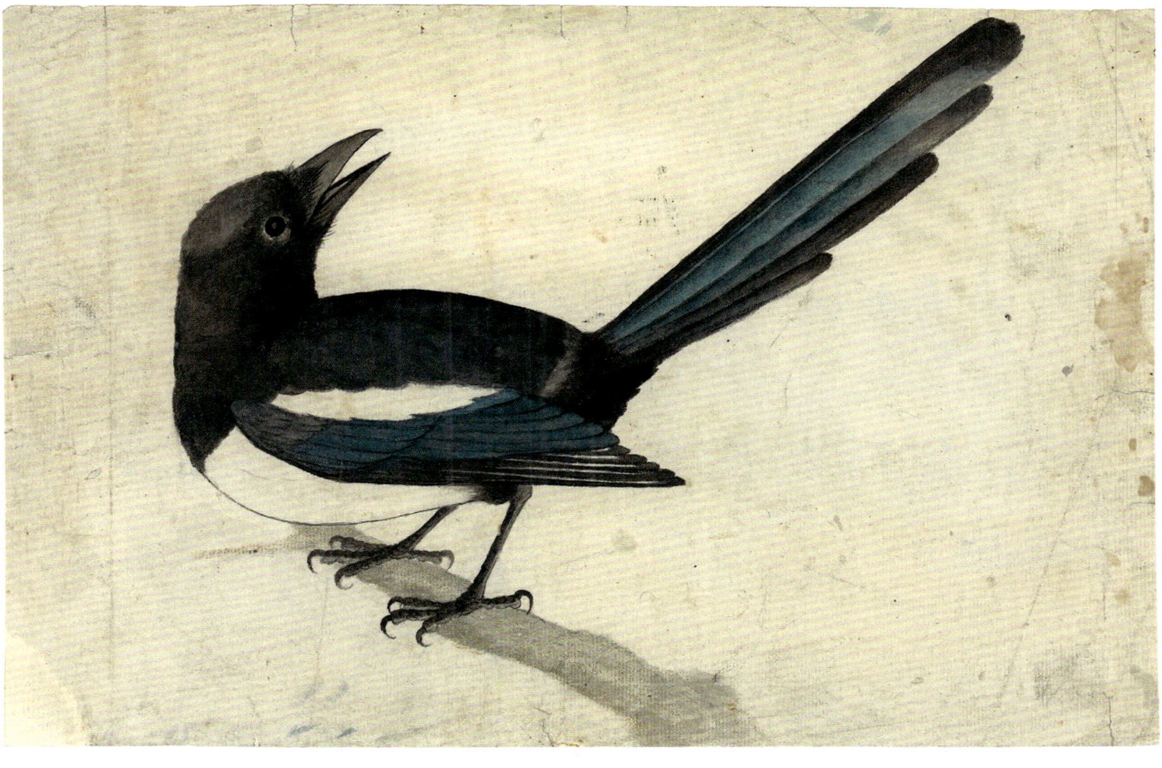

John Bewick, *Magpie*, ca. 1777–1782, watercolor, graphite, and pen and black ink. Yale Center for British Art, Paul Mellon Collection

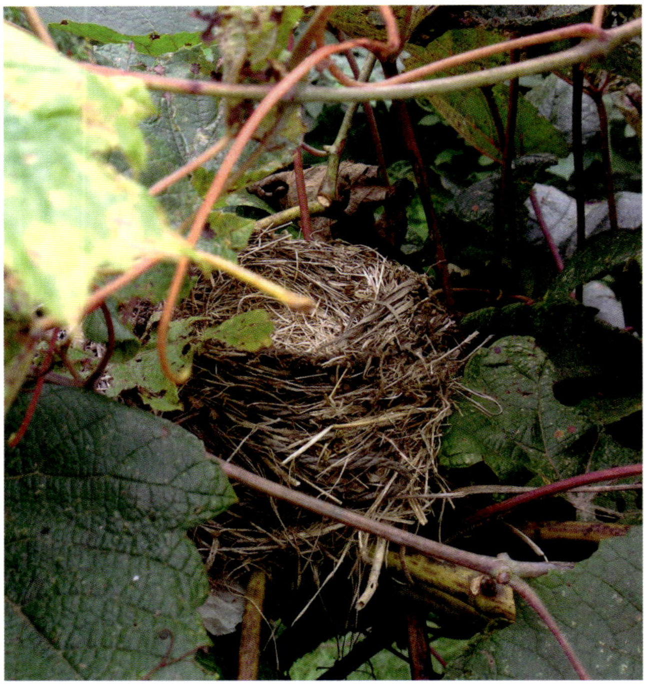

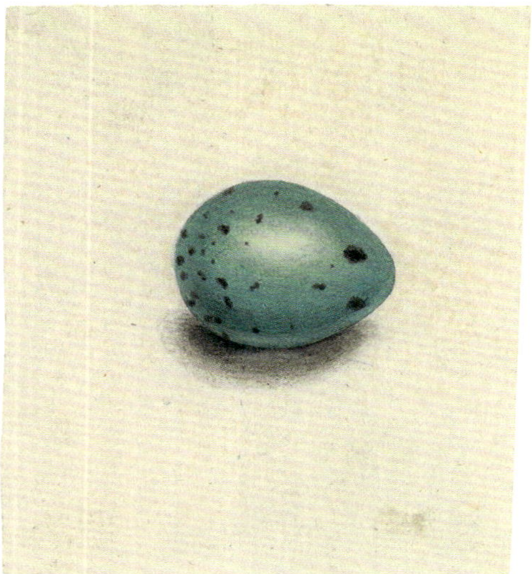

James Sowerby, *Bird egg, Song Thrush (Turdus philomelos)*, ca. 1790, watercolor over graphite. Yale Center for British Art, Paul Mellon Collection

ABOVE: Bird's nest in a summer garden, in Chester, Connecticut, 2013

James Bolton, Self-Taught Naturalist

The most notable publications by the naturalist/artist James Bolton (1735–1799) were his pioneering books on English ferns and fungi, and he is acknowledged as one of the preeminent British specialists on the latter, of which he discovered many new species. His watercolors, however, evoke his *Harmonia Ruralis*, which he dedicated "To the British Ladies, to Naturalists, and to all such as admire the Beauty or Melody of the Feathered Warblers," offering it as a handy field guide to those interested in identifying local species. The lively scenes of birds in natural settings, often with their eggs, or colorful groups of plants, insects, and shells, such as that on the cover of this book, follow in the footsteps of such naturalists/artists as Mark Catesby and Maria Sybilla Merian, who set a style for depicting flora and fauna engaged with their environments. But the drawings also emulate those same artists' willingness sometimes to intermingle flora and fauna from disparate regions, such as mixing shells from the Indo-Pacific region with insects and butterflies from North and South America, to create aesthetically pleasing productions that owe as much to the naturalists' cabinet as they do to observations of nature—in this case the museum or cabinet of the noted naturalist Anna Blackburne, who likely commissioned these images from Bolton. Just as many of the artists in the current exhibition have created imaginary worlds for their creatures to inhabit, so Bolton in these drawings created a world that brings together flora and fauna in a way that they would never have mingled in nature, while still depicting them faithfully and exactingly enough to serve as recognizable and accurate examples of their type.

Lisa Ford
HISTORIAN

OPPOSITE: James Bolton, ?House Wren (*Troglodytes troglodytes*) and eggs with a specimen from the rose family (*Rosaceae*), Peacock butterflies (*Nymphalis io*), both closed and open, and butterfly chrysalis, larva caterpillar, daddy longlegs, spider (*Phalangida)* egg case, and snout beetle, from the natural history cabinet of Anna Blackburne, ca. 1768, watercolor and gouache over graphite on parchment. Yale Center for British Art, Paul Mellon Fund, in honor of Jane and Richard C. Levin, President of Yale University (1993–2013)

OVERLEAF: Peter Brown, *European Goldfinches (Carduelis carduelis) in a Snare*, 1766?, gouache on parchment. Yale Center for British Art, Paul Mellon Collection

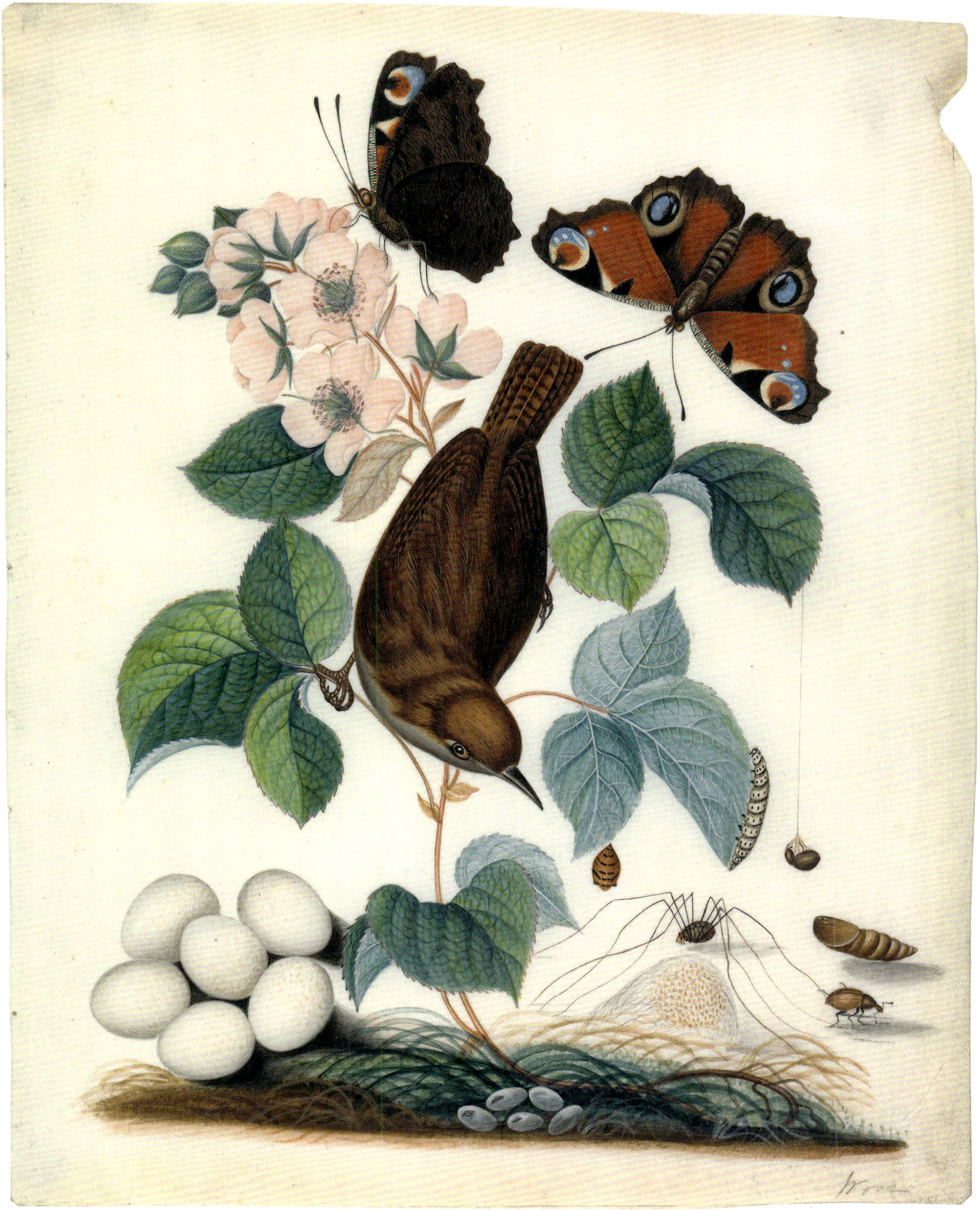

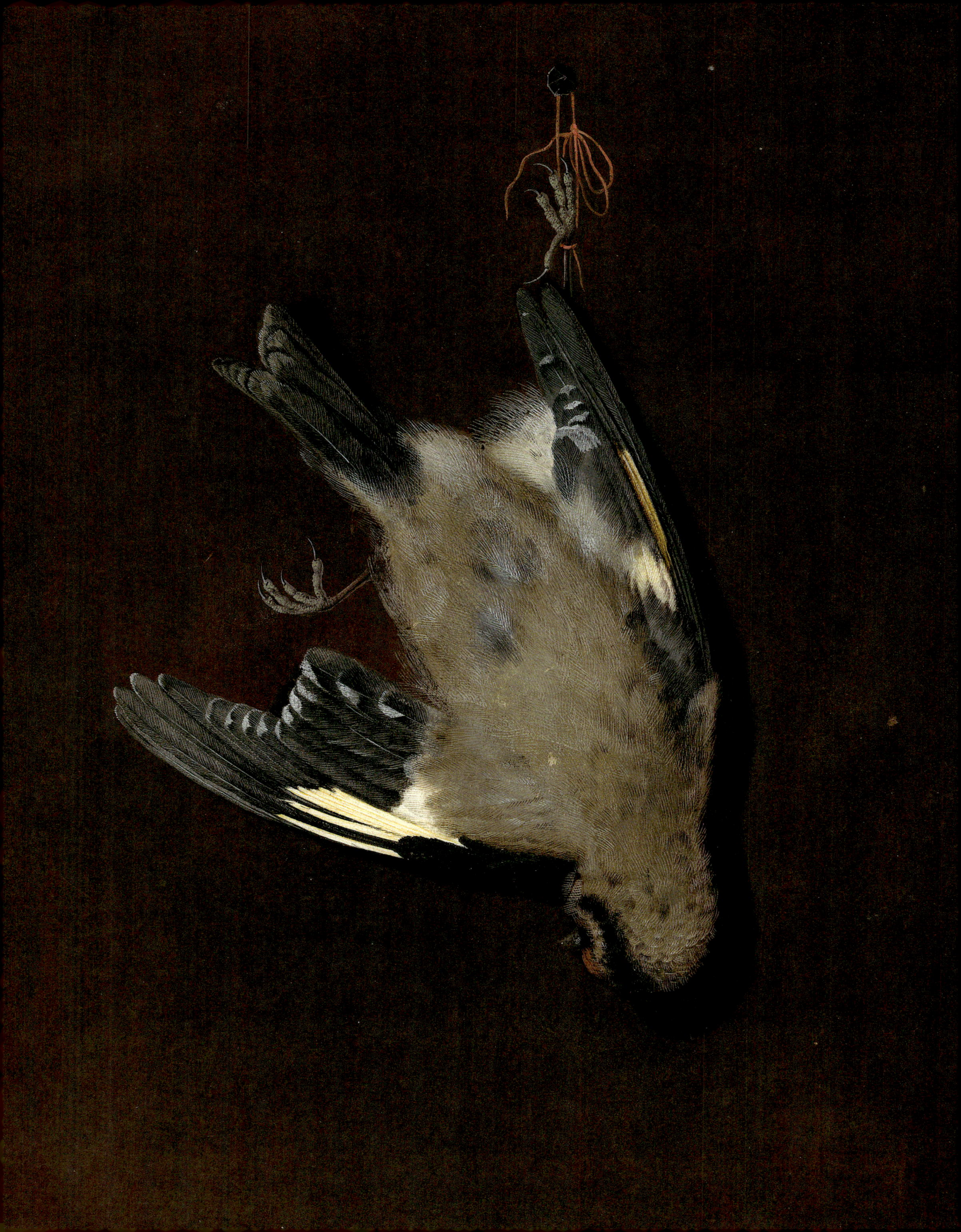

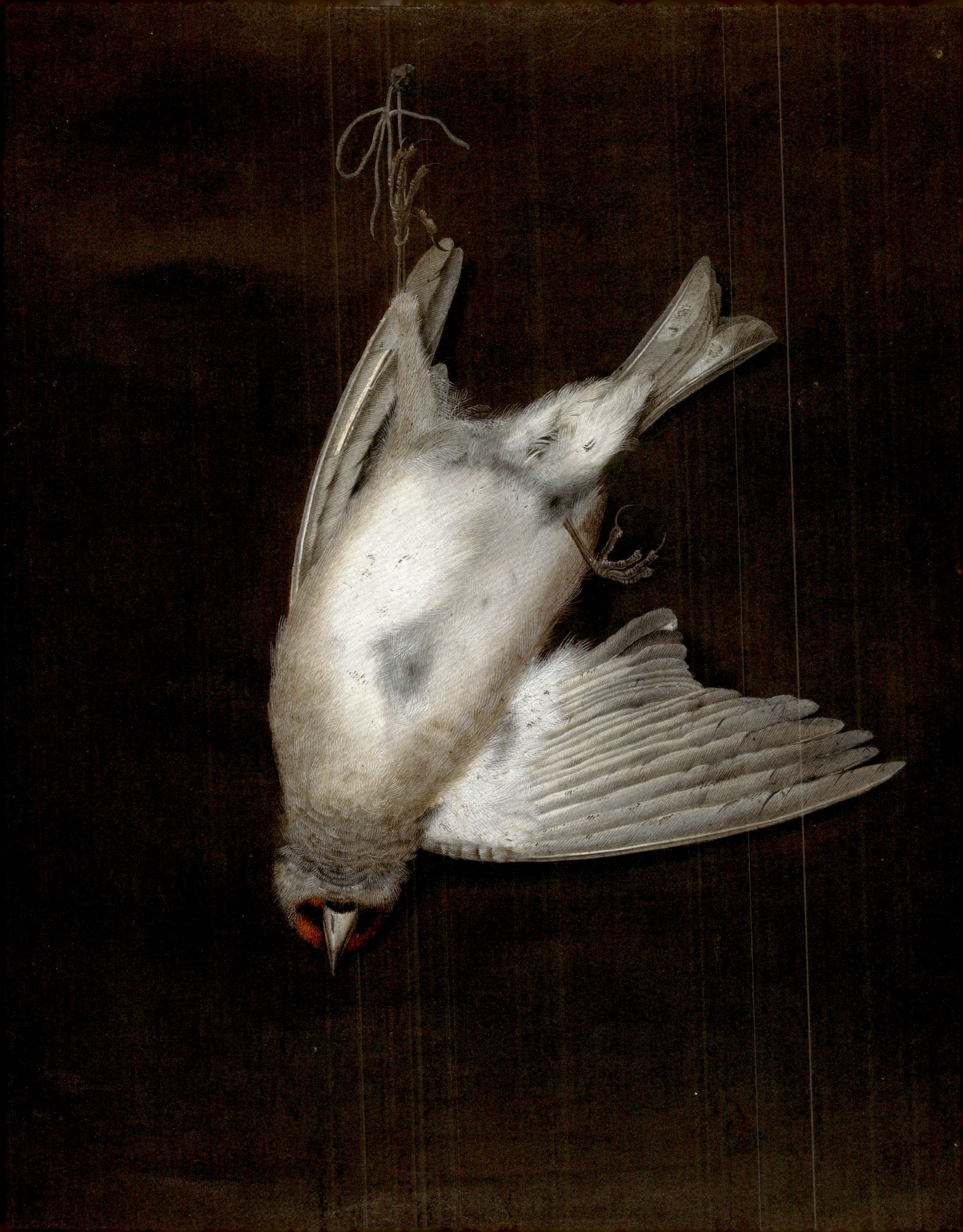

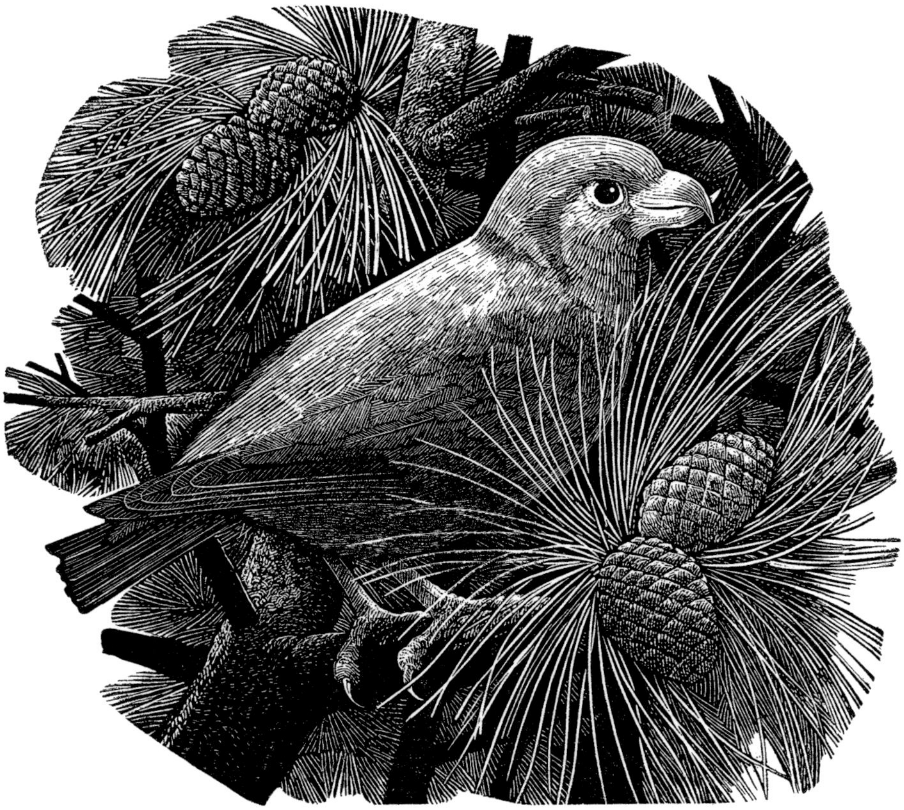

Ian Stephens, *Crossbill*, 2009, wood engraving. Collection of David Burnett

OPPOSITE: James Bolton, Red Crossbill (*Loxia curvirostra*), male, with Scots Pine (*Pinus sylvestris* L.), from the natural history cabinet of Anna Blackburne, ca. 1768, watercolor and gouache over graphite on parchment. Yale Center for British Art, Paul Mellon Fund, in honor of Jane and Richard C. Levin, President of Yale University (1993–2013)

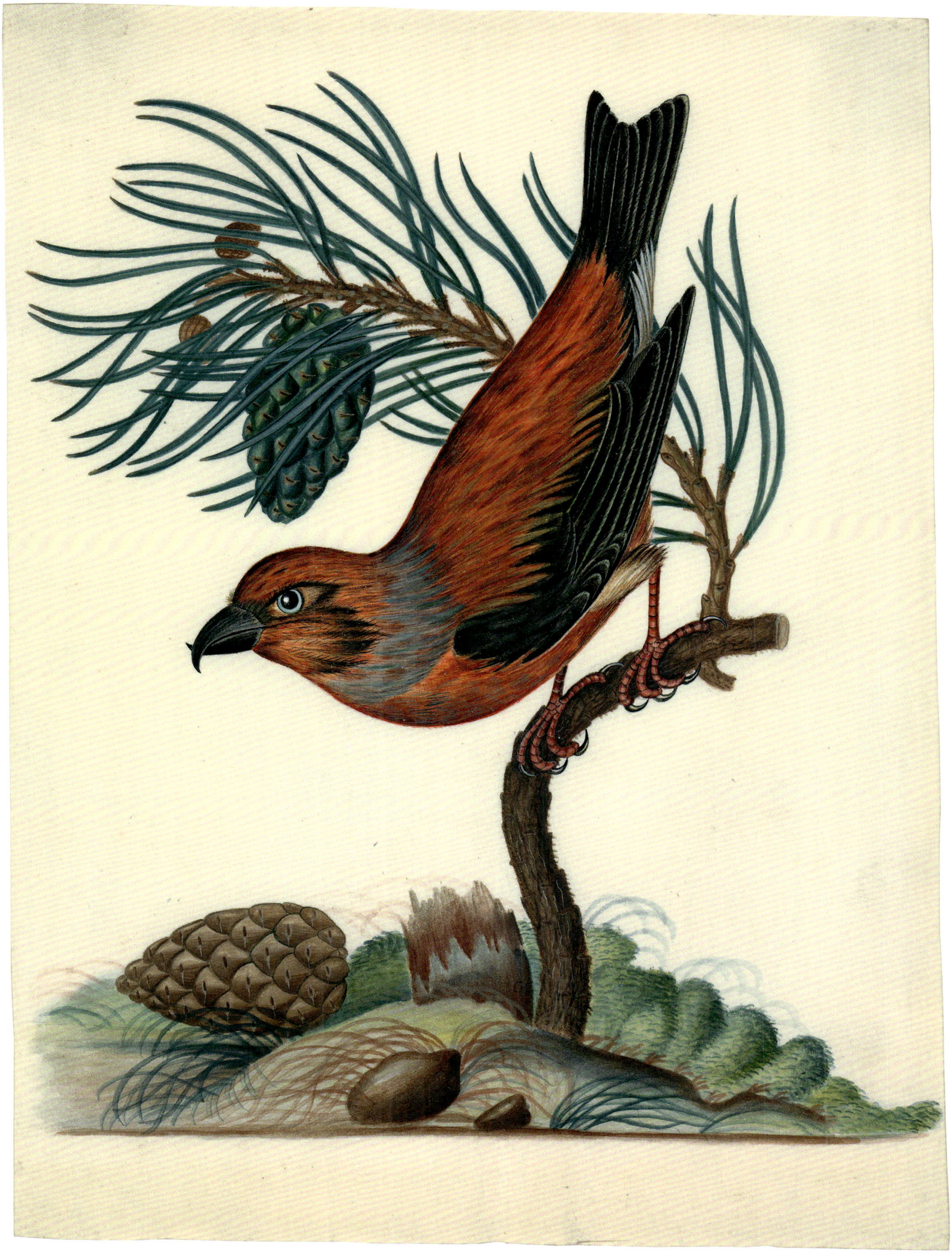

The most unusual birds I ever observed in these parts were a pair of *hoopoes* (*upupa*), which came several years ago in the summer, and frequented an ornamented piece of ground, which joins to my garden, for some weeks. They used to march about in a stately manner, feeding in the walks, many times in the day; and seemed disposed to breed in my outlet; but were frighted and persecuted by idle boys, who would never let them be at rest.

Gilbert White
The Natural History and Antiquities of Selborne, London: T. Bensley, 1789, p. 30

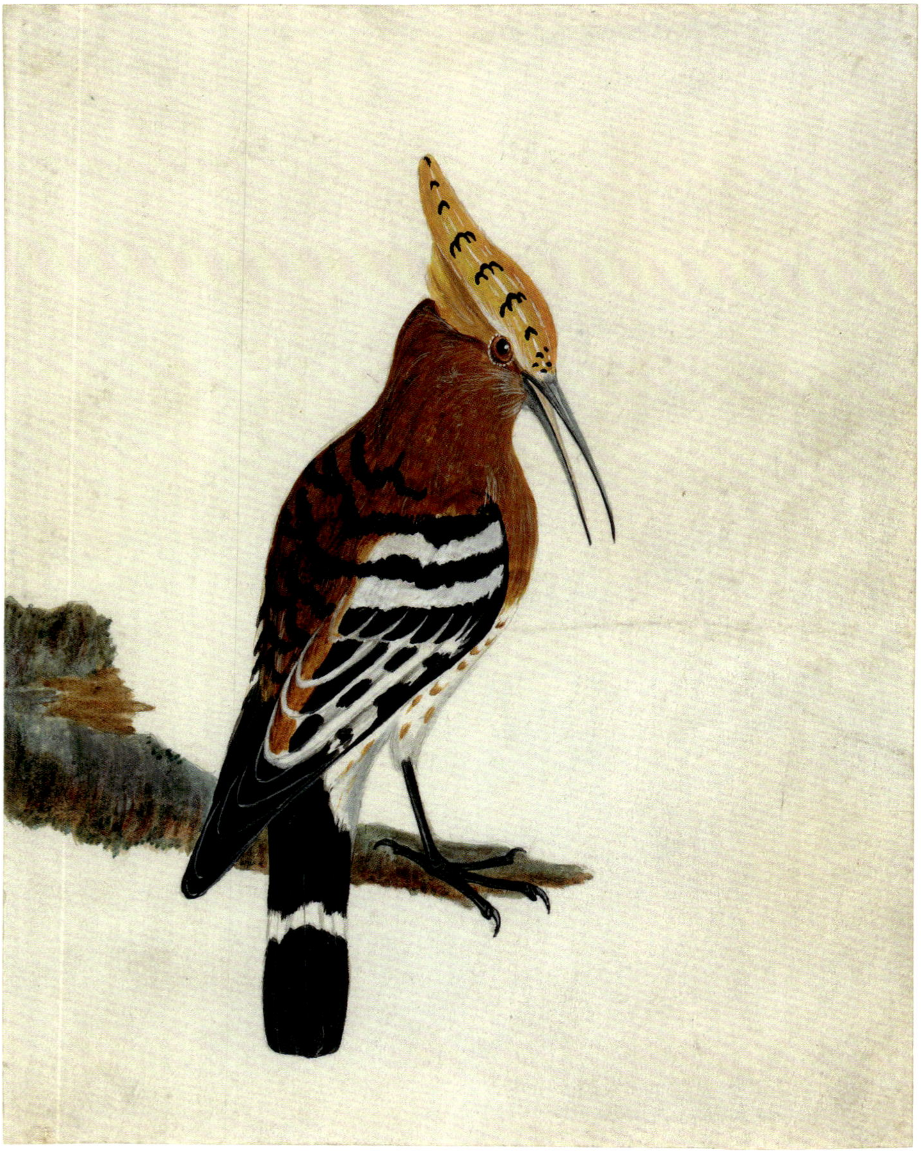

William Lewin, *Common Hoopoe*, ca. 1789, watercolor and gouache over graphite. Yale Center for British Art, Paul Mellon Collection

OPPOSITE: Workshop of Tulsi Das Nimbark, *Hoopoe on a Branch*, Jaipur, India, 2009, gouache on printed leaf of text. Yale Center for British Art, Gift of Holly Shaffer

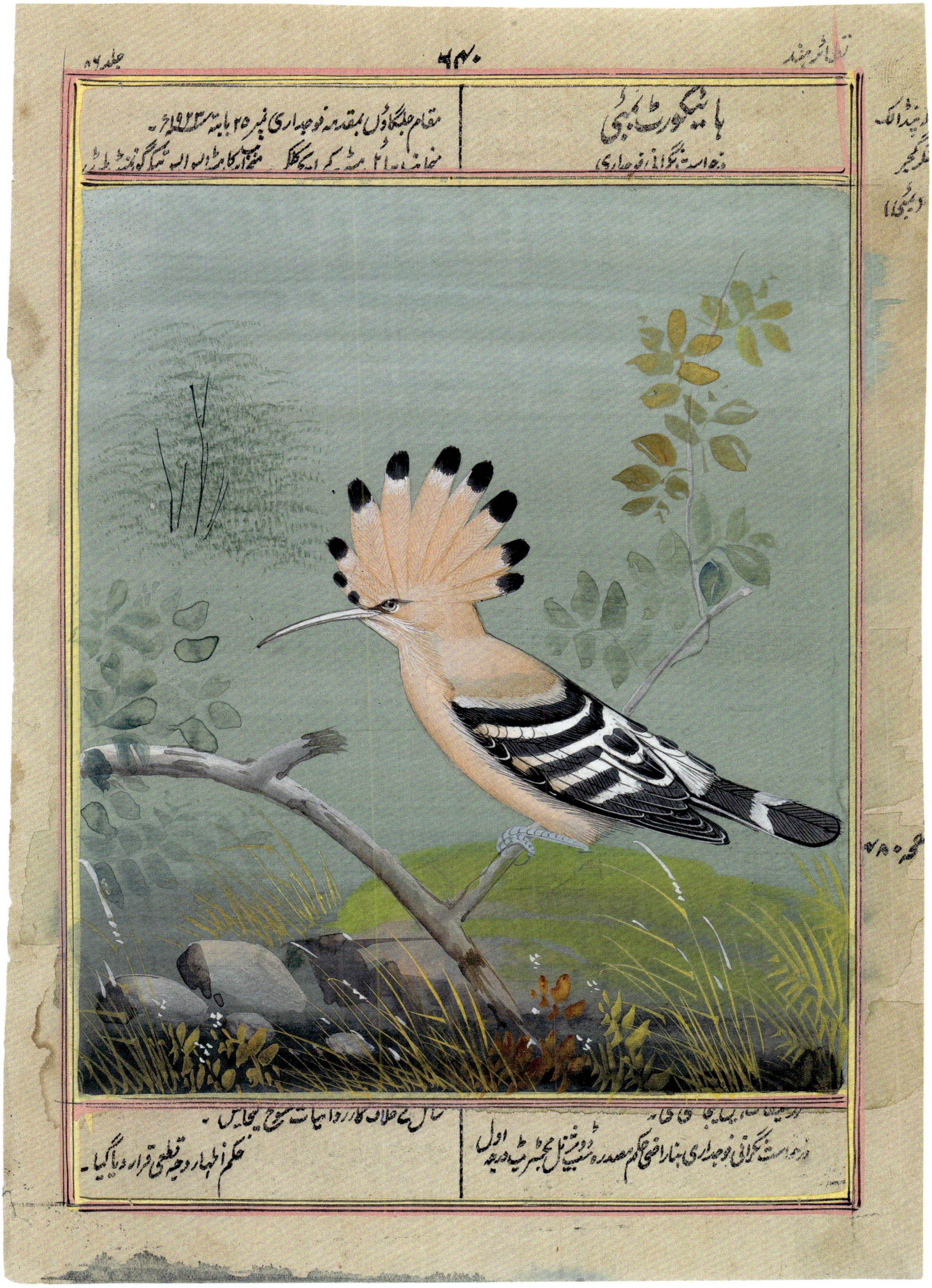
ہائیکورٹ بمبئی

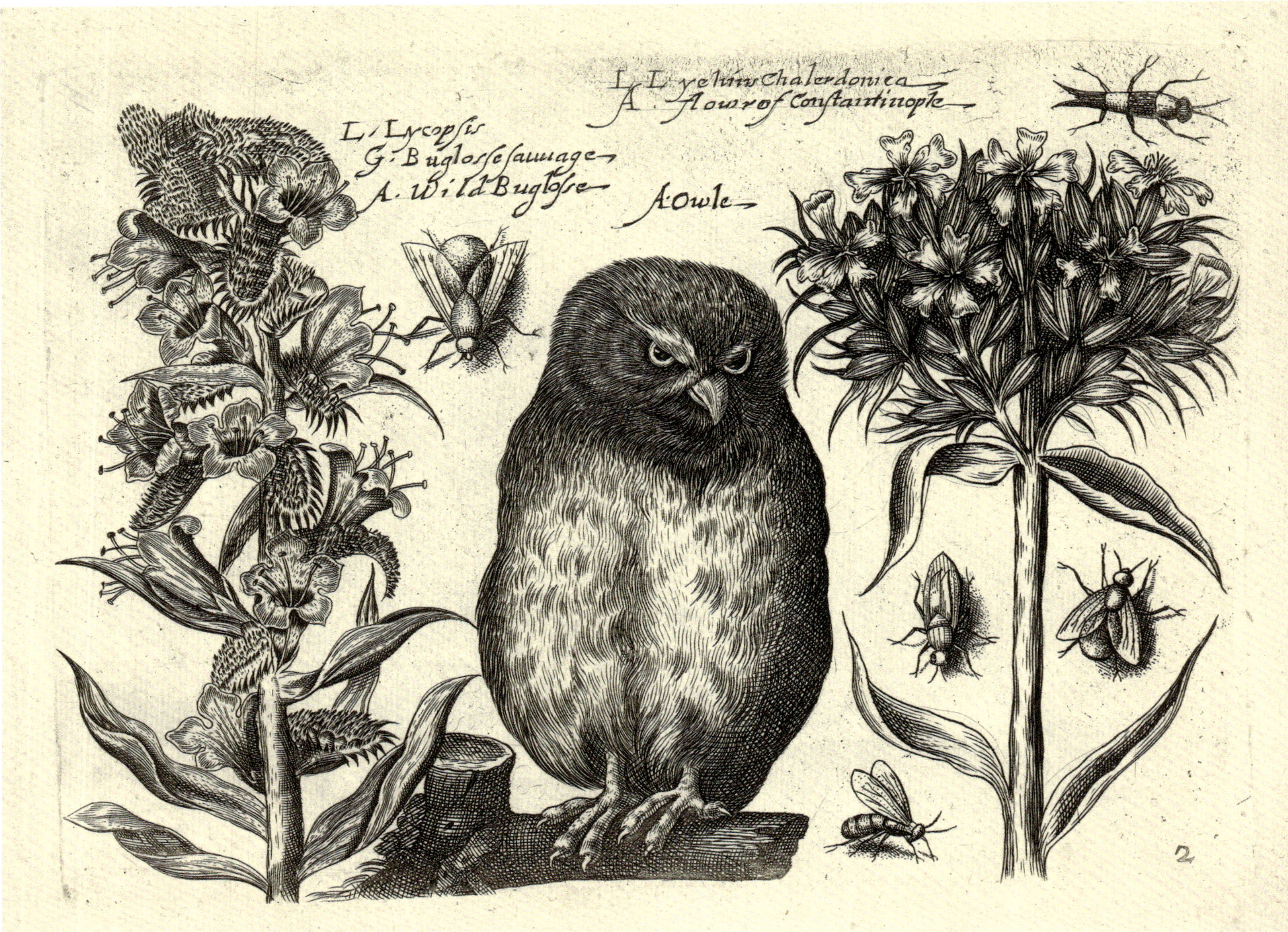

John Dunstall, *The Therd Booke of Flowers Fruits Beastes Birds and Flies Exactly Drawne*, London: Peter Stent, 1661, etching. Yale Center for British Art, Paul Mellon Collection

The owls throughout [Bewick's *History of British Birds*] are stupendous in feather cutting. [The white owl] is one of the two finest.

John Ruskin
pen and ink annotation in his own copy of Bewick's book

OPPOSITE: Thomas Bewick, *History of British Birds*, Newcastle upon Tyne: S. Hodgson, 1797 [i.e. 1798]–1804. Yale Center for British Art, Paul Mellon Collection

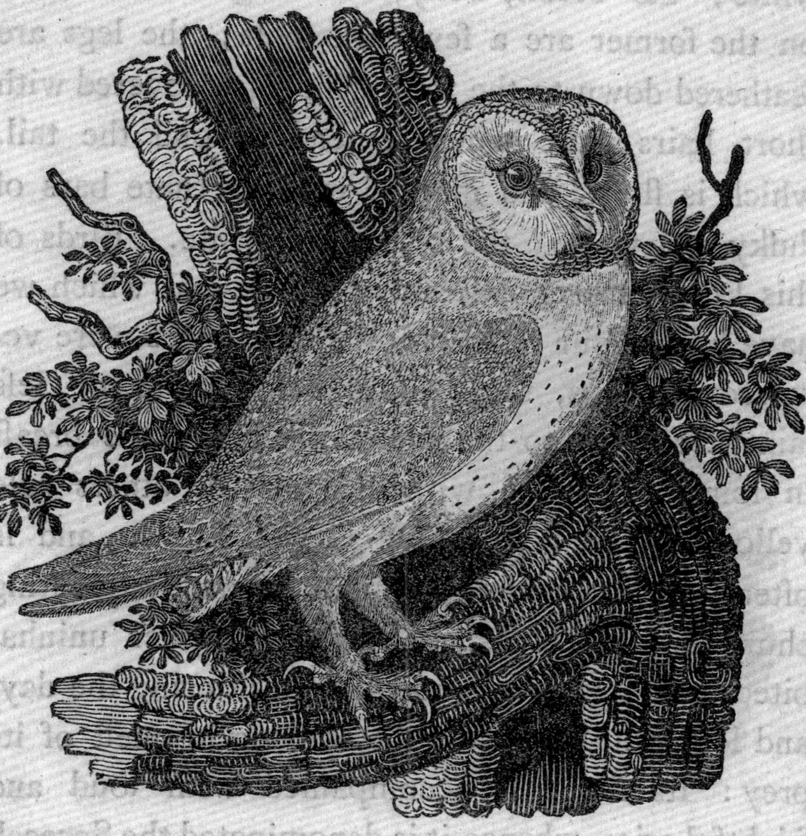

THE WHITE OWL.

BARN OWL, CHURCH OWL, GILLIHOWLET, OR SCREECH OWL.

(*Strix Flammea*, Lin.—*L'Effraie, ou la Frefaie*, Buff.)

Length fourteen inches: Bill pale horn colour; eyes dark; the radiated circle round the eye is compofed of feathers of the moft delicate foftnefs, and perfectly white; the head, back, and wings, are of a pale cheftnut, beautifully powdered with very fine gray and brown fpots, intermixed with

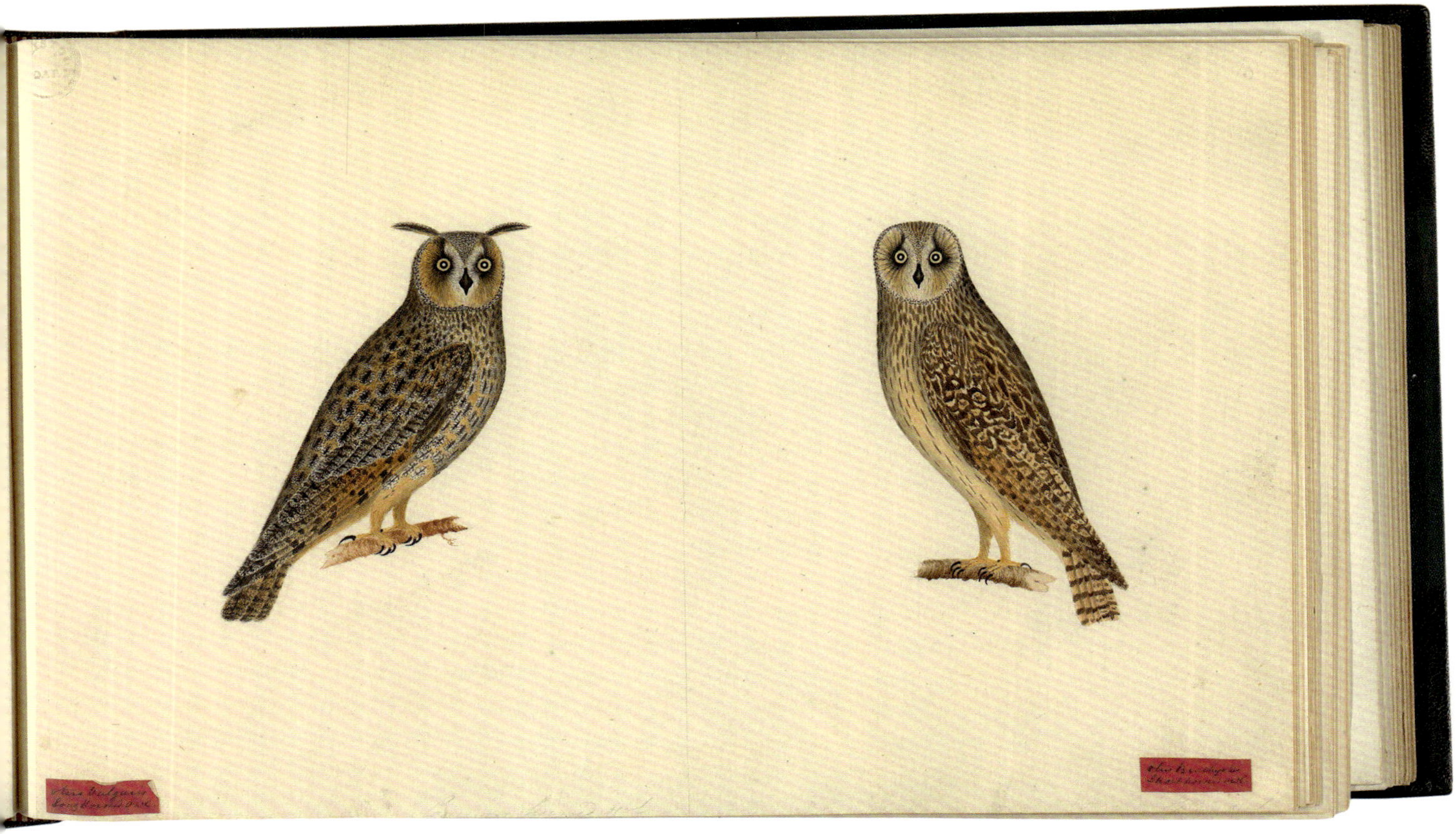

Album of Watercolors of British Birds, ca. 1820–1830, watercolor with pen and ink labels. Yale Center for British Art, Paul Mellon Fund

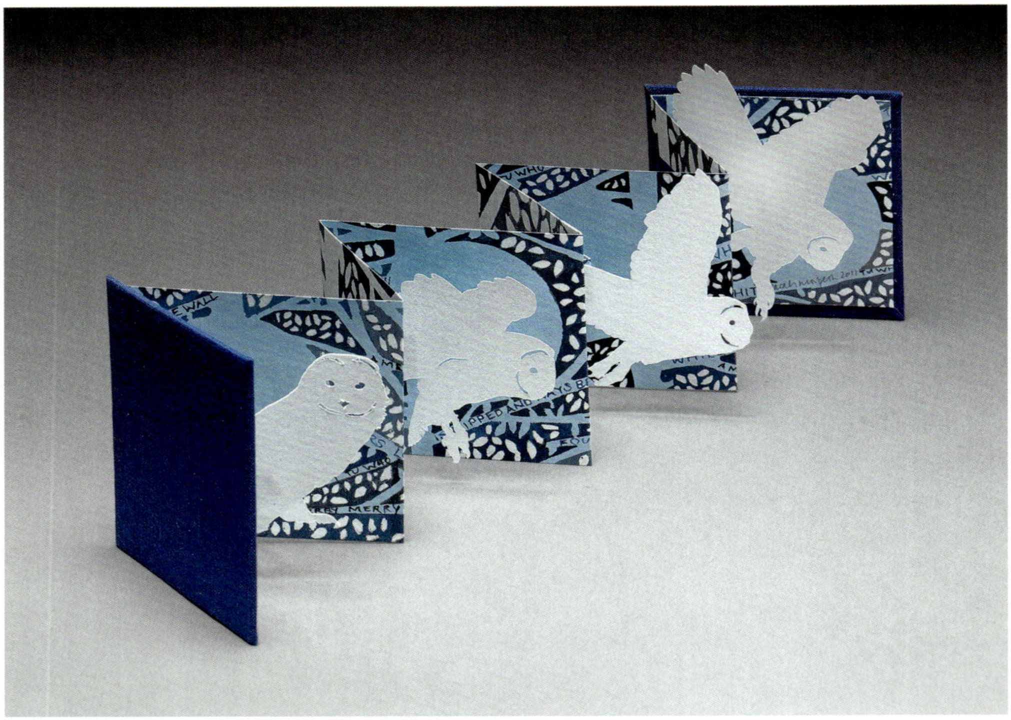

Sarah Morpeth, *Owl*, Elsdon, Northumberland, 2011, cut and folded paper and digital print. Yale Center for British Art, Friends of British Art Fund

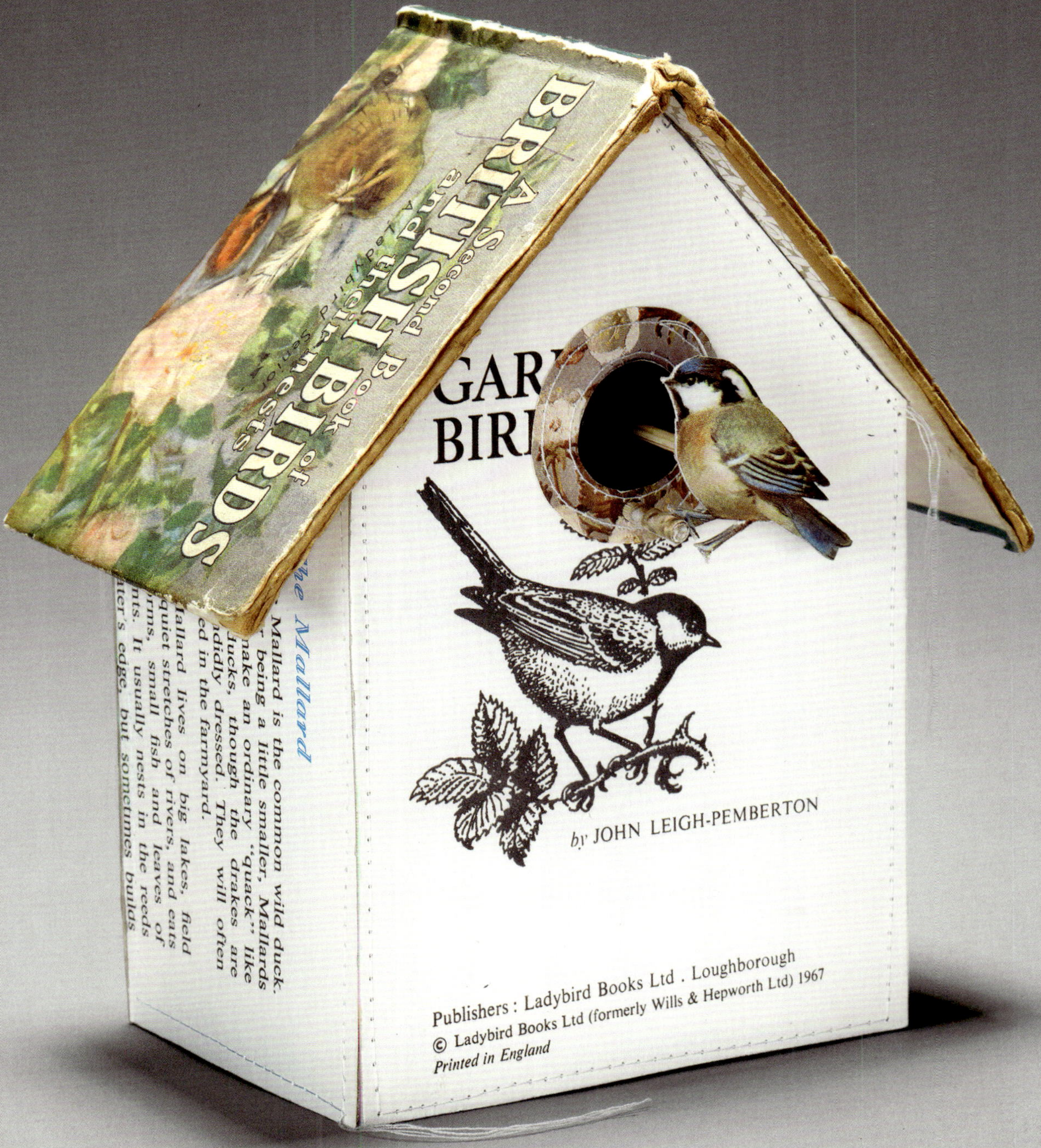

Jennifer Collier, *Bird House*, Stafford, England, 2013, altered book, cut paper, wood, and thread. Yale Center for British Art, Friends of British Art Fund

Butterflies, Moths & Caterpillars

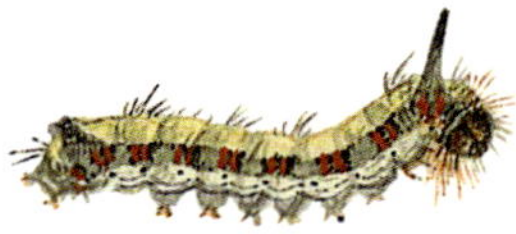

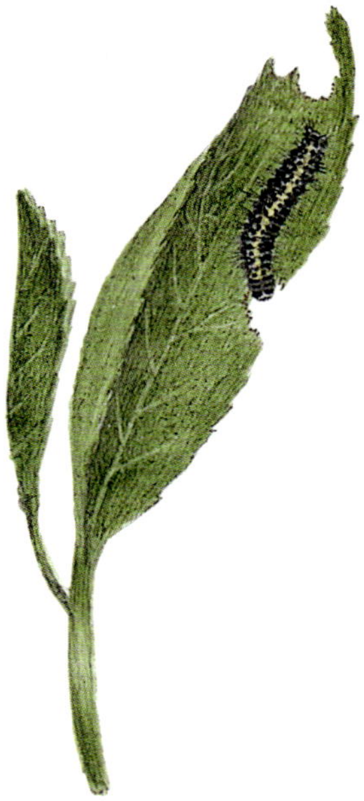

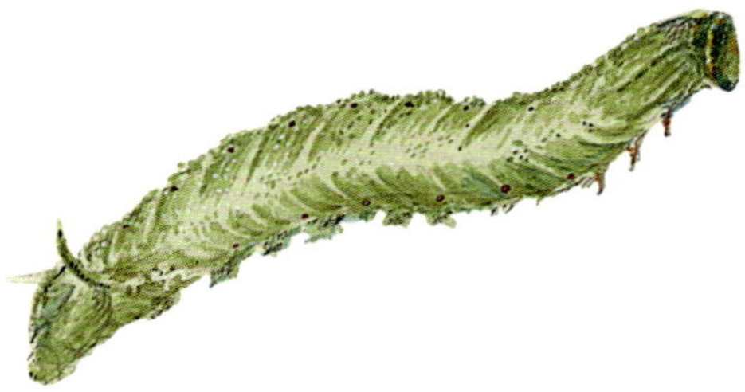

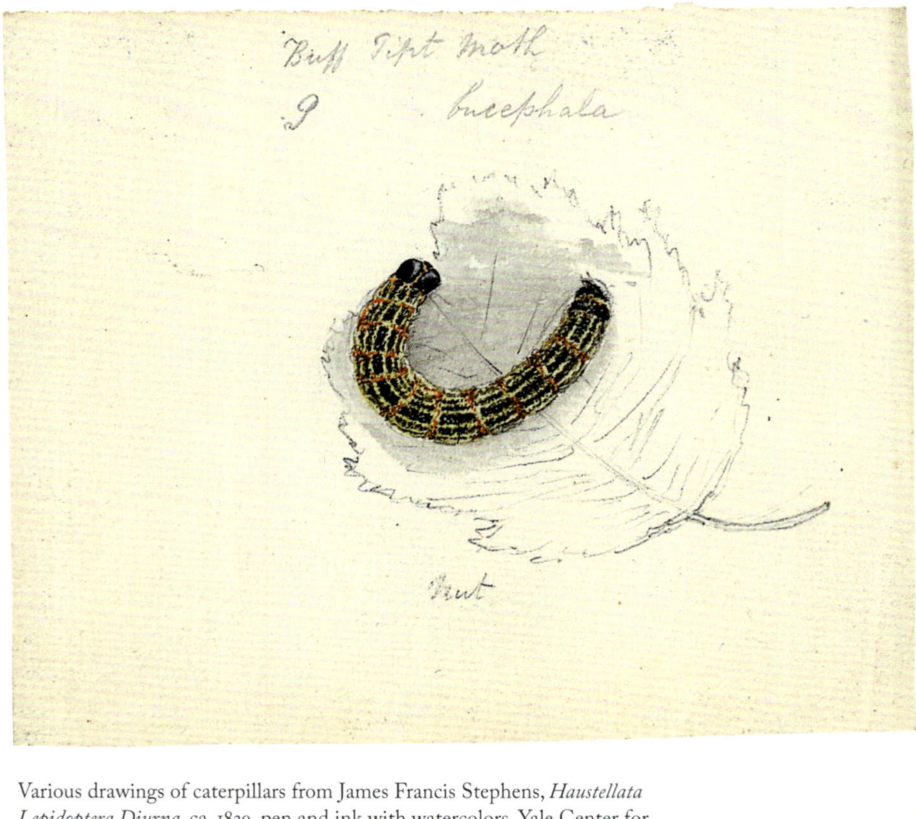

Various drawings of caterpillars from James Francis Stephens, *Haustellata Lepidoptera Diurna*, ca. 1839, pen and ink with watercolors. Yale Center for British Art, Paul Mellon Fund

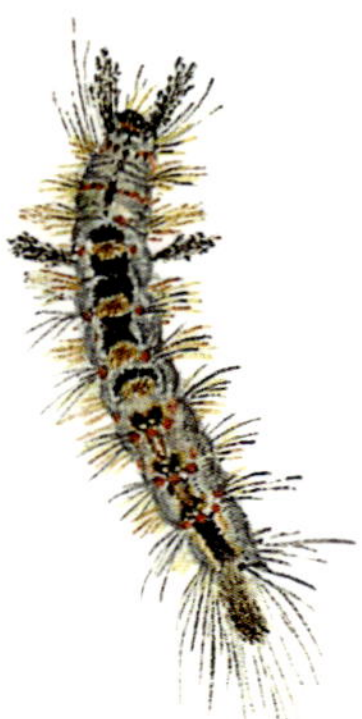

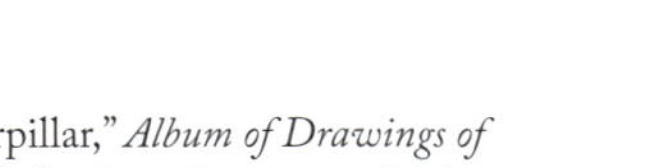

OPPOSITE: "The Great Tyger Moth & Caterpillar," *Album of Drawings of English Moths, Butterflies, Flowers, and Mollusks*, 1805–1822, pen and ink and watercolor. Yale Center for British Art, Paul Mellon Fund

The Great Tyger Moth & Caterpillar,
it will eat many plants, but the favorite food of this Caterpillar is the stinging Nettle it is commonly found in May & June

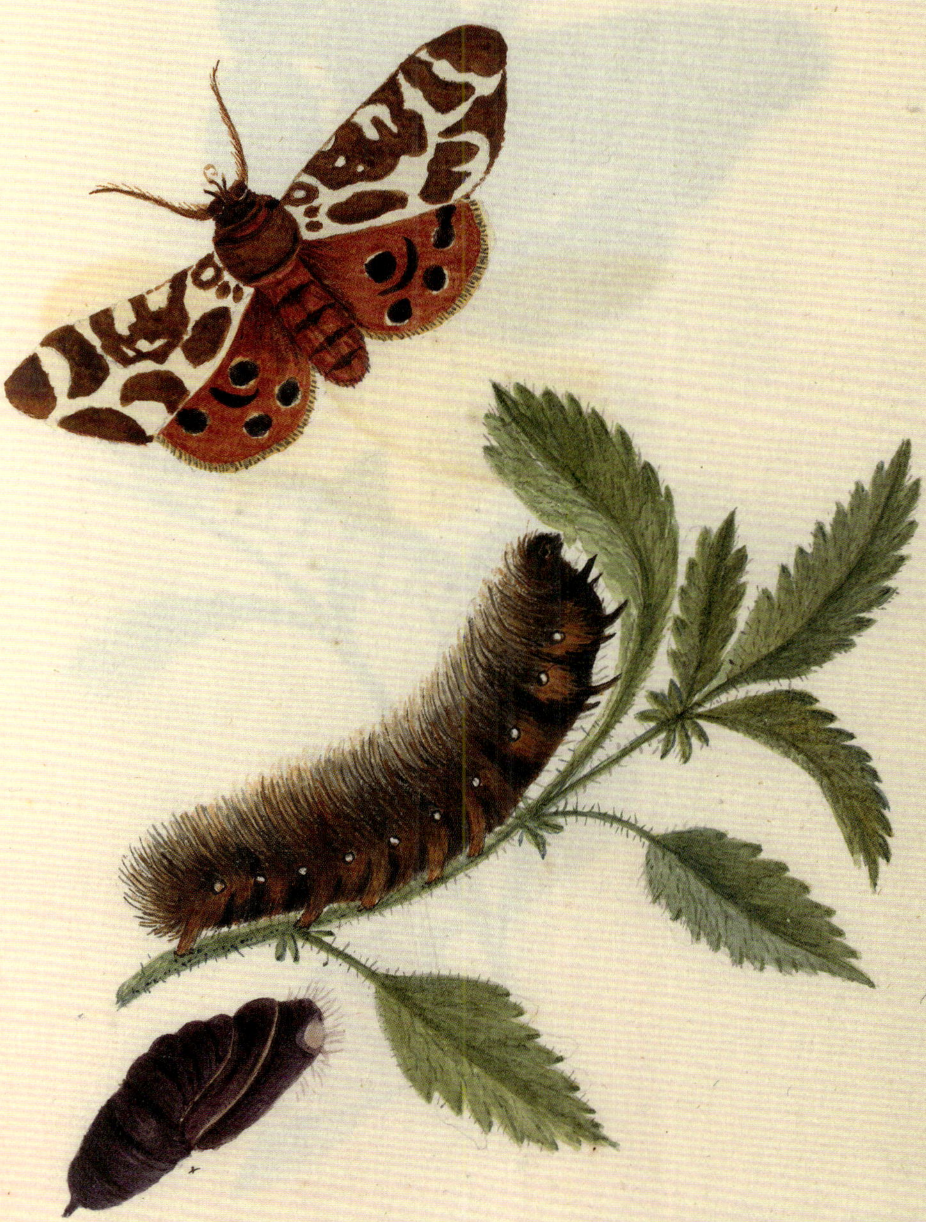

* The Chrysalis after the Moth was flown

The Caterpillar from which this drawing was taken spun in a Glass in June, & the Moth came out in August, the same year

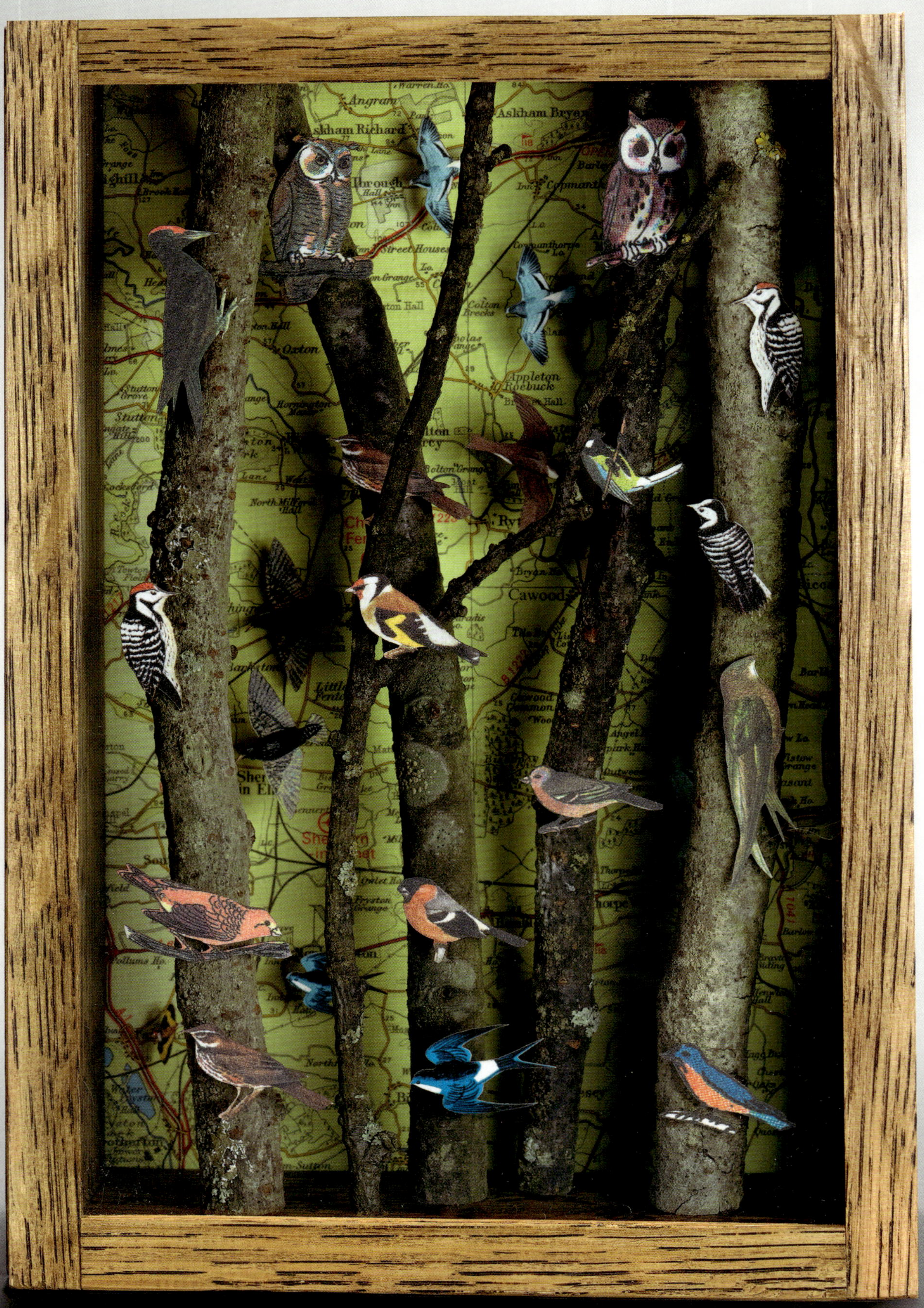
Angram
Oxton
Appleton Roebuck
Cawood

Categorizing Nature

I spent my childhood exploring nature in my grandparents' large garden, getting down in the soil and observing in detail. I learned a lot about nature from their point of view as gardeners. They treated nature as a battleground, attempting to eradicate the advancing "weeds and pests." If sides had to be taken, I think even then I was on the side of those unwelcome plants and creatures.

These creatures and many others that I had never seen in real life were all laid out in illustrated plates in the books on nature I had as a child and still collect today. They have had as much of a lasting effect on me as nature at first hand.

Fascinated by the human activity of categorizing species, I became interested in the cases of Lepidoptera amassed during the Victorian era. In these, the moths and butterflies are typically pinned and labeled in neat rows. In my own work I imagined these creatures coming back to life and trying to escape from the case. The result is a frenzied mass of moths hitting the glass as they compete with each other toward the light.

My *Moth Collection* boxes also work for me in a formal way; the boxes have to be full of moths to achieve an allover continuous surface in the manner of an abstract expressionist action painting, such as in Jackson Pollock's work.

The woodland boxes (such as *Bishop Wood*) are partly inspired by the museum cases made to show off the nineteenth-century taxidermy collections that I saw as a child. In these, a model of a natural scene is created and staged with an example of each species. I found these both strangely natural and unnatural. I also saw them in a poetic way, imagining how the creatures might live in harmony away from the human gaze.

The maps serve as a pictorial background showing the landscape laid out, often with woodland represented by tiny tree symbols, and therefore a landscape can be imagined from a map. Maps are also just a human record, the land recorded and explained for our purpose in the way that species have been labeled, categorized, and laid out in a book plate. I am more interested in human recordings of nature than trying to recreate nature.

John Dilnot
ARTIST

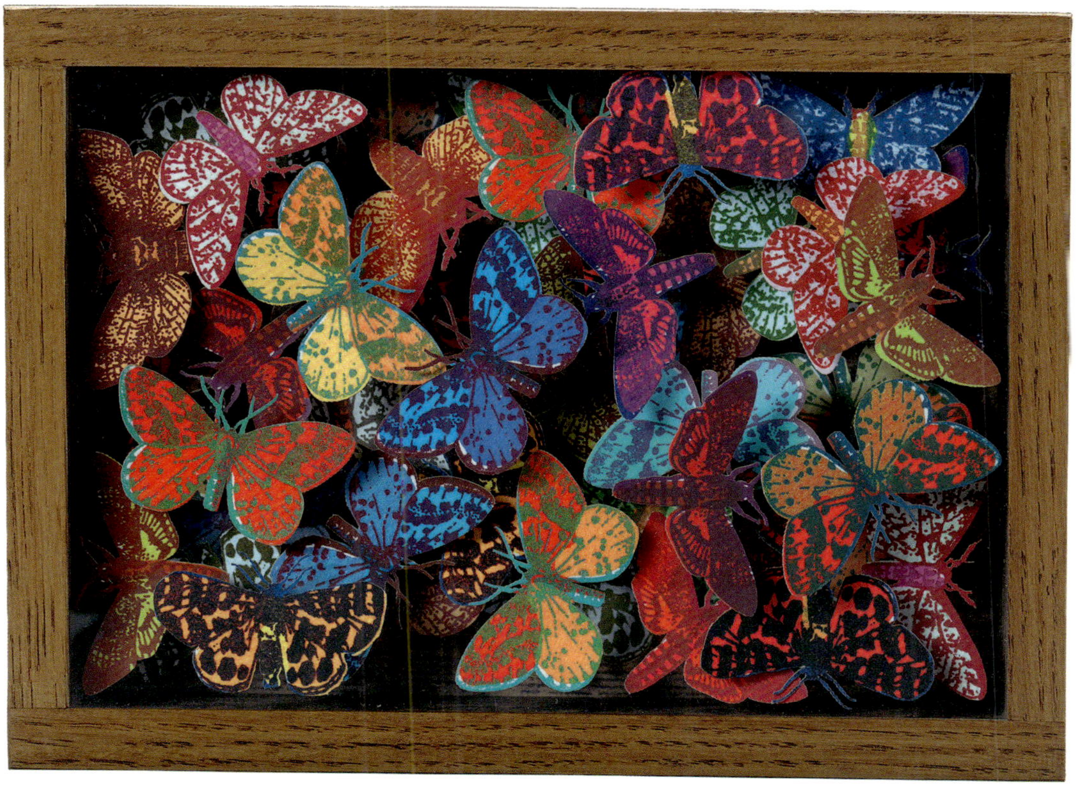

OPPOSITE: John Dilnot, *Bishop Wood*, Brighton, England, 2013, sticks, printed map, and cut paper in glass-topped wooden box. Yale Center for British Art, Friends of British Art Fund

John Dilnot, *Moth Collection*, Brighton, England, 2013, screen print and cut paper in glass-topped wooden box. Yale Center for British Art, Friends of British Art Fund

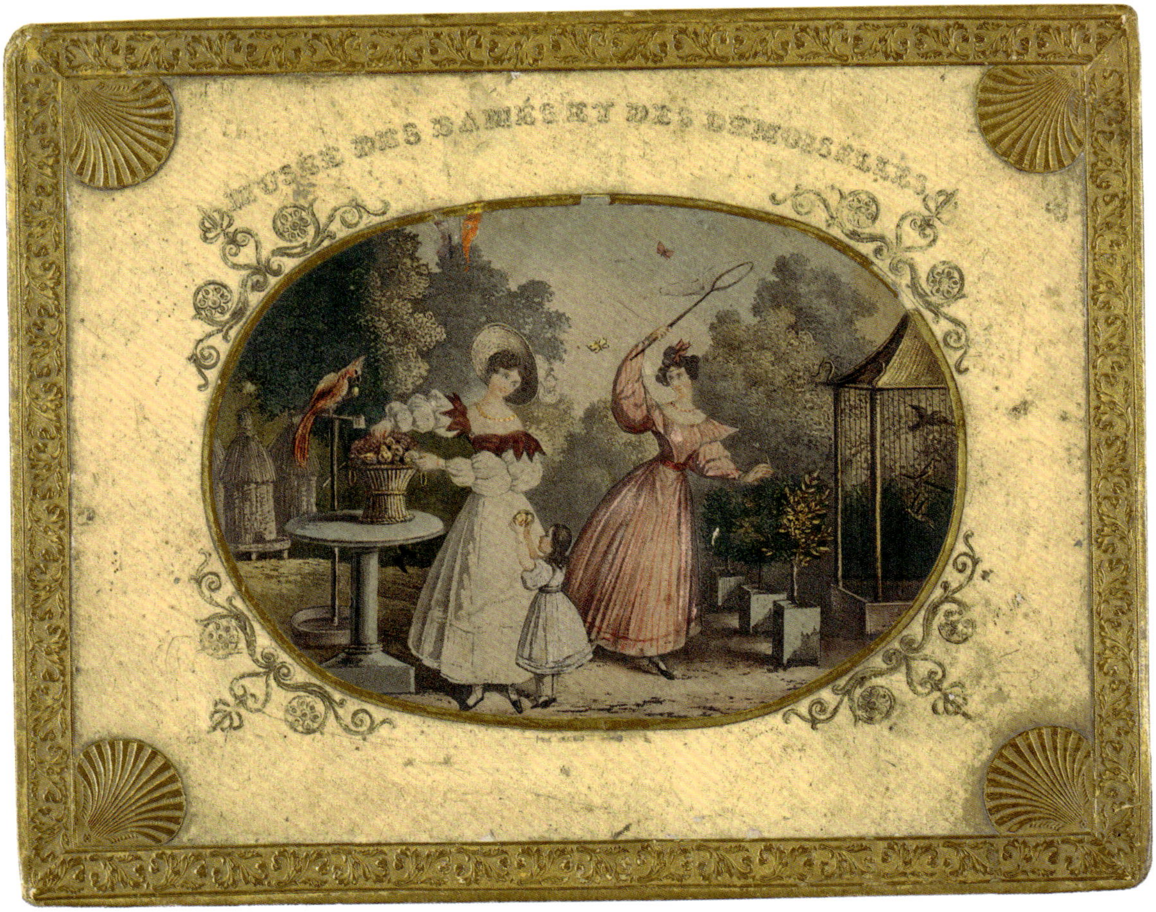

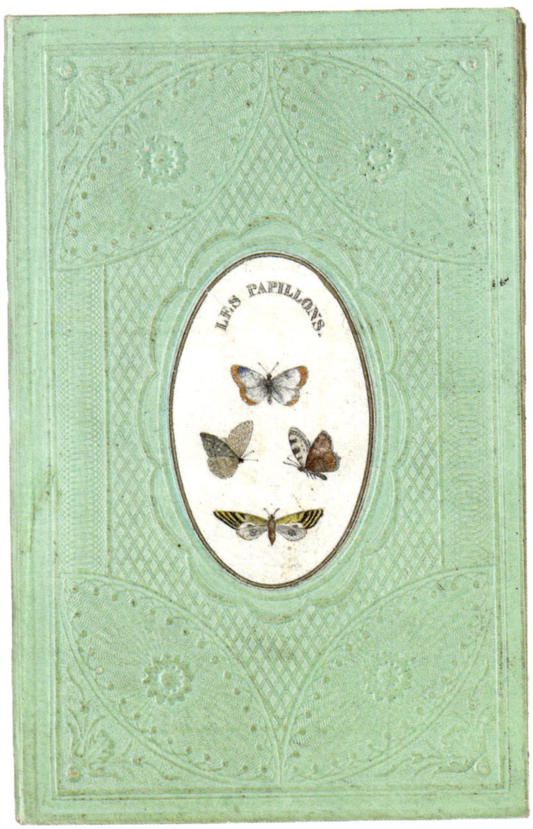

Box lid and one of six books from *Musée des dames et des demoiselles: les oiseaux, les insectes, les papillons, les fleurs, les fruits, les minéraux et les coquillages*, Paris: Chez Marcilly Ainé, ca. 1825. Yale Center for British Art, Paul Mellon Collection

OPPOSITE: *Design for Embroidered Waistcoat*, Britain or France, ca. 1845, watercolor and gouache on black coated paper. Yale Center for British Art, Paul Mellon Collection

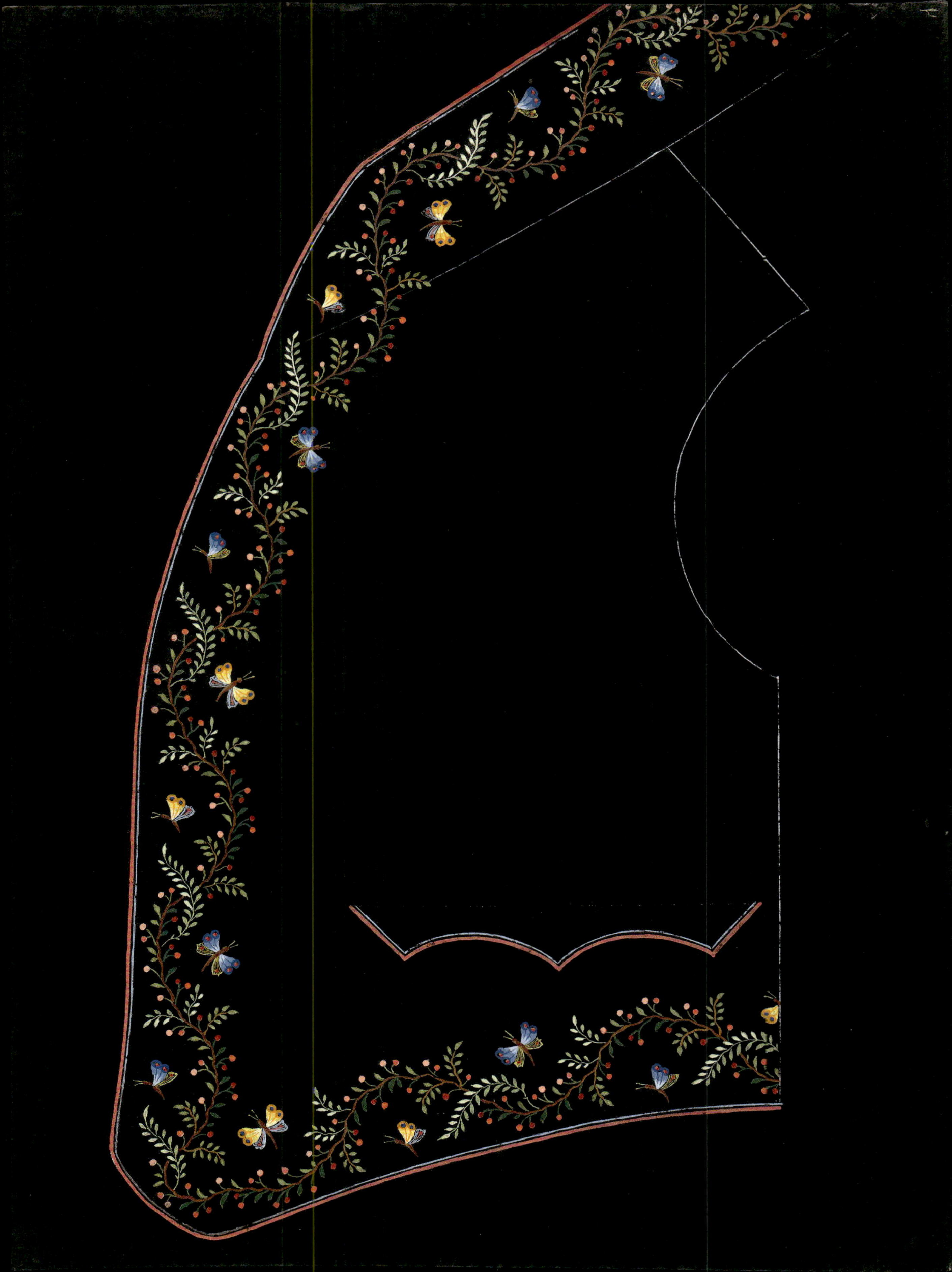

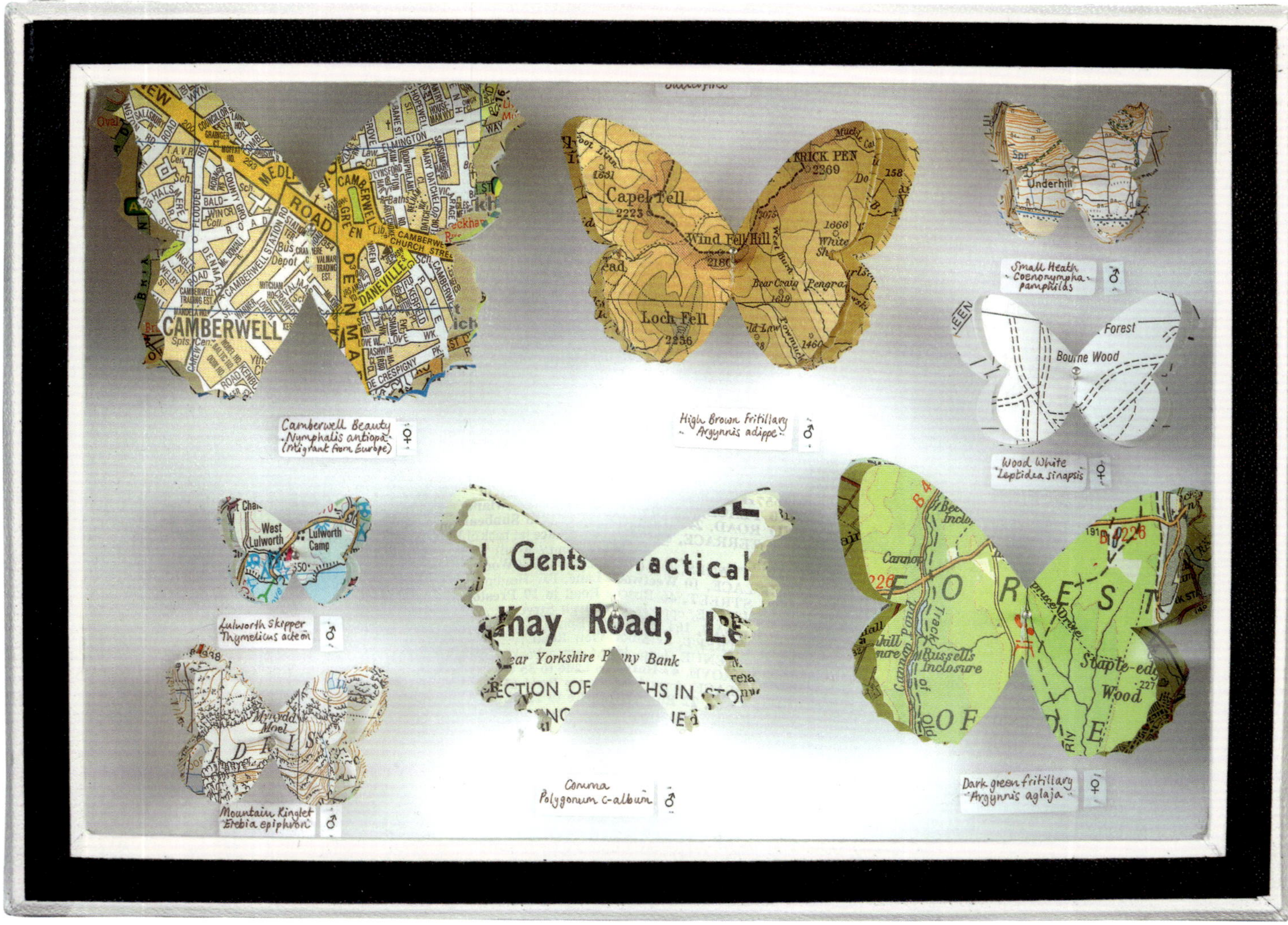

Tracey Bush, *British Butterflies: Museum Box, no. 1*, London, 2004, hand-cut paper specimens mounted in glass-topped box. Yale Center for British Art, Friends of British Art Fund

I create artworks that are informed by cultures of collecting the natural world. My entomological boxes contain collections of butterflies and moths, hand cut from paper ephemera and held together with a bookbinder's pamphlet stitch. There is usually a quirky link between the paper material used for the butterfly and its common name. For example, the *Camberwell Beauty* [top left] is cut from the Camberwell district in the *London A–Z* street map—although these butterflies are no longer found in that locale.

Tracey Bush
November 2013

In August and September, some of our finest Butterflies make their appearance. . . . We give the place of honour, where it is deserved, to the Camberwell beauty (*Vanessa antiopa*), so common on the Continent and so rare in this country. It is a Butterfly which no one who has once seen can possibly mistake, and a glance will immediately detect it even when on the wing. This splendid Butterfly has its rich chocolate-coloured wings edged with a black band, a row of blue spots, and a broad band of white, which runs completely along the outer edges of the wings.

John George Wood
The Calendar of the Months; or, The Young Naturalist Abroad, London: George Routledge & Sons, 1873, p. 112

OPPOSITE: "Camberwell Beauty" in *Album of Drawings of English Moths, Butterflies, Flowers, and Mollusks*, 1805–1822, pen and ink and watercolor. Yale Center for British Art, Paul Mellon Fund

Found in every part of Europe

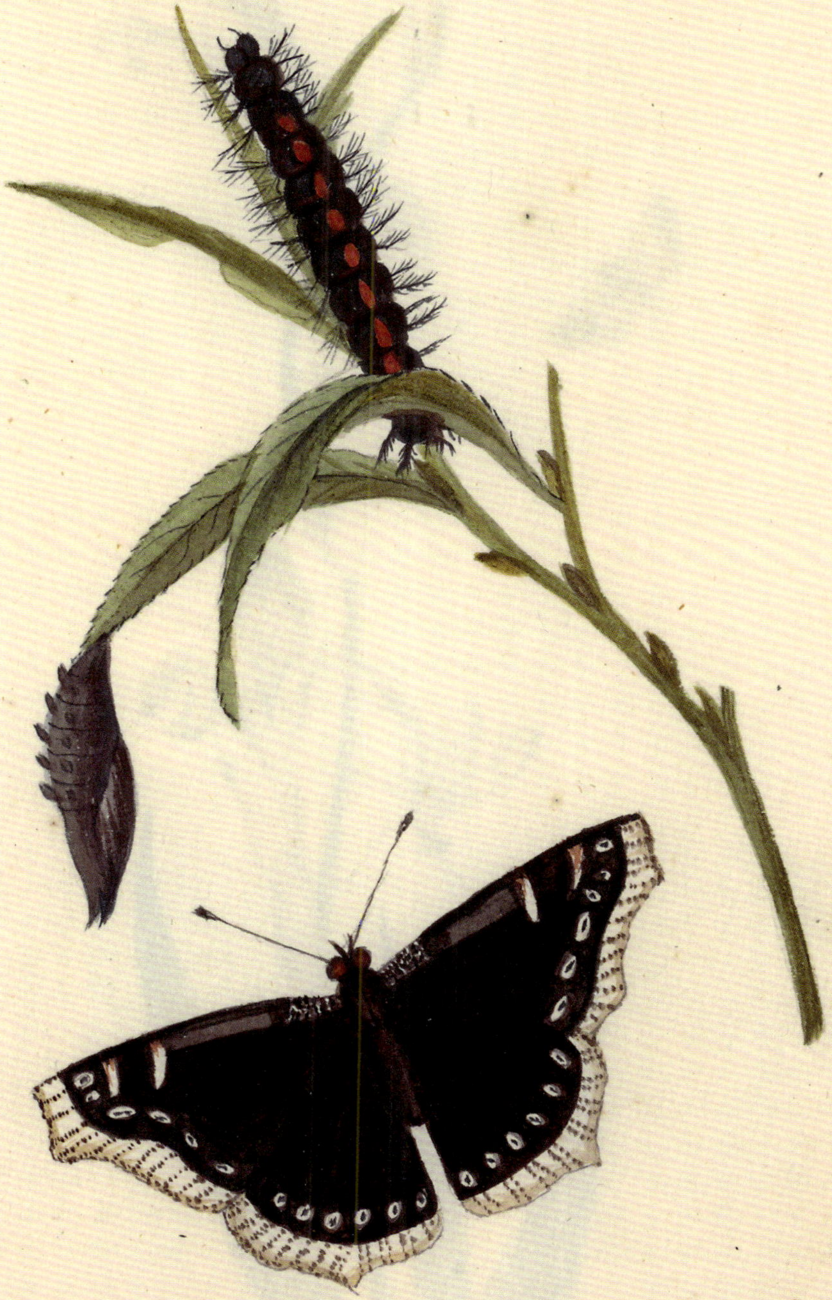

Camberwell Beauty
or
Grand surprise Butterfly.

8

in 1793 they were common near London
JE 1821

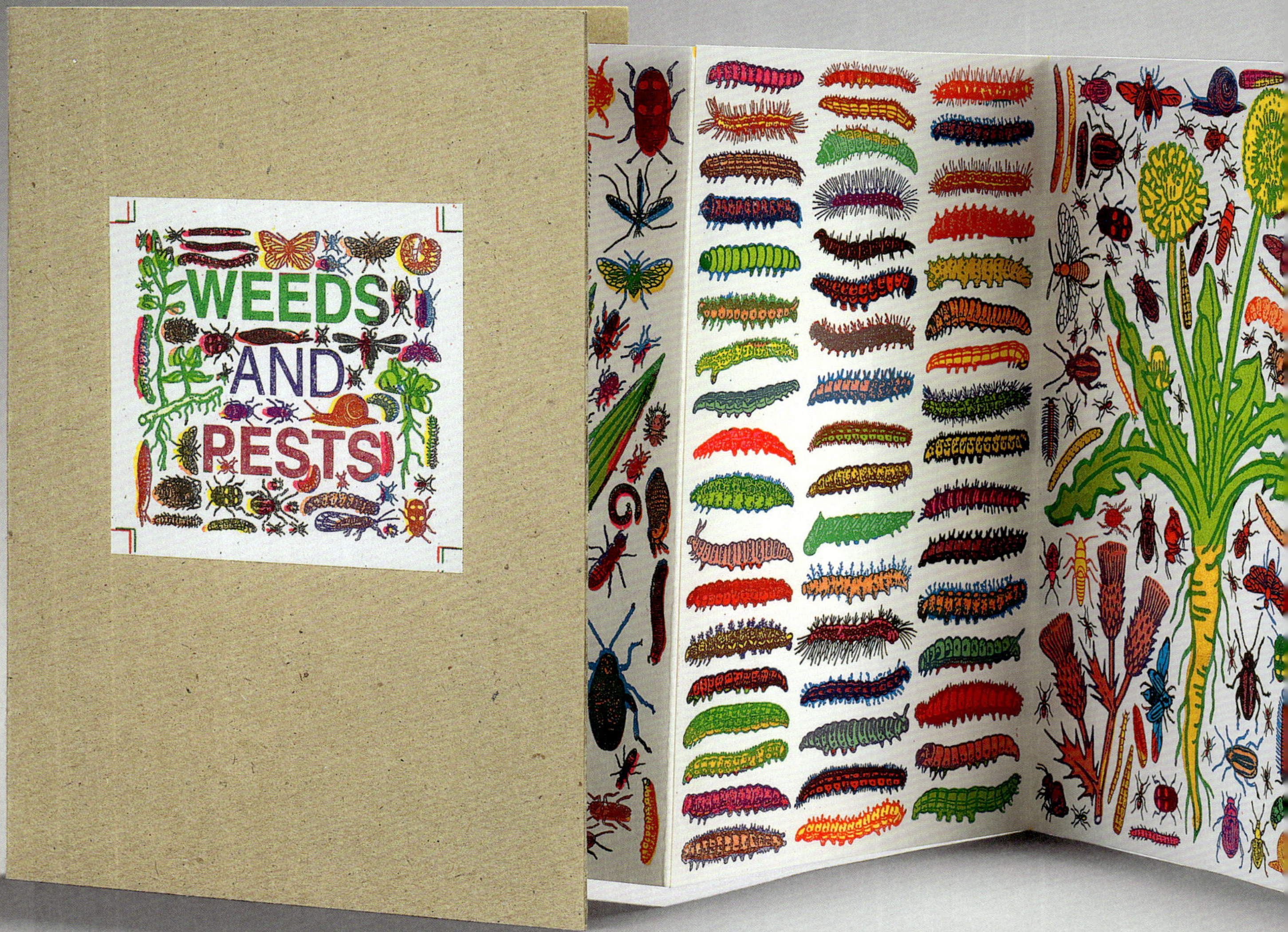
WEEDS
AND
PESTS

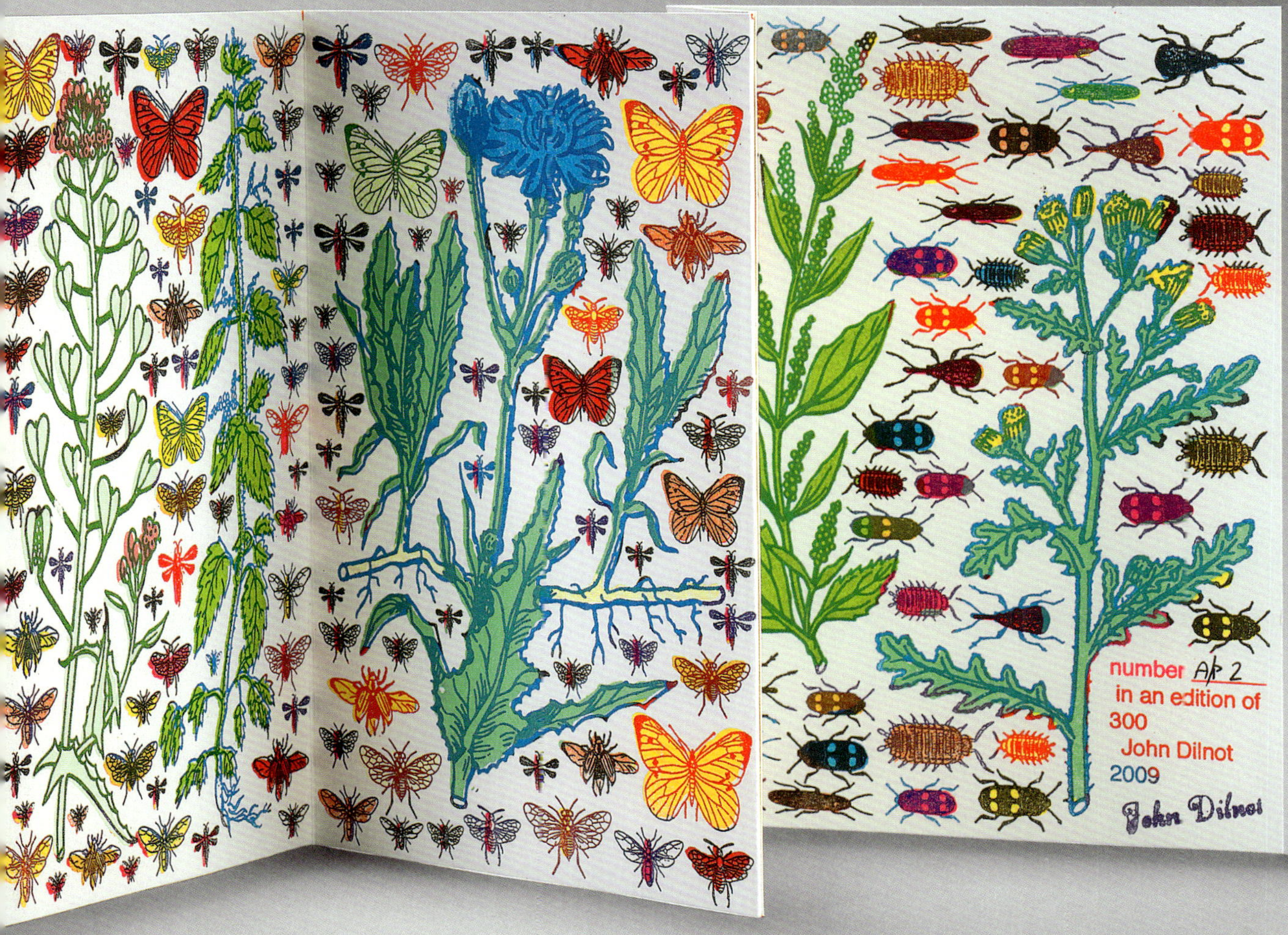

John Dilnot, *Weeds and Pests*, Brighton, 2009, screen print.
Yale Center for British Art, Friends of British Art Fund

Beetles, Bees & Spiders

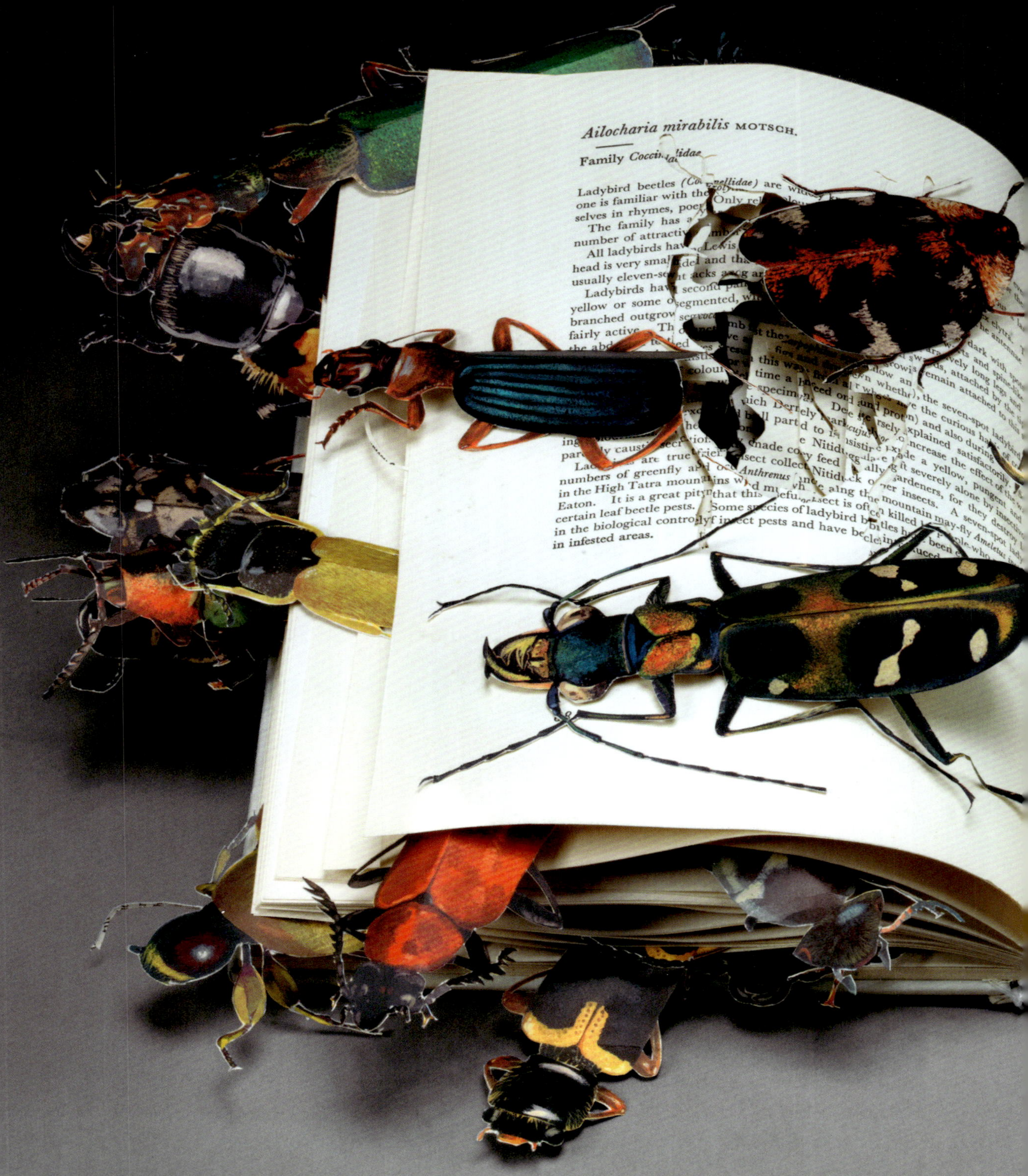

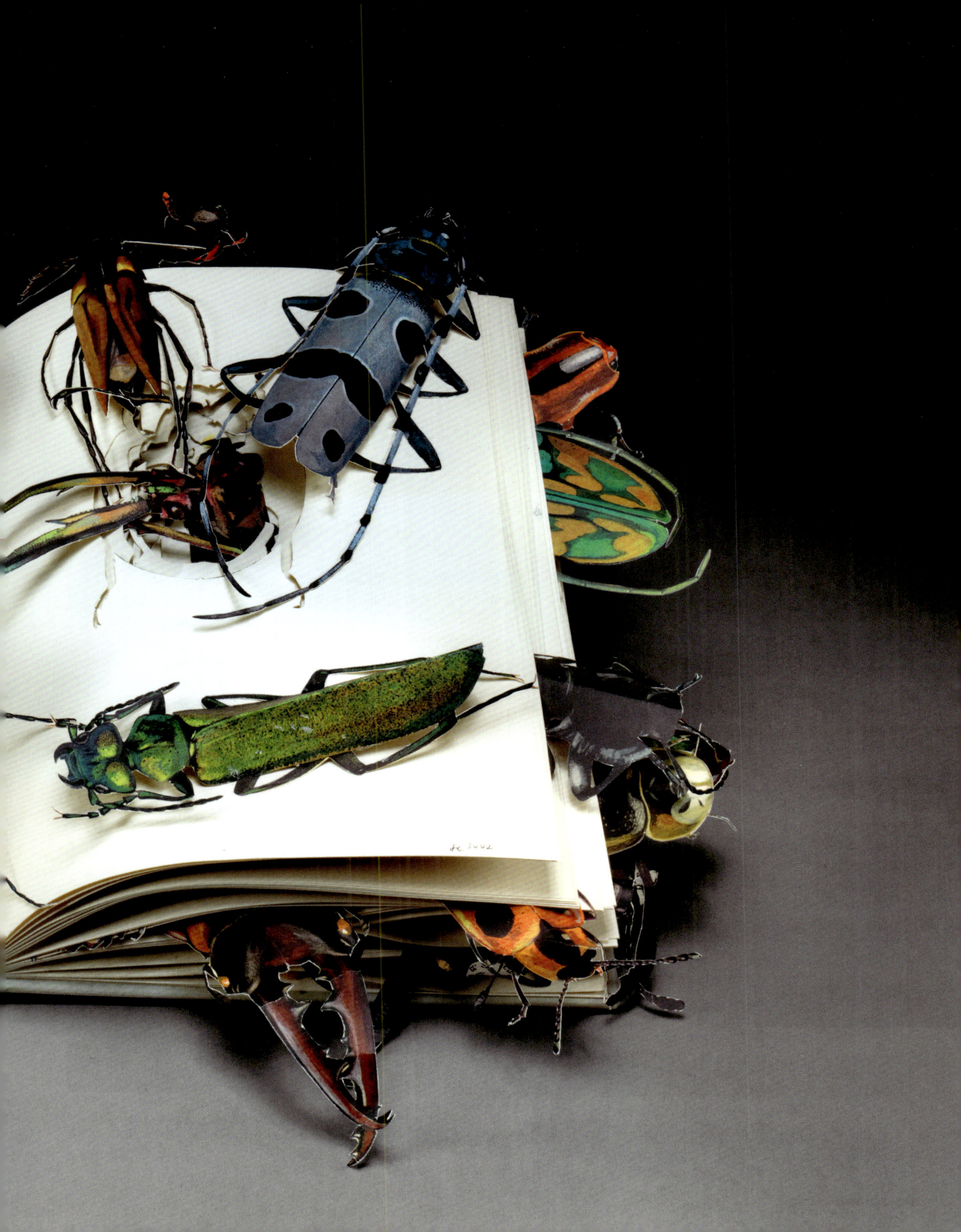

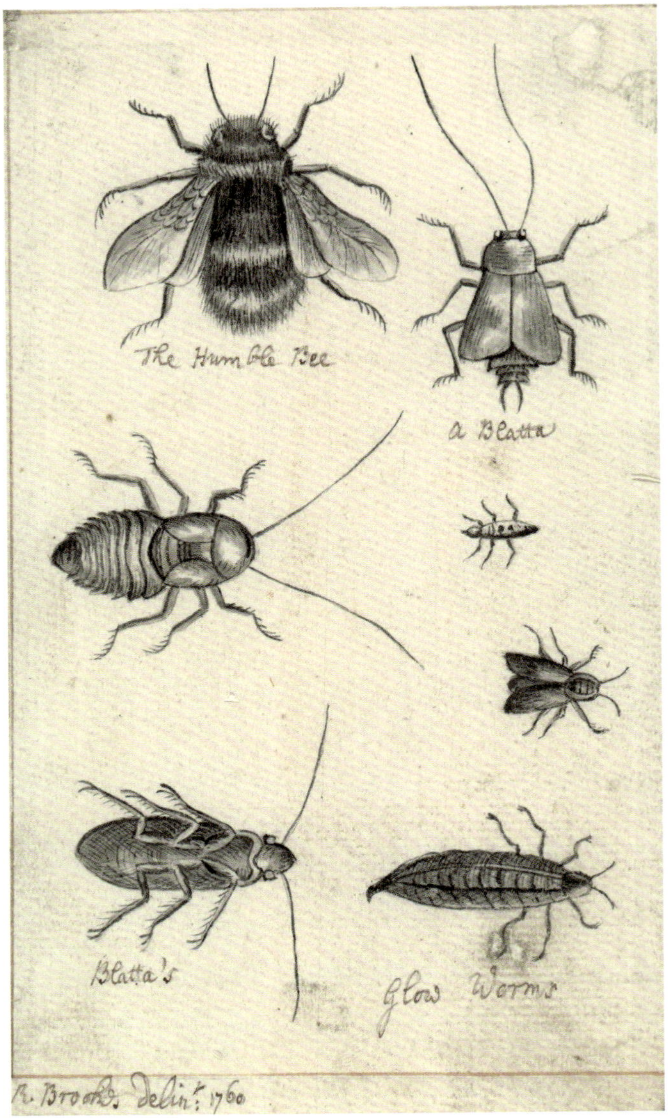

Richard Brookes, *The Humble Bee and Insects*, 1760, pen and black ink, graphite, and gray wash. Yale Center for British Art, Paul Mellon Collection

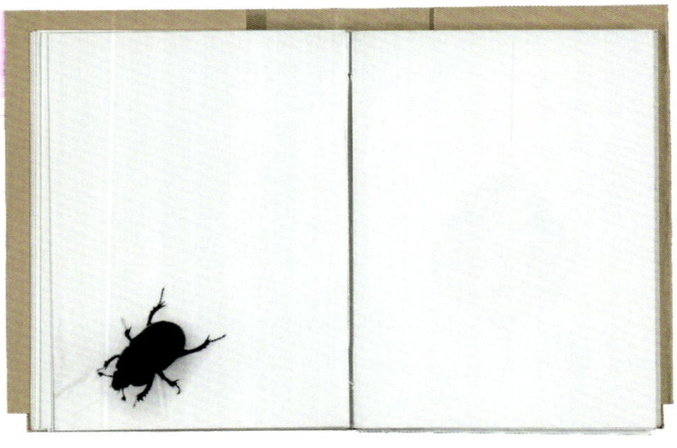

Helen Douglas, *Field Notes*, Yarrow, Scotland: Weproductions, 2006, digital print. Yale Center for British Art, Friends of British Art Fund

OPPOSITE: *Commonplace Book, with Drawings of Insects*, ca. 1830, pen and ink with watercolor. Yale Center for British Art, Paul Mellon Fund

PREVIOUS PAGES: Julie Cockburn, *Beetles Book*, 2002, altered book. Yale Center for British Art, Friends of British Art Fund

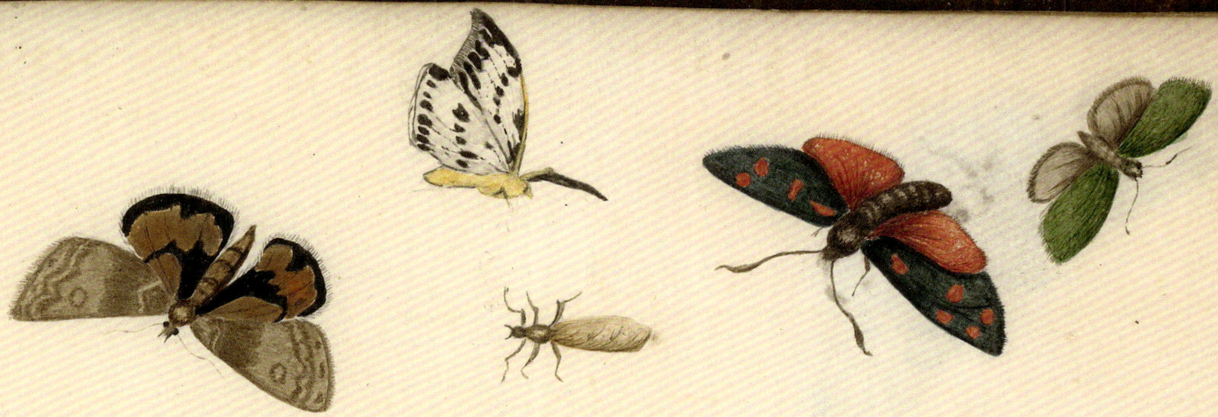

Come take up your Hats, and away let us haste
To the Butterfly's Ball and the Grasshopper's feast,
The trumpeter Gadfly has summon'd the crew,
And the revels are now only waiting for you;

On the smooth shaven grass, by the side of a Wood,
Beneath a broad Oak, which for ages has stood;
See the children of earth, and the tenants of air
To an ev'ning's amusement together repair.

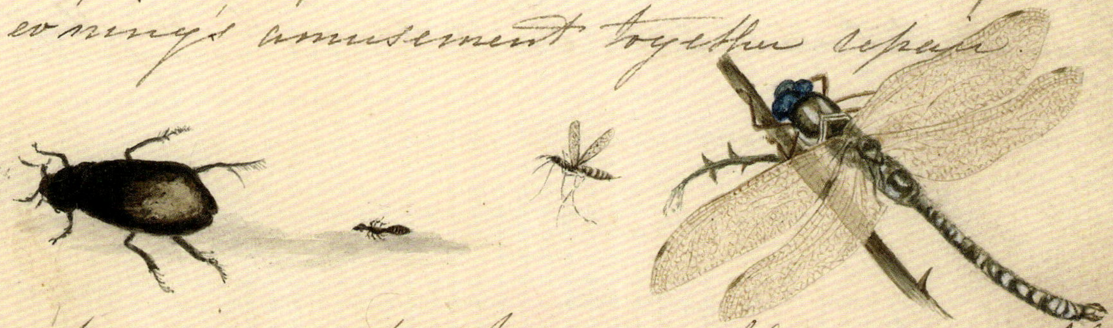

And there came the Beetle, so blind and so black,
Who carried the Emmet, his friend, on his back,
And there came the Gnat, and the Dragon fly too,
And all their relations, green, yellow, and blue;

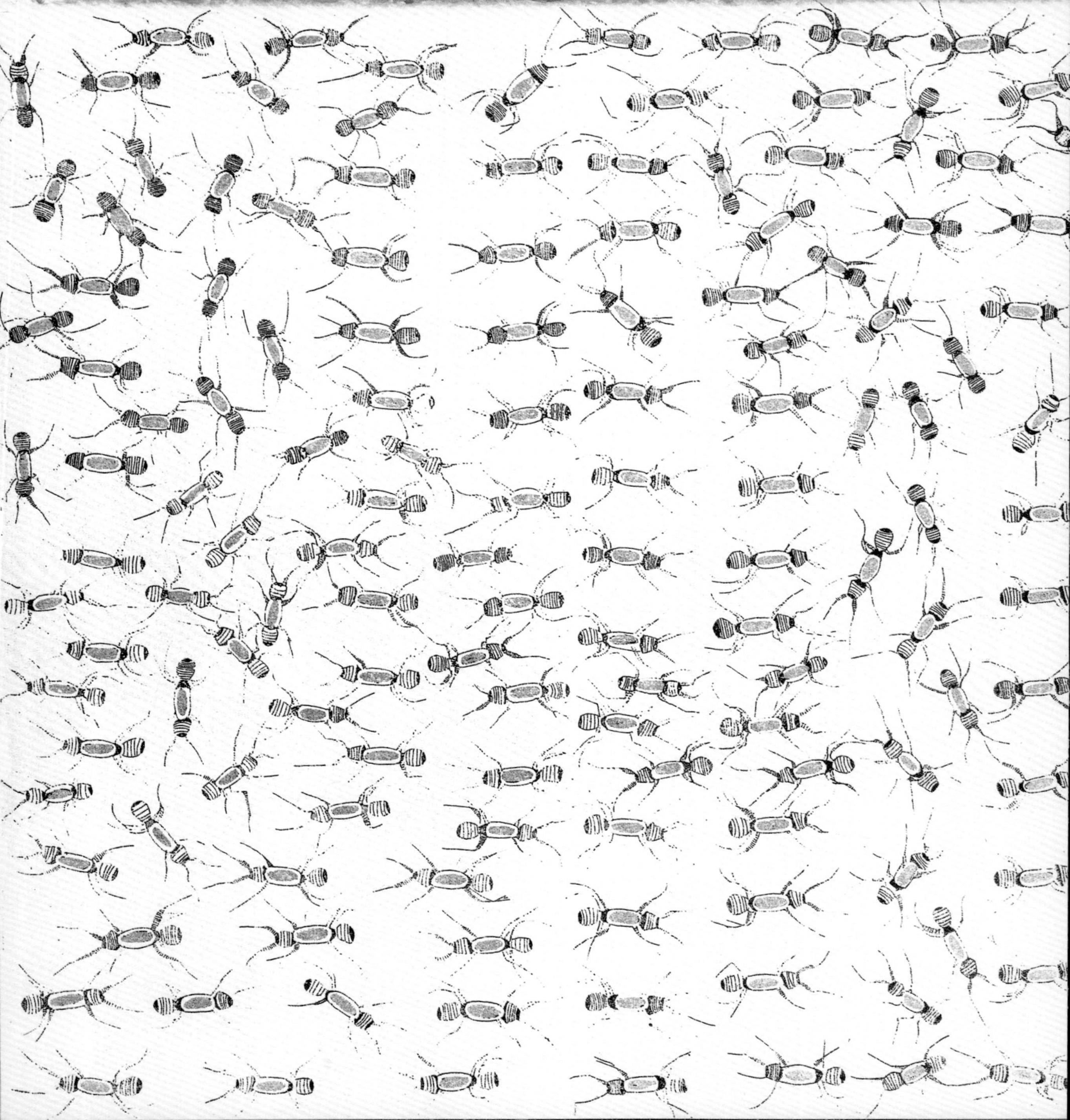

Ant

Stick insect

Millipede

Moth

Crab spider

Beetle

Slug

Catapillar

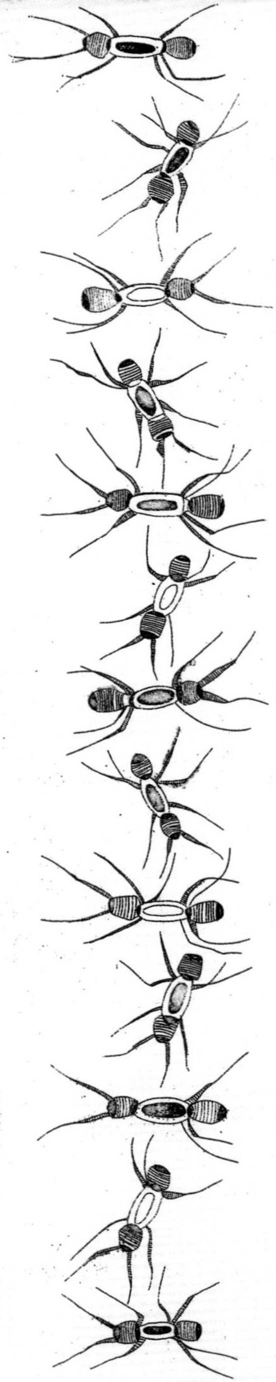

Tsungo

Bandauha

Jongolo

Boho

Juijui

Miigogwe

Kivonyongwei

Bulu

grow to be prickles, which is the lesse to be marvelled at, because there be Mise in *Egypt* (as *Pliny* saith) which have hair like Hedge-hogs. It hath none of these prickles on the belly, and therefore, when the skin is off, it is in all parts like a Hog.

Albertus. His stones are inward and cleave to his loins like as a birds, he hath two holes under his tail, to eject his excrements, which no creature living hath besides him. His meat is Apples, Wormes, or Grapes; When he findeth apples or grapes on the earth, he rowleth himself upon them, untill he have filled all his prickles, and then carryeth them home to his den, never bearing above one in his mouth. And if it fortune that one of them fall off by the way, he likewise shaketh off all the residue, and walloweth upon them afresh, untill they be all setled upon his back again, so forth he goeth, making a noise like a cart wheele. And if he have any young ones in his nest, they pull off his load wherewithal he is loaded, eating thereof what they please, and laying up the residue for the time to come.

When they are nourished at home in houses and brought up tame, they drink both milk and Wine: But there is an Herb (called *Potomagiton*) whereof if they tast, they die presently.

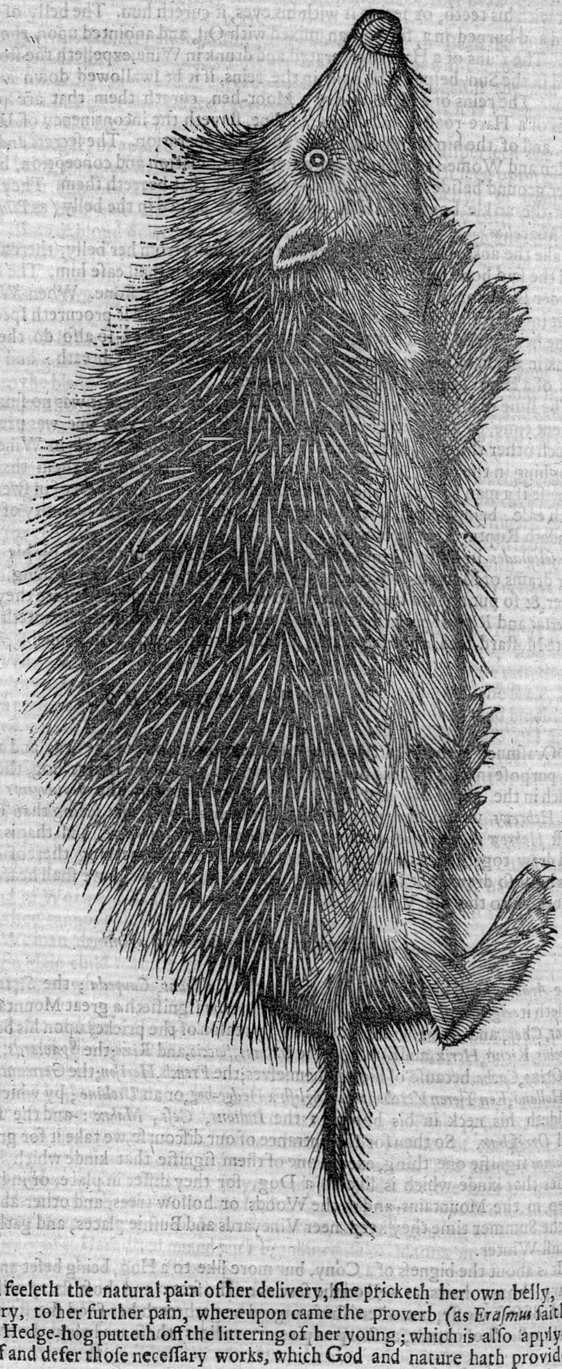

Their copulation. When they are in carnall copulation they stand upright, and are not joyned like other beasts, for they imbrace one another, standing belly to belly: but the prickly thornes upon their backs will not suffer them to have copulation like Dogs or Swine, and for this cause they are a very little while in copulation, because they cannot stand long together upon their hinder legs. When the female is to bring forth her young ones, and feeleth the natural pain of her delivery, she pricketh her own belly, to delay and put off her misery, to her further pain, whereupon came the proverb (as *Erasmus* saith) *Echinus partum differt*, the Hedge-hog putteth off the littering of her young; which is also applyed against them which put off and defer those necessary works, which God and nature hath provided them

Mammals

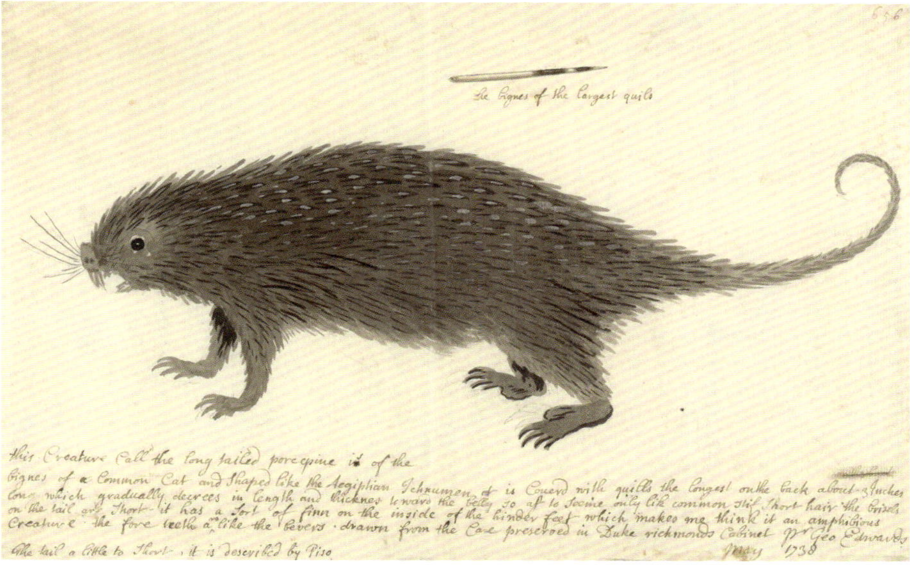

George Edwards, *The Long-Tailed Porcupine*, 1738, pen and gray and brown ink, brown wash, watercolor, and graphite. Yale Center for British Art, Paul Mellon Collection

Notwithstanding all the persecutions from prejudice and wantonness to which the hedgehog (*Erinaceus europaeus*) is exposed, it is yet common with us; sleeping by day in a bed of leaves and moss, under the cover of a very thick bramble or furze-bush, and at times in some hollow stump of a tree. It creeps out in the summer evenings; and, running about with more agility than its dull appearance promises, feeds on dew-worms and beetles, which it finds among the herbage, but retires with trepidation at the approach of man. In the autumn, crabs, haws, and the common fruits of the hedge, constitute its diet. In the winter, covering itself deeply in moss and leaves, it sleeps during the severe weather; and, when drawn out from its bed, scarcely any thing of the creature is to be observed, it exhibiting only a ball of leaves, which it seems to attach to its spines by repeatedly rolling itself round in its nest.

John Leonard Knapp

The Journal of a Naturalist, London: John Murray, 1829, pp. 96–97

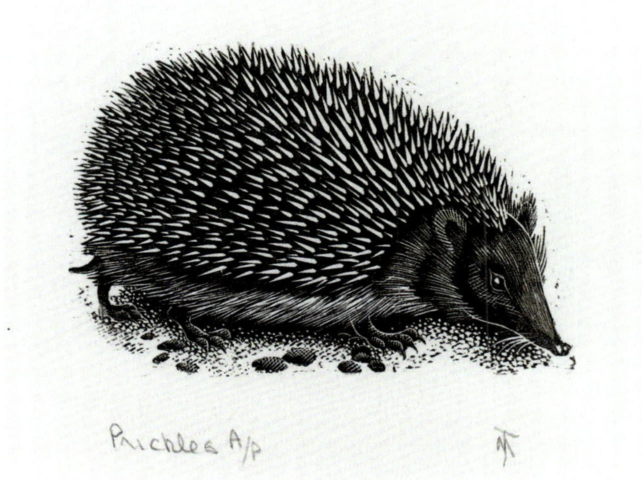

Sister Margaret Tournour, *Prickles*, ca. 1995, wood engraving. Collection of David Burnett

OPPOSITE: Edward Topsell, *The History of Four-Footed Beasts and Serpents*, London: E. Cotes, for G. Sawbridge, 1658, wood cut. Yale Center for British Art, Paul Mellon Collection

PREVIOUS PAGES:
Mandy Bonnell, *Antmothbeetlemillipedespider*, with poems by Gabriel Gbadamosi, London: EMH Arts, 2006, wood engraving. Yale Center for British Art, Friends of British Art Fund

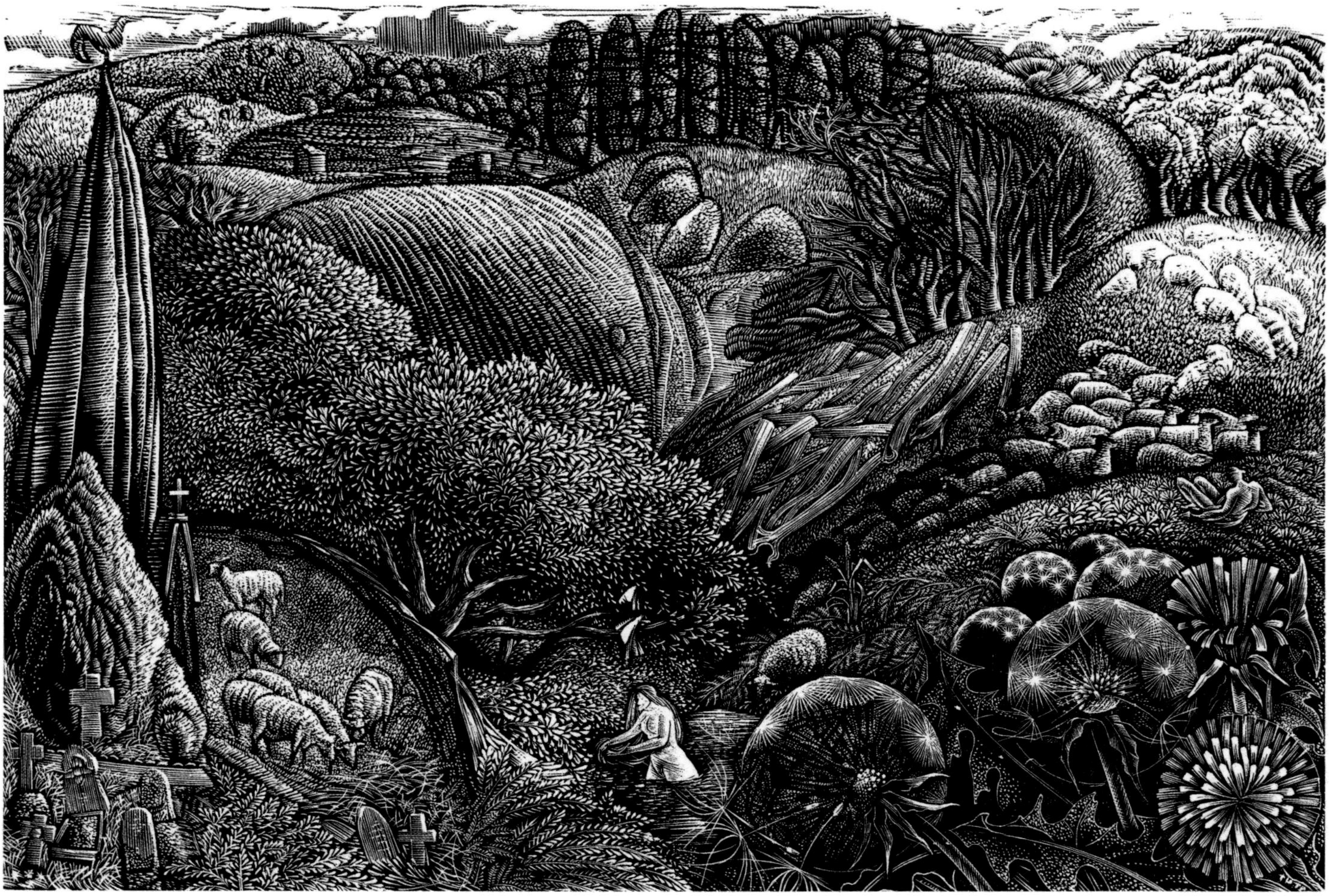

George Tute, *Bather in Landscape*, ca. 1979, wood engraving.
Collection of David Burnett

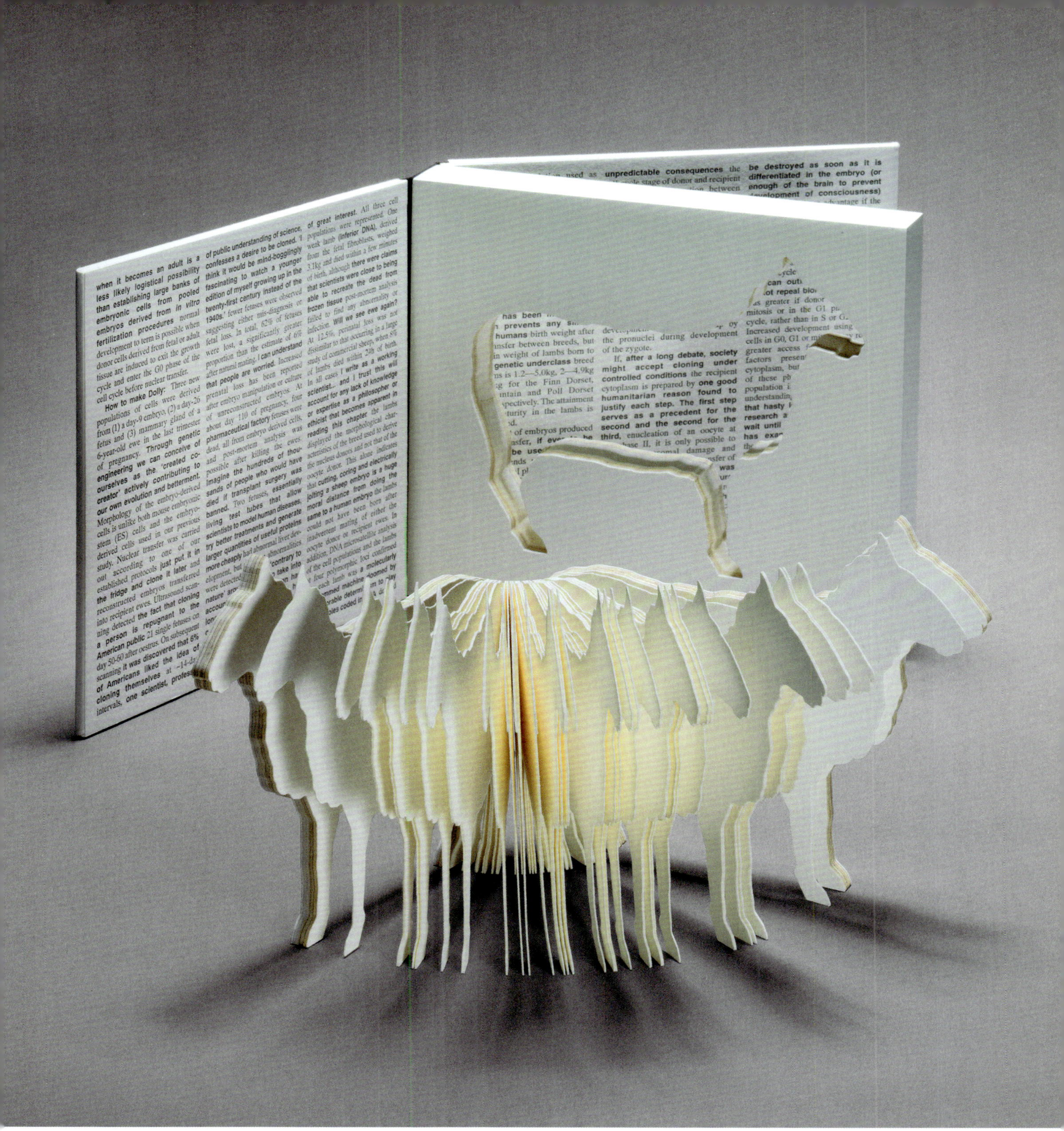

Karen Bleitz, *Dolly*, London: Circle Press, 1997, fret-sawed painted wood and paper. Yale Center for British Art, Friends of British Art Fund

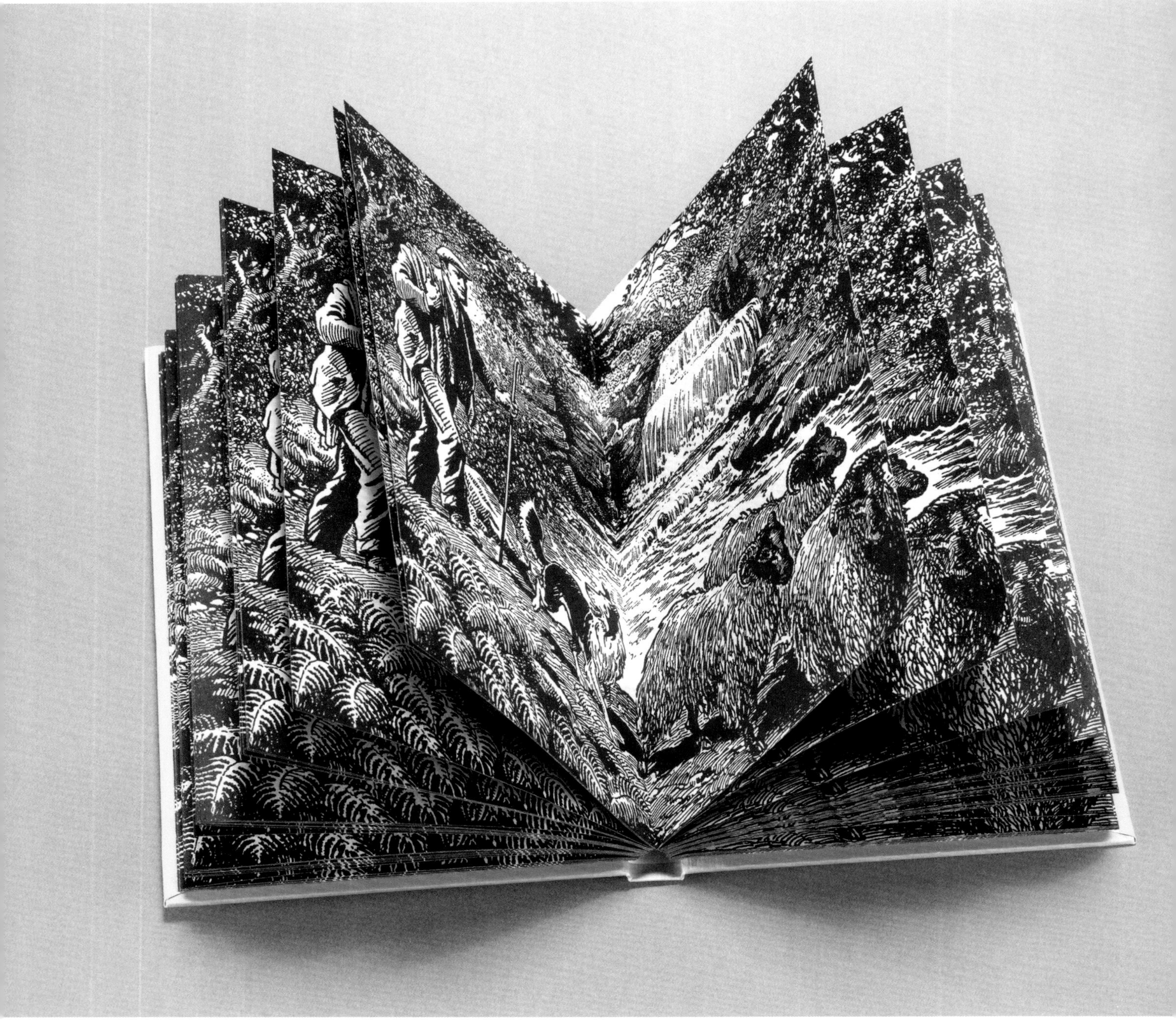

Colin Sackett, *Black Bob*, London: Coracle Press, 1989, line block engraving. Yale Center for British Art, Friends of British Art Fund

Reading *Black Bob*

It doesn't dictate or imply a pace, it has all sorts of paces, but it's also completely static. . . .[1]

The completely white book, the monochrome or black book, silent and abstract, intends not the small scale reading of words, sentences, and pages—the detail and accumulation of content—but a larger-scale, maybe distant reading.[2] Here the minimalist "blank" book is redeployed, the commandeered subject matter repeated throughout, the dense black impression of letterpress line-block balanced with unprinted white stock.[3] From the first to the last spread there is no change, and so this emphatic demonstration of direction—the page after page rightward movement of shepherd, dog, sheep, and the flow of the river—is identical to the narrative of a book with blank stock as its subject.[4]

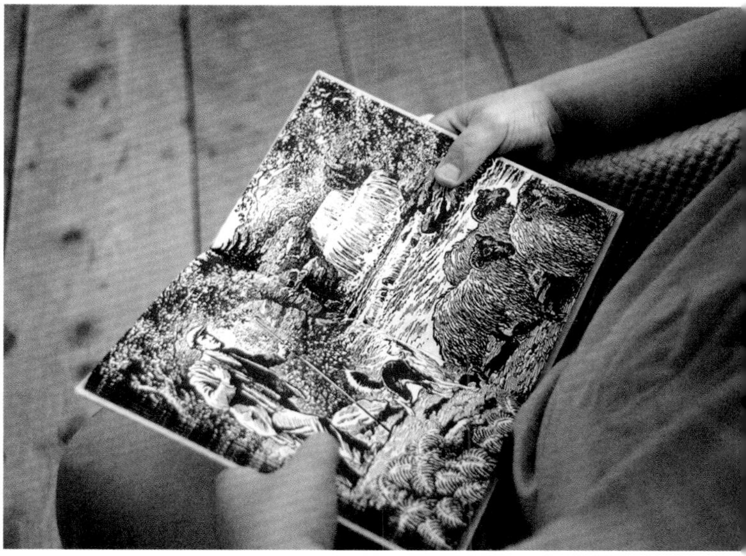

Colin Sackett
ARTIST AND PUBLISHER

1. "*Black Bob*, on the face of it, is an illustrated book whose single illustration is repeated throughout, from opening to closing page. It runs over the whole construction of the book, the margins, the gutters, the endpapers, the cut edges of the leaves. Only the outer boards of the binding are exempt, serving at once as cover and title page. When the book is turned over, we find that the back cover is a mirror image of the front." (critic and writer John Bevis, correspondence with the artist, November 2013)
2. Herman de Vries's *Wit = Weiss* (i.e., White), for example, was the first determinedly blank book, and instrumental in the existence of *Black Bob*. "The first version was published in 1962 and had an edition of 5. The second version was published by Hansjörg Mayer in Stuttgart in 1967 in an edition of 500. The front and back were printed as well as the first page where there was an introduction by the poet J. C. van Schagen. But the following 250 pages were blank. The title and the names of the author, editor and preface author have been printed in a white band that surrounds the book, which is completely white this time. On the inside of the band, there are the words: 'in this book, where all contradictions have been removed and the contents of which are suitable for everyone, where we have moved beyond all problems and where nothing is cut from reality or is unilateral, we can say that this is a positive contribution to the summary of the whole.'" (Accessed November 22, 2013, http://www.hermandevries.org/publications_1960-1969.php)
3. "The pastoral procession is shown heading left to right, the direction in which one turns the pages and, and yet the figures go nowhere. They are stuck among the repetitions, as somebody said, like a jammed gramophone record. However, the sound that emerges here is not a shriek or a judder or a hole in sound but a constant true note pitched in a natural key that has no need to expire. We have a transfixed moment among tumbling streams and ferny paths, heading expectantly across the page yet forever held in anticipation of what awaits. The event that both is and is not there is so totally obvious that it seems like a wonder, tremulously maintained. It raises questions and at the same makes evident that the irrelevant questions are ones we have imported." (J. C. C. Mays, *Rare Sighting in the Haldon Hills*, Exeter, 2009; see http://www.artcornwall.org/features/Colin_Sackett.htm)
4. "Despite the insistence of the repeated image, of shepherd, dog, sheep and river, we suspect that it has not been chosen for its content, its message, or even as a signifier of its—perhaps nostalgic—found origin. More likely it is the uniform overall tone of the woodblock-like pen and ink drawing that makes it work in this context. The more we look at it the more it tends away from image and detail towards print and pattern, a record of the repeated impressions. The book becomes a mechanism for turning a figurative image into an abstraction, its narrative to a silence." (Bevis, correspondence with the artist, November 2013)

Gardens

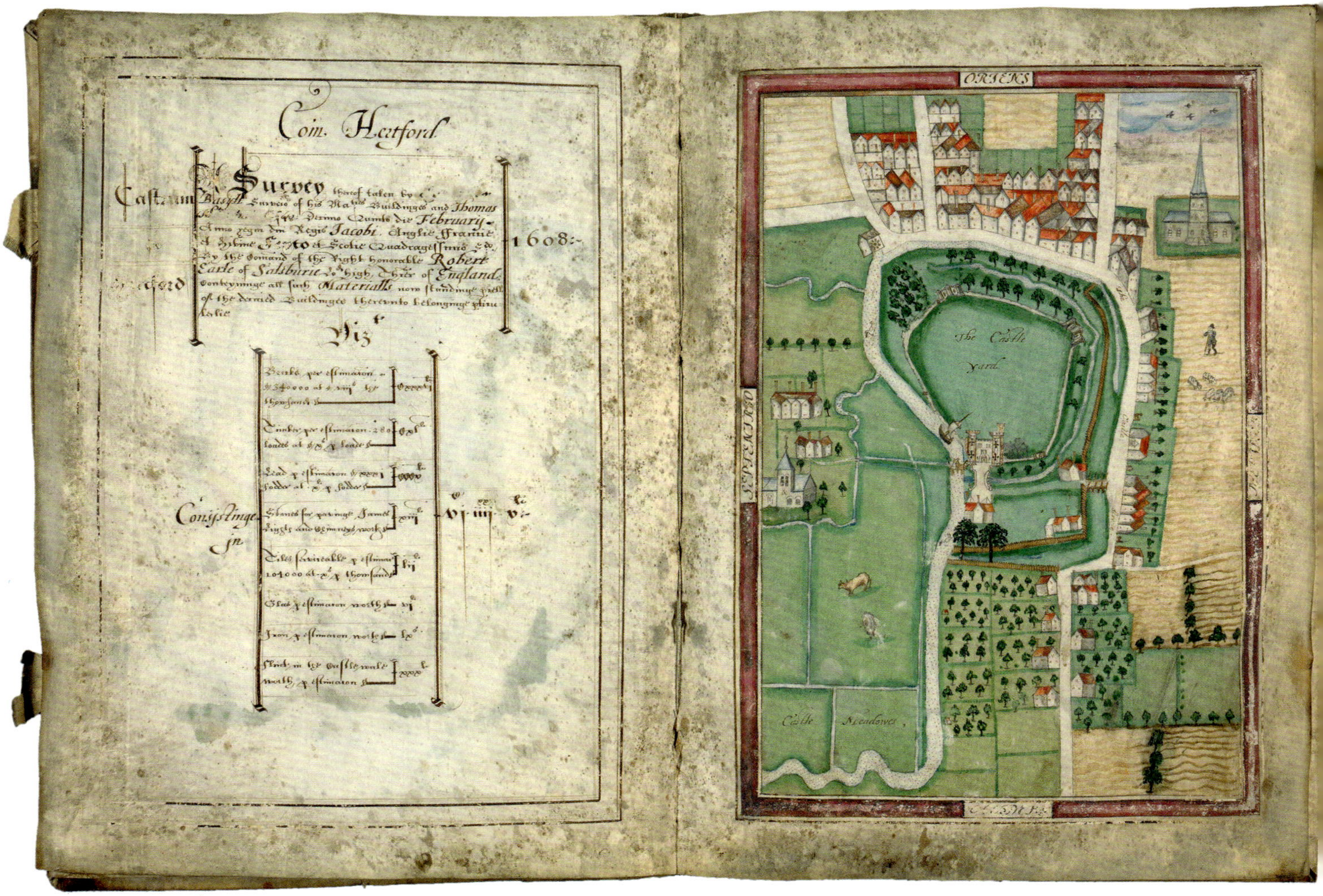

Simon Basil, *Survey of Hertford Castle*, 1608, pen and ink, watercolor, and gouache on parchment. Yale Center for British Art, Paul Mellon Collection

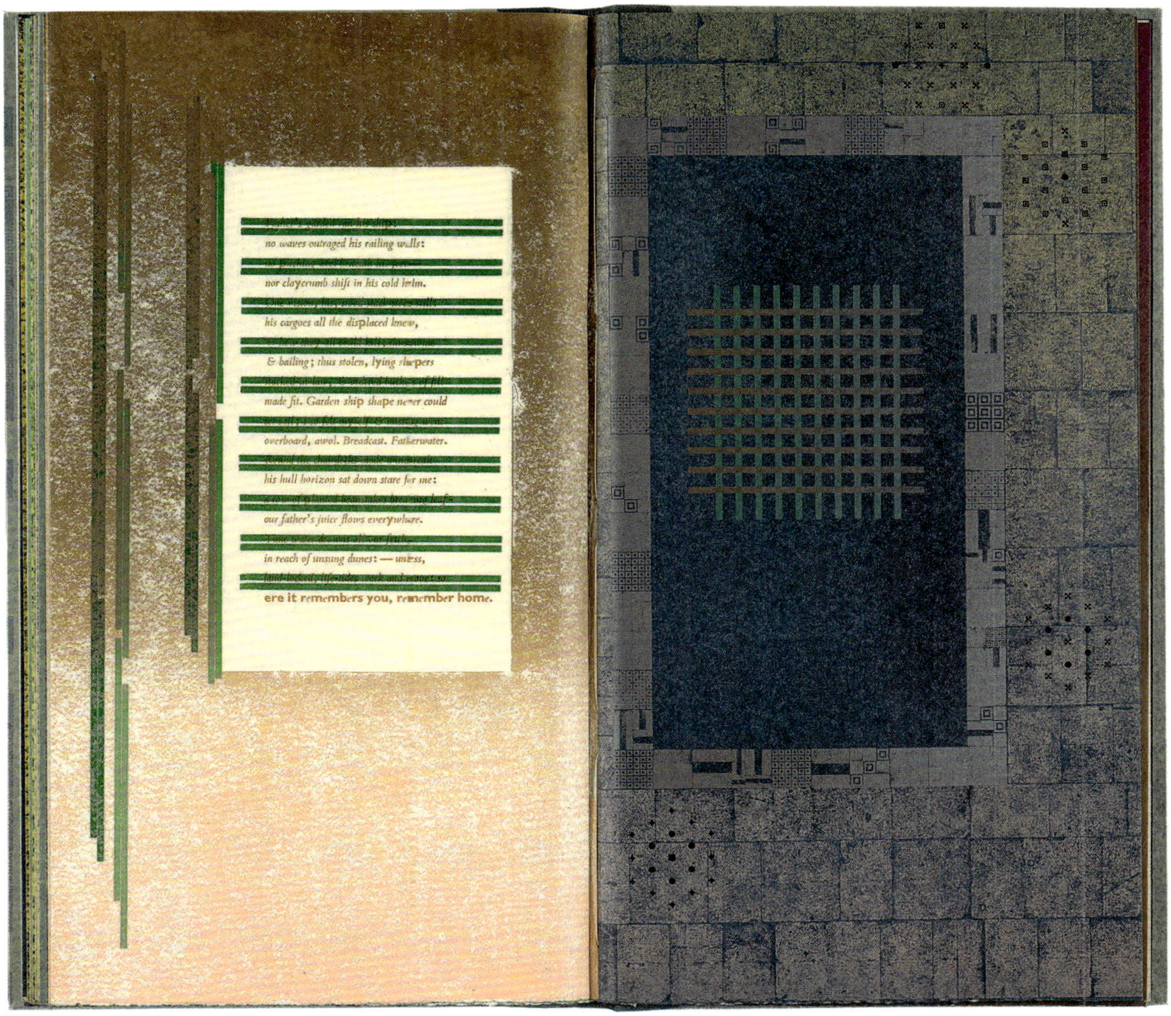

Ken Campbell, *Father's Garden*, Oxford, 1989, letterpress, printed in multi-color and metallic inks. Yale Center for British Art, Friends of British Art Fund

An Allotment Garden

My grandfather had an allotment garden near the coal mine in county Durham where he spent his working life. He spoke a local mining dialect called "Pitmatic." He wore a flat cap, a "gansey," and a pair of "aad byuts." And, when he'd run out of "baccy" he'd smoke chrysanthemum leaves in his pipe. My grandfather kept bees, handling the hives with his bare hands. He thrived on the stings as an aid to longevity. In winter he took the queen bee to bed with him in a matchbox, tucked in his pajama pocket. As a boy I wanted to be like him: no, I wanted to be him. Now, aged fifty-four, I am. I have an allotment near my home—a short bike ride away. I grow vegetables: potatoes, cabbage, sprouts, beans, tomatoes, beetroot, and leeks. I also have a few flowers: marigolds, hollyhocks, and Michaelmas daisies. I go there to be alone. I also go there for company: the friendly robin who sits on my fork handle and pilfers worms, and my battered scarecrow sporting a pair of 3-D sunglasses, natty combat trousers, and my son's old school tie. Sometimes I say hello to my neighbor in the adjacent plot. We talk "veg." I like to dig and plant, and get my hands dirty. I also like to sit on my oak bench and think. I contemplate past triumphs and failures: this year's magnificent row of savoy cabbages and last year's tomatoes that collapsed with blight. I philosophize about my enemies: the mice that nibbled my broad bean seeds no sooner than I'd stuck them in the earth, the slugs that decimated my infant courgette plants, and the band of squirrels who made off across the railway line with my sweet corn as I looked on helplessly. At dusk, I pick up my plastic bag of swag: a bunch of spinach, a handful of runner beans, and a few sticks of rhubarb. I don my flat cap, put my pair of muddy "aad byuts" back in the shed, pull on my "gansey," and "gan yem" on my bike.

Martin Postle
ART HISTORIAN AND GARDENER

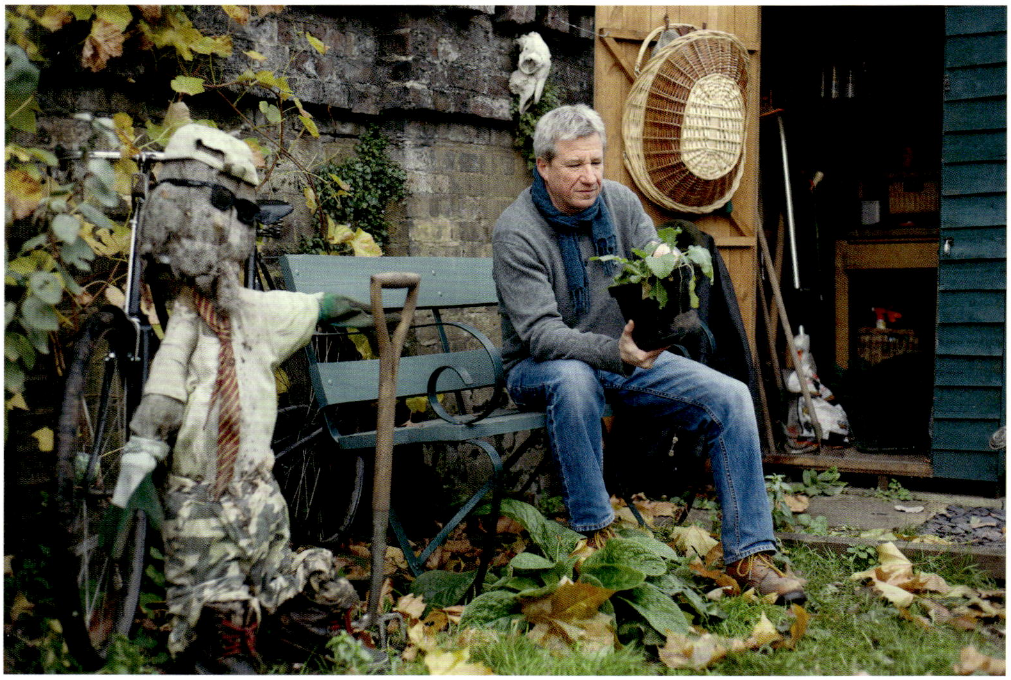

Martin Postle in his allotment garden in London, autumn 2013

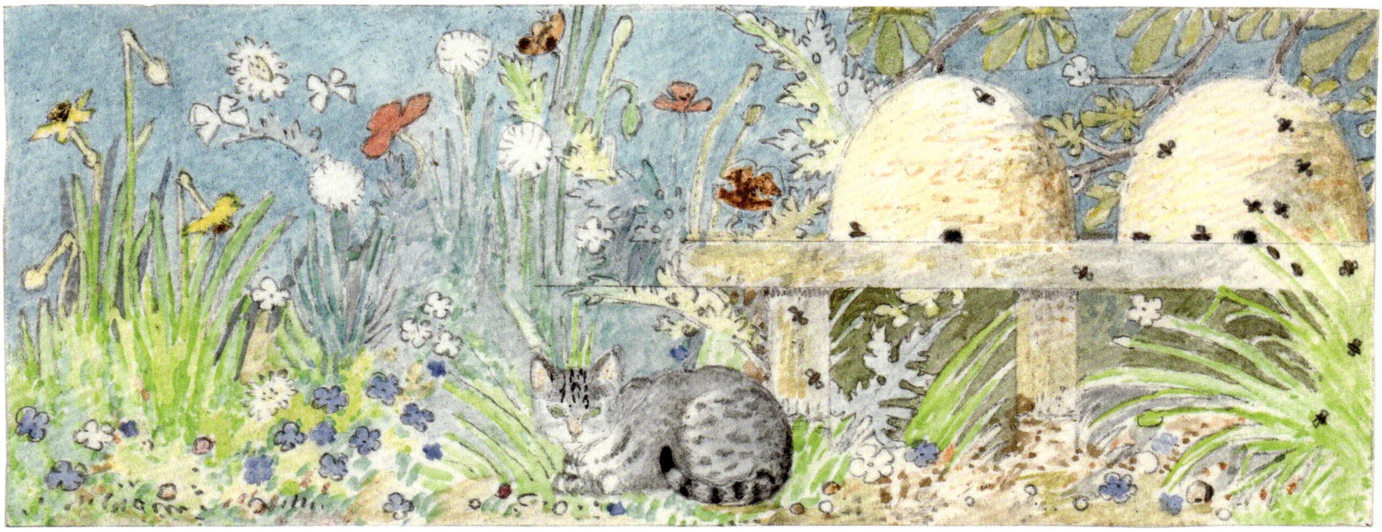

Mary Leslie, *Sketchbook*, ca. 1890, watercolor, graphite, and pen and ink. Yale Center for British Art, Paul Mellon Collection

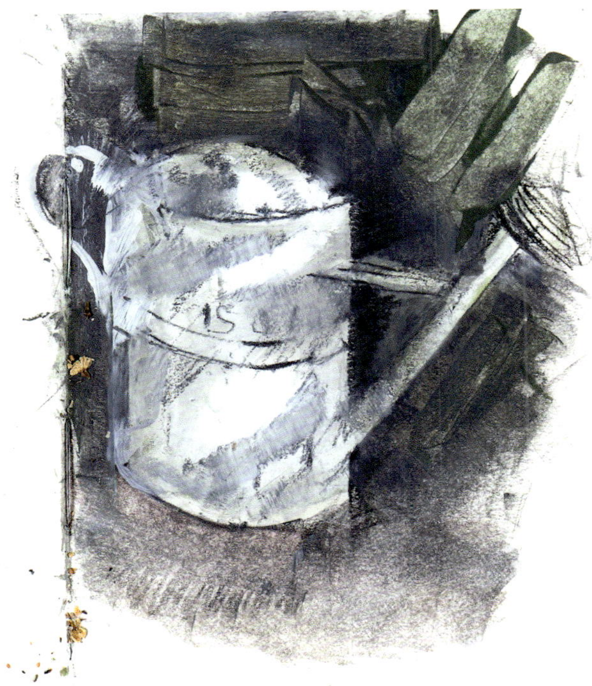

Eileen Hogan, "Kettle Drum" (detail) from *Sketchbooks Recording Little Sparta, Scotland*, 1997–2013, graphite, pen and ink, and watercolor. Collection of the artist

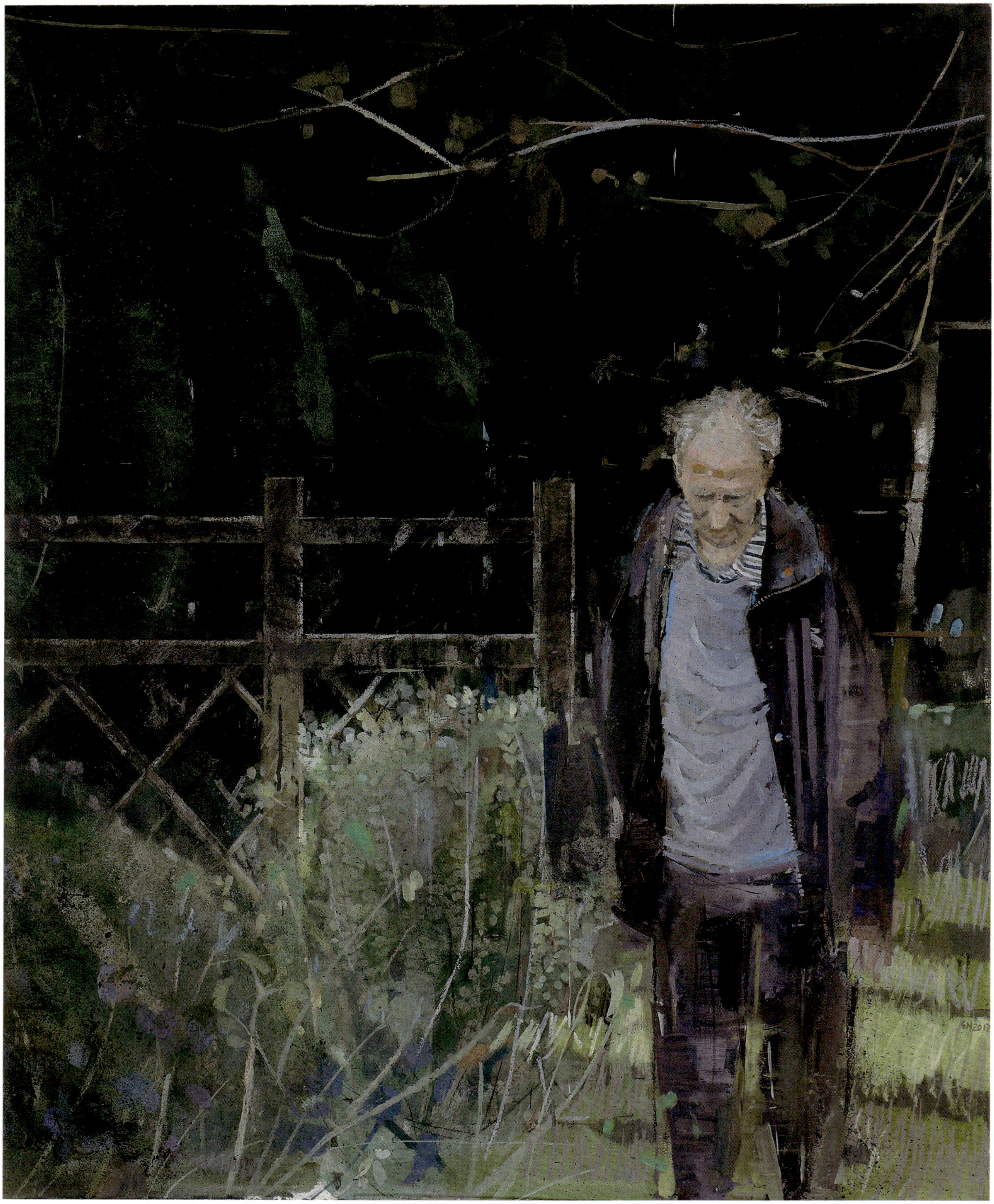

Eileen Hogan, *Ian Hamilton Finlay*, 2012, oil and charcoal on paper, mounted on card. Collection of the artist

Little Sparta

The inspiration for this series of paintings and drawings, made between 1998 and 2013, has been Ian Hamilton Finlay's garden Little Sparta, near Dunsyre in the Borders area of Scotland. Finlay and his wife Sue began the garden in 1966, carving it out of the Pentland Hills. It is a poet's creation, in which language, sculpture, and historical allusion interweave with the landscape and its plantings.

When I visited the garden for the first time, in 1997, I never expected that it would become my subject matter for fifteen years. I have usually taken my source material from within London, where I was born. I am not a landscape painter, and although green spaces have often been my subject matter, they usually have been bounded by architecture.

Walking, drawing, and making notes are always the starting point of my work. In recent years, a walk from home to my studio introduced me to the gardens of Bryanston, Portman, Manchester, and Montagu squares—the sites of my *Four Squares* series of paintings. I gained access to these private spaces, painting from the inside rather than outside looking in. The wildness of the hills and farmland surrounding Little Sparta could hardly contrast more with those urban backdrops, but the same notion of enclosure underlay my impulse to paint there.

Finlay allowed me to paint wherever and whenever I wanted. His presence seeped into several of my paintings, as did an unexpected trip to the Coffee Spot in the nearby town of Biggar, where I drew him in the midst of the café's more conventional clientele, an occasion which led to a series of painted portrait heads.

Finlay died in 2006, but I have kept returning to the garden, witnessing different seasons and dramatically different weathers, gradually absorbing its complexity. The goose hut and Lochan Eck have been the focus of my most recent work, but the main subjects have been the trio of white beehives in the section of the garden known as the English Parkland. The structures echo past concerns in my work, when I have repeatedly been drawn to the play of light dappling white surfaces in green surroundings.

Eileen Hogan
ARTIST

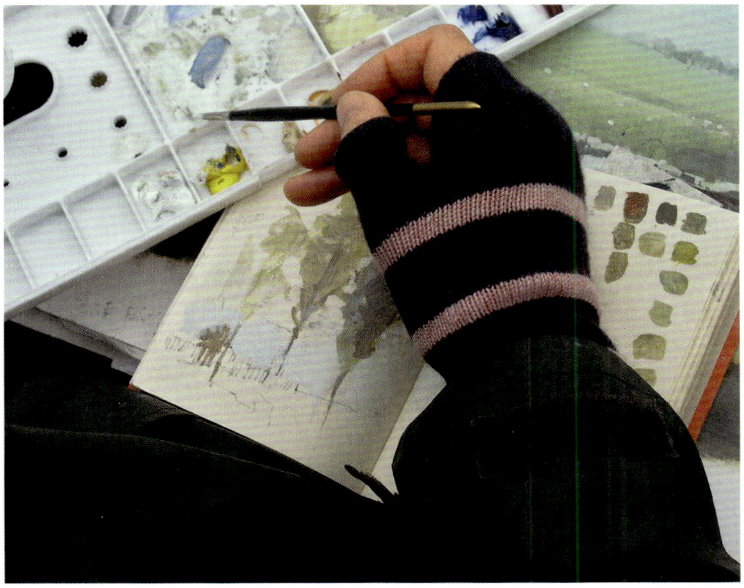

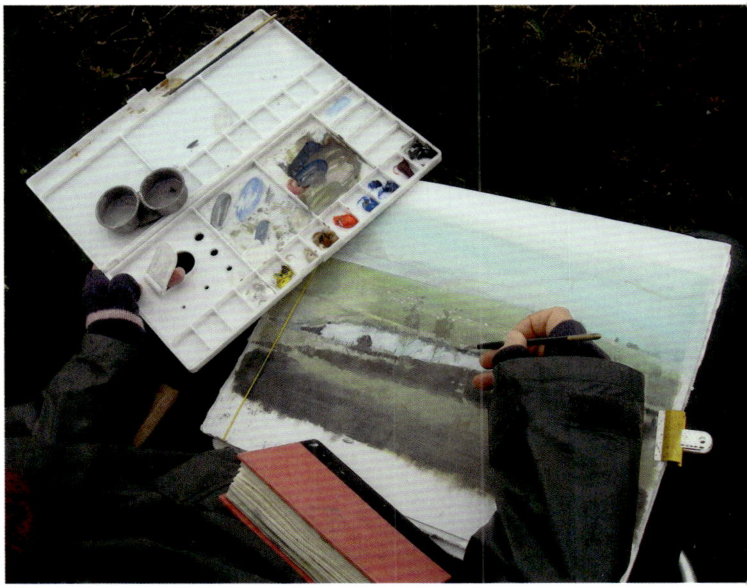

Eileen Hogan painting at Little Sparta, summer 2013

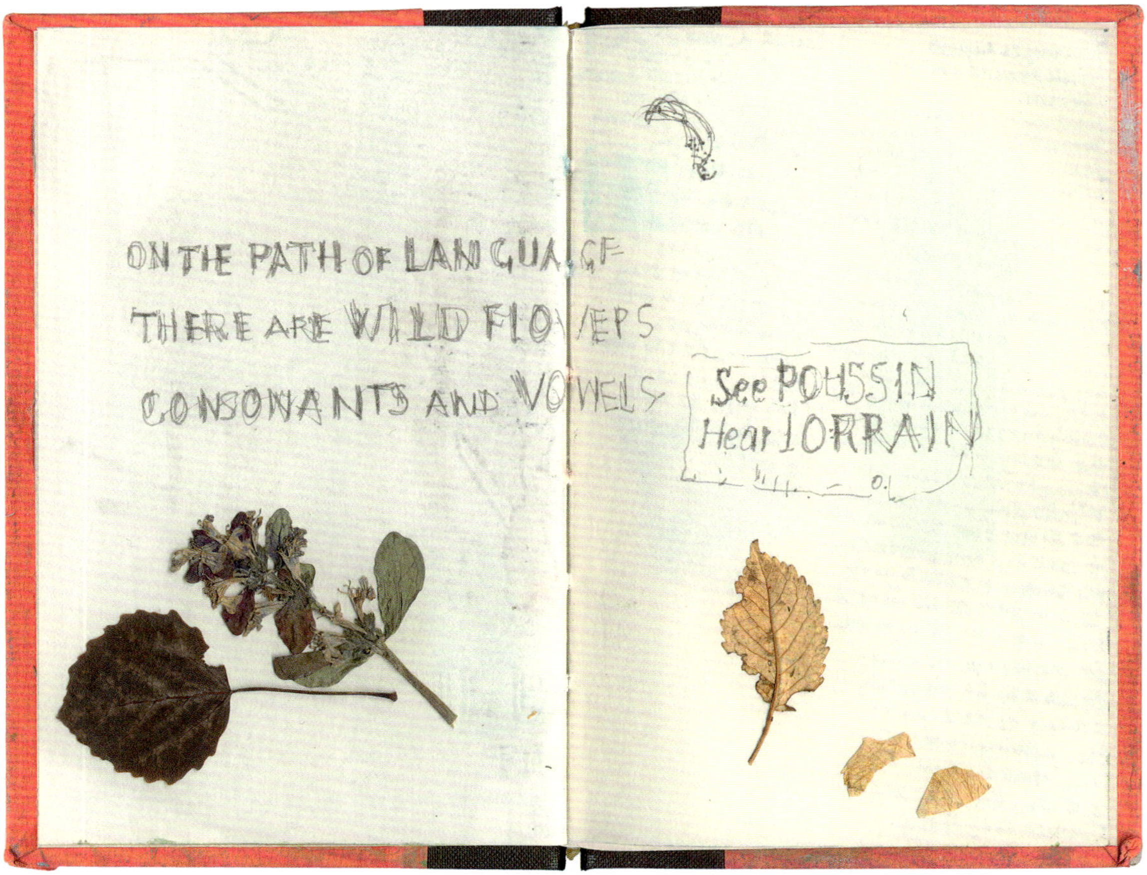

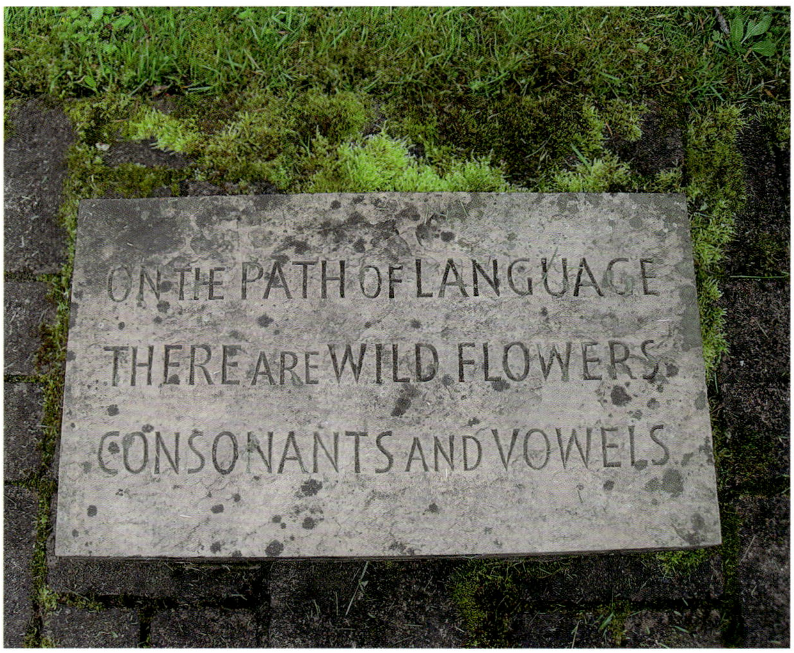

Stone tablet, Little Sparta

Walking and stopping to draw are always the beginning for me, scribbling and making notes in my sketchbook. I find out what connects me to a place, or even what I think, through drawing. The next stage is to find the positions where I will work and, once decided, I always work from the same spot. Sitting there for long periods is important, not necessarily because of what I produce, but because staring and observing become a sort of ritual and make the place mine. Memory is important, even directly working from nature involves memory, memories of seeing the garden in other lights and in other seasons, memories of Ian Hamilton Finlay, and of other people.

Eileen Hogan
March 2014

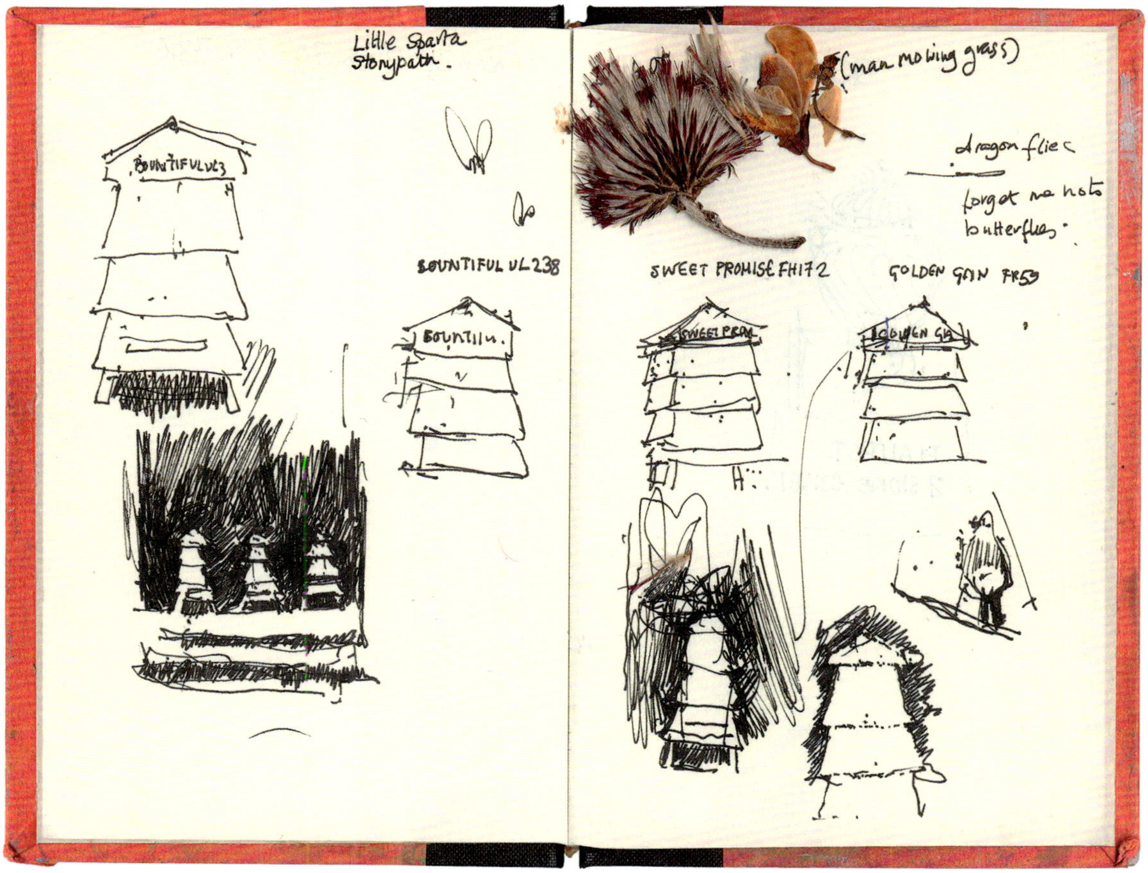

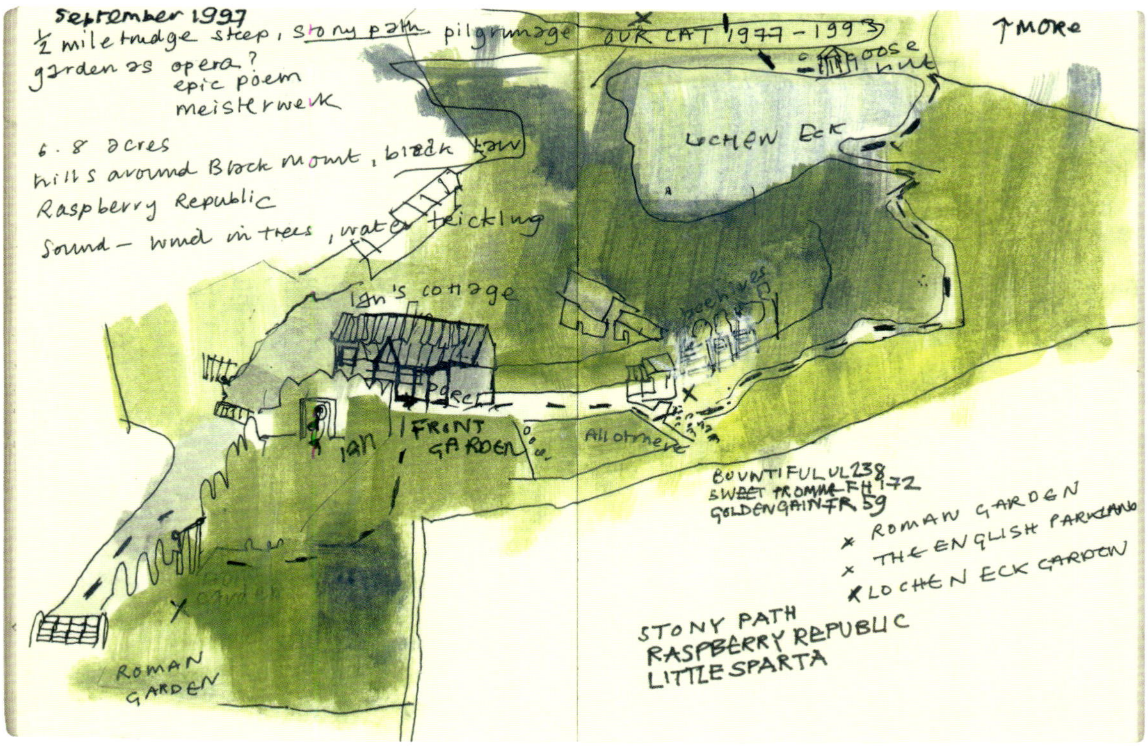

OPPOSITE ABOVE AND THIS PAGE: Eileen Hogan, *Sketchbooks Recording Little Sparta, Scotland*, 1997–2013, graphite, pen and ink, and watercolor, with natural specimens. Collection of the artist

Countryside Walks

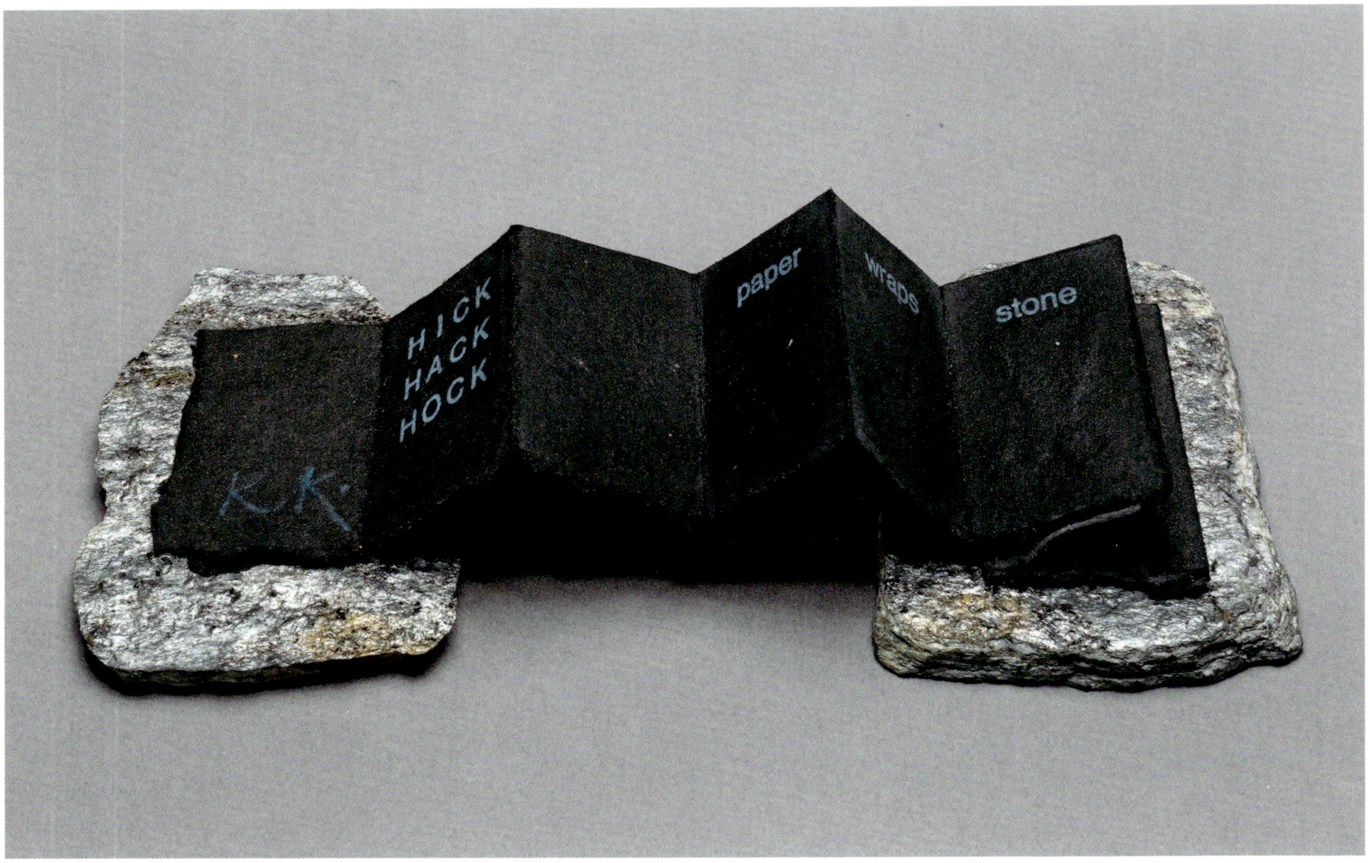

Ron King, *Hick, Hack, Hock*, London: Circle Press, 1995, stone and blind intaglio relief. Yale Center for British Art, Friends of British Art Fund

Sketching in the Lake District

As anyone who has ever visited the Lake District knows, the beauty of the landscape is shaped by the spectacularly unpredictable weather. The clouds rolling in to obscure the peaks pose a real danger to the unprepared walker but do little more than frustrate the artist. The view that was there no longer is, and drawings must adjust to the new reality.

In these circumstances the challenge of working within the fixed sketchbook format (as opposed to loose sheets) is part of the thrill—failed pages can't be torn out or hidden; the linear narrative of the book rules. And very often it is under pressure that barely adequate sketches can be resuscitated such that they can justify their page within the sketchbook and survive to capture their small part of the story.

Rebecca Salter
ARTIST

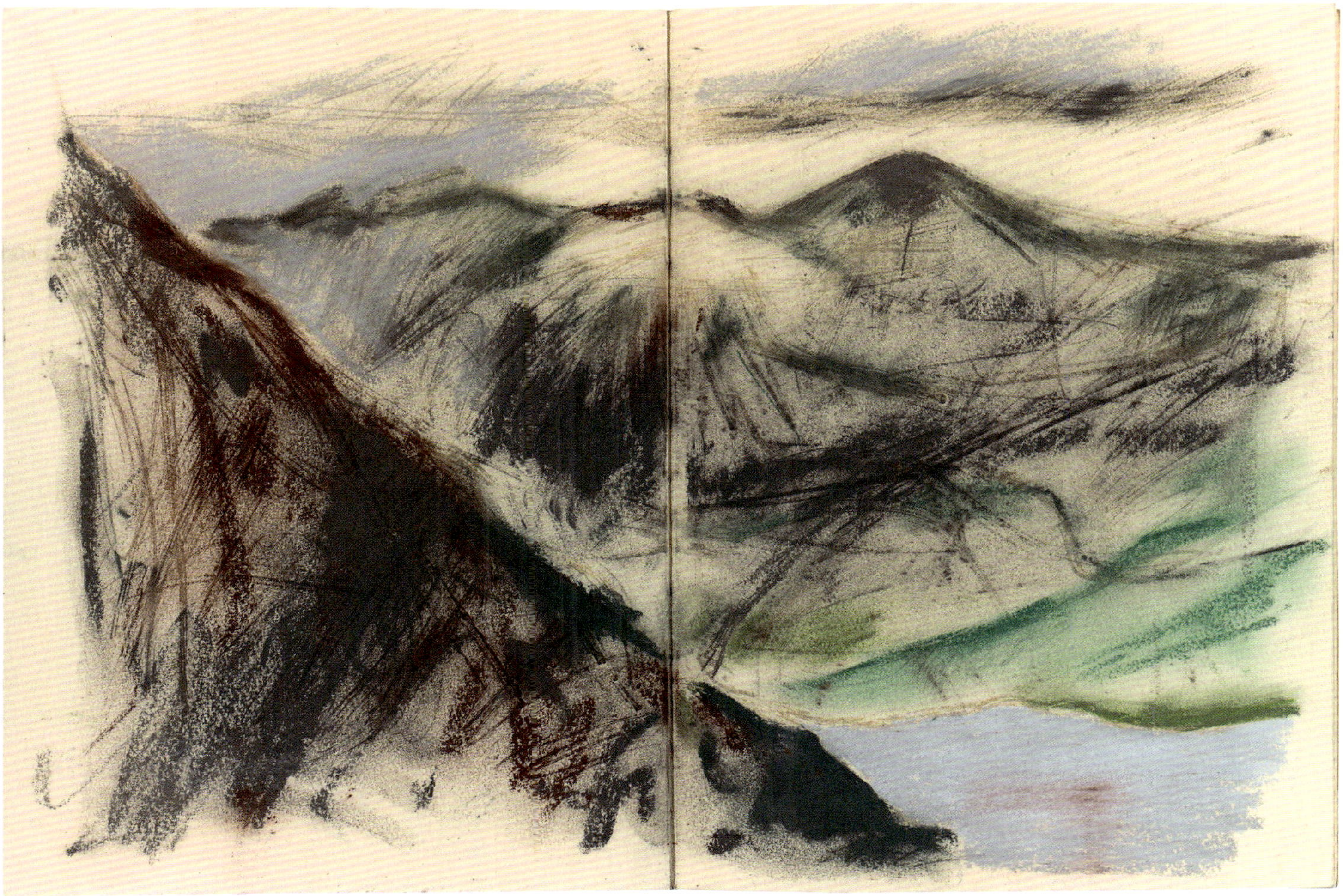

Rebecca Salter, *Lake District Sketchbook*, 1990, pastel. Yale Center for British Art, Gift of the artist in memory of her father Samuel Salter (1922–2005)

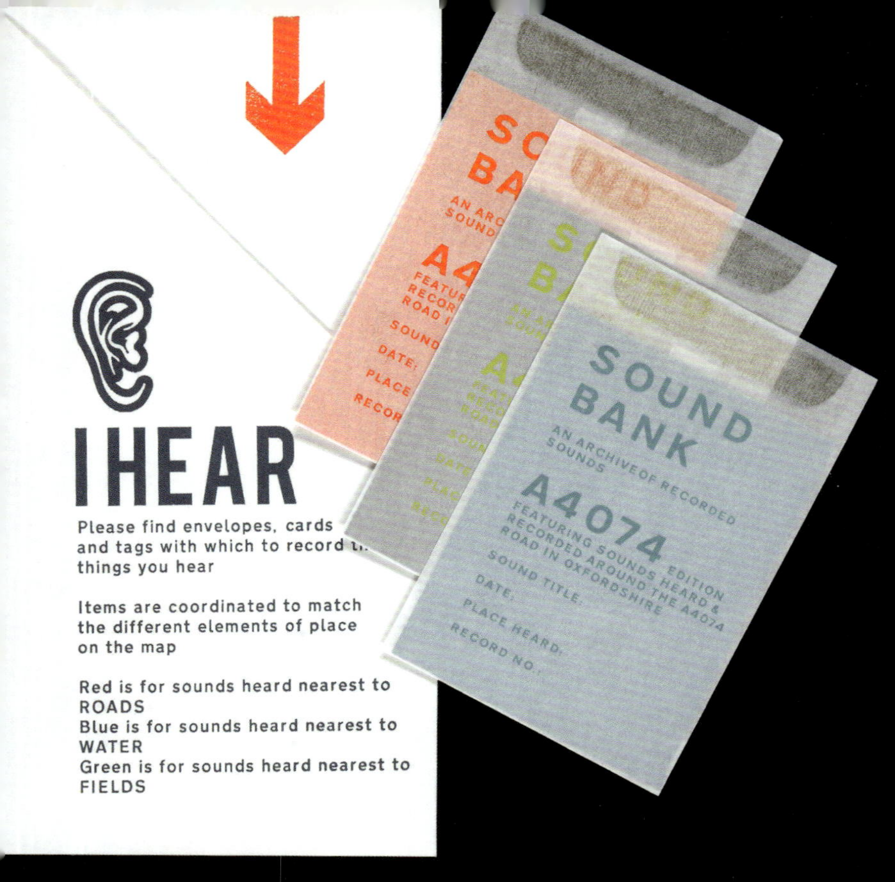

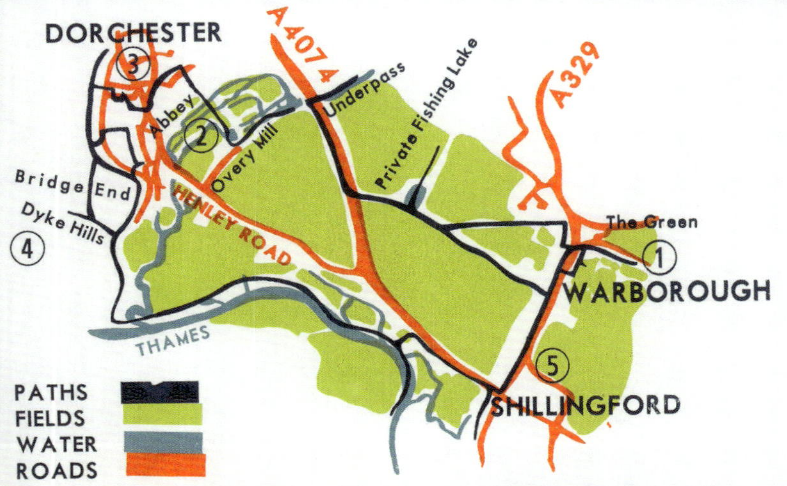

(1) From The Green, Warborough, head to the main thoroughfare and find the footpath opposite the church that leads out across the fields towards the A4074. In these fields you will pass a large stone on your left and a private fishing lake on your right. At the far corner of this field, search for an underpass running beneath the A4074.
Go through into a field with views ahead to Wittenham Clumps.

(2) When the path veers right onto a road, follow this road and a curved brick wall around and go over the stile that sits between two gates. Continue on past Overy Mill and through a meadow Cross a bridge and continue on the path ahead, until you reach a footpath sign and a junction with a road to your left. Take this route into Dorchester, following Manor Farm Road to a T-junction. At the T-junction, turn left and walk until you reach the main thoroughfare.

(3) Dorchester Abbey is down the road to your left. To continue on your route, turn right up the main street and then left down Malthouse Lane, which is just beyond the Co-operative shop. At the end of Malthouse Lane, search for a narrow path which leads up to the right, between the houses. Follow this until you reach another lane. Turn left and walk until you pass some allotments. Look for a footpath sign to the right, marked Days Lock 1/2 and follow this towards Dyke Hills.

(4) At the foot of Dyke Hills, turn left and walk to a gate in the corner of the field. Go through this and follow the path down through a meadow to the Thames. Go along the Thames Path, and look out for the second fork in the path up to your left. You will know it is the correct turn when you see a house near the Thames up in front of you, and a stile leading onto the A4074 to your left. Go over this stile, cross the road, and walk up the pavement into Shillingford.

(5) Turn left up Warborough Road (opposite the Kingfisher Inn Hotel) and find the path on the right of the road. Walk back up to Warborough.

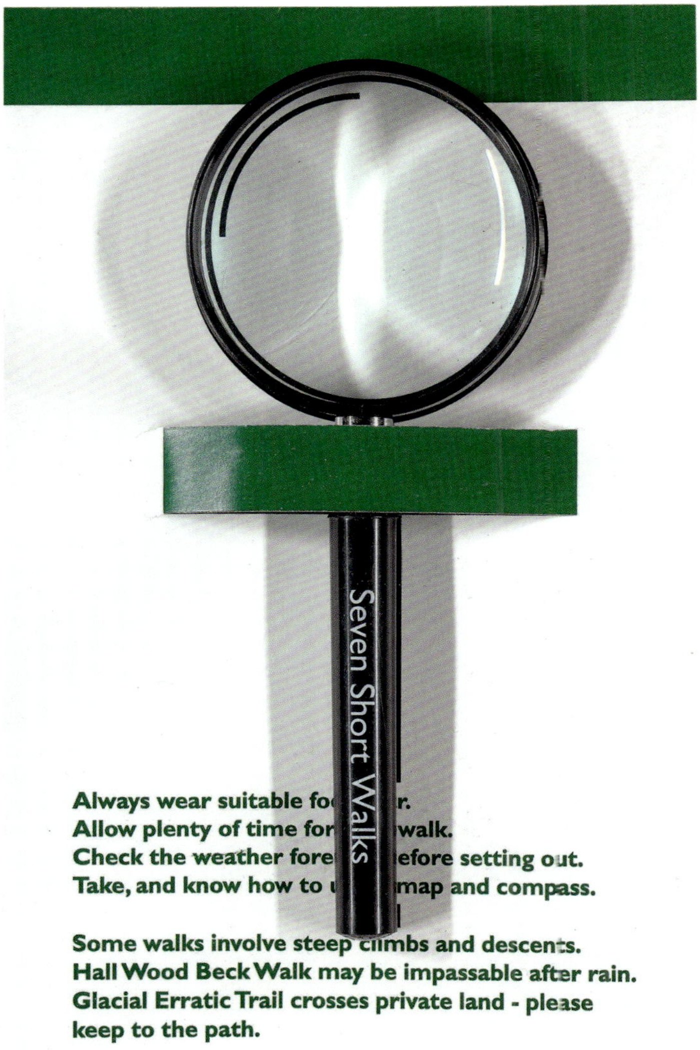

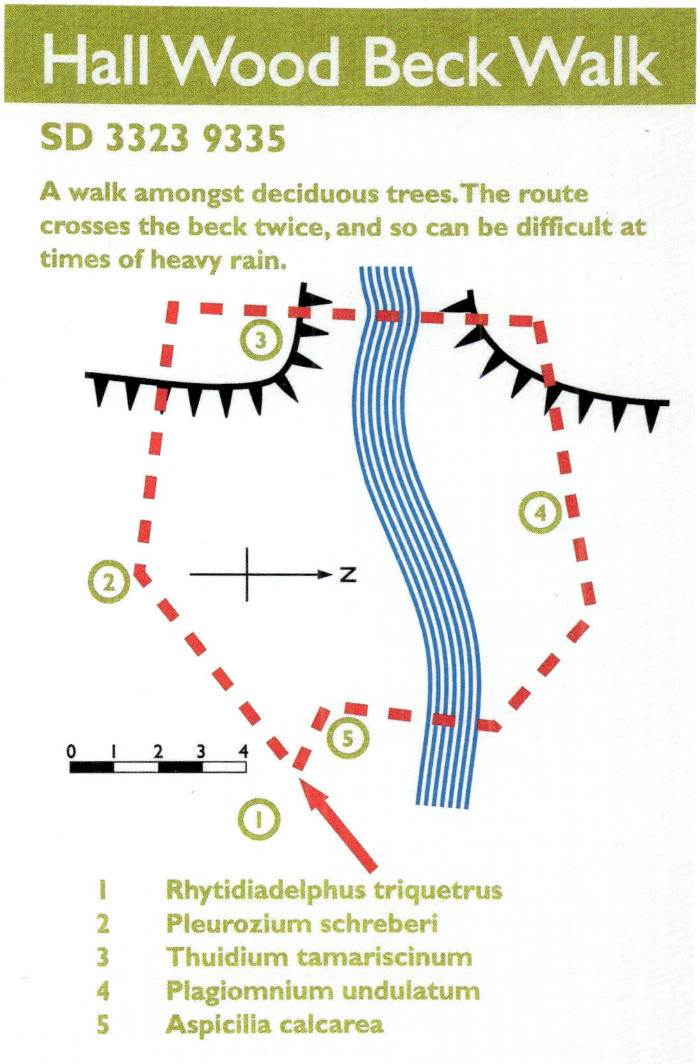

Stuart Mugridge, *Seven Short Walks*, Ambleside, England: Grizedale Books, 2002, screen prints with magnifying glass. Yale Center for British Art, Friends of British Art Fund

OPPOSITE: Felicity Ford, *I Hear: Warborough, Overy, Dorchester, Shillingford: A Circular Walk Exploring the Landscape and the Ambient Sounds Surrounding a Section of the A4074 Road*, 2010, screen prints. Yale Center for British Art, Friends of British Art Fund

Seaside Walks

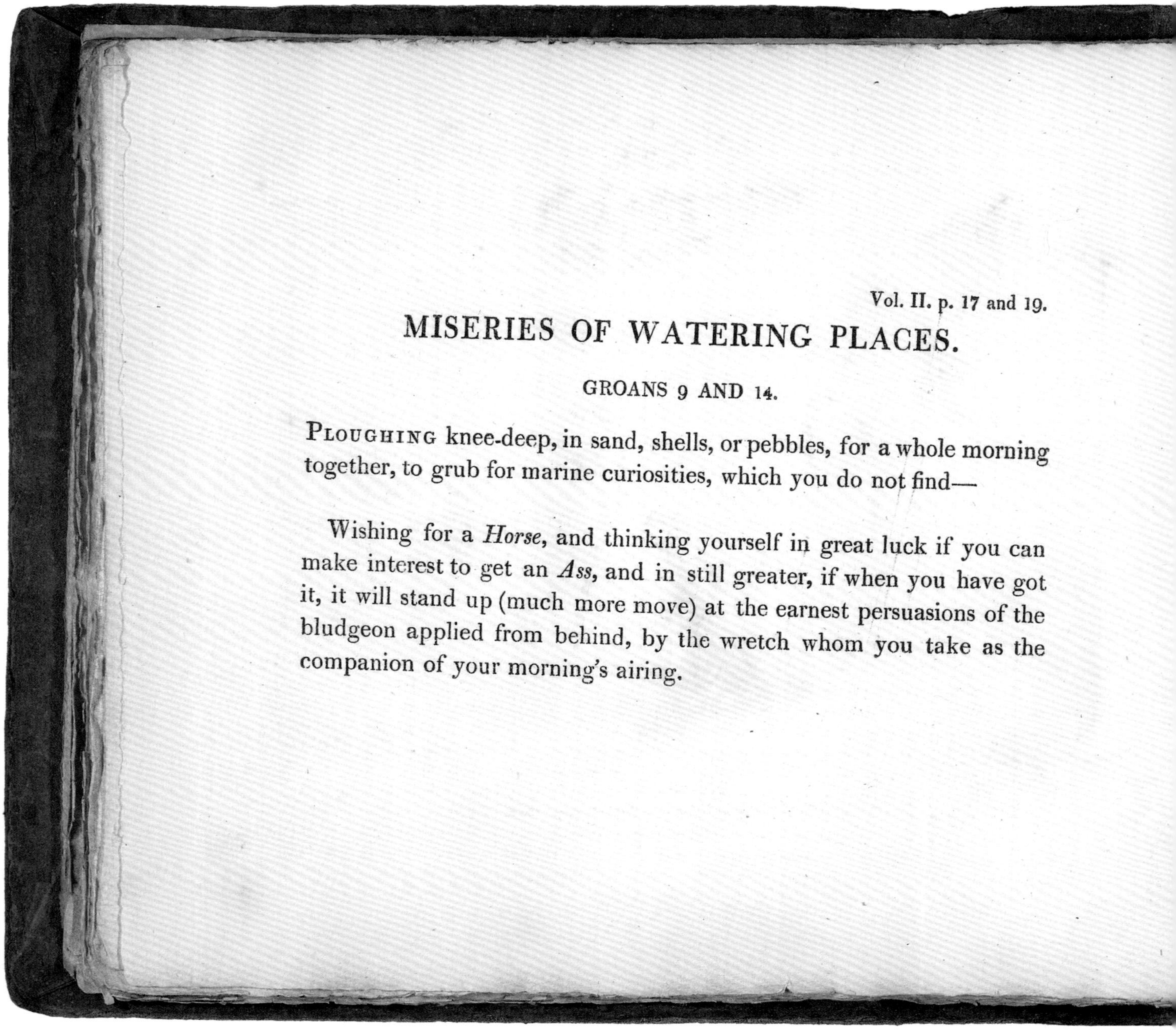

Vol. II. p. 17 and 19.

MISERIES OF WATERING PLACES.

GROANS 9 AND 14.

PLOUGHING knee-deep, in sand, shells, or pebbles, for a whole morning together, to grub for marine curiosities, which you do not find—

Wishing for a *Horse*, and thinking yourself in great luck if you can make interest to get an *Ass*, and in still greater, if when you have got it, it will stand up (much more move) at the earnest persuasions of the bludgeon applied from behind, by the wretch whom you take as the companion of your morning's airing.

John Atkinson, *Sixteen Scenes Taken from the Miseries of Human Life*, London: William Miller, 1807, hand-colored etching. Yale Center for British Art, Paul Mellon Collection

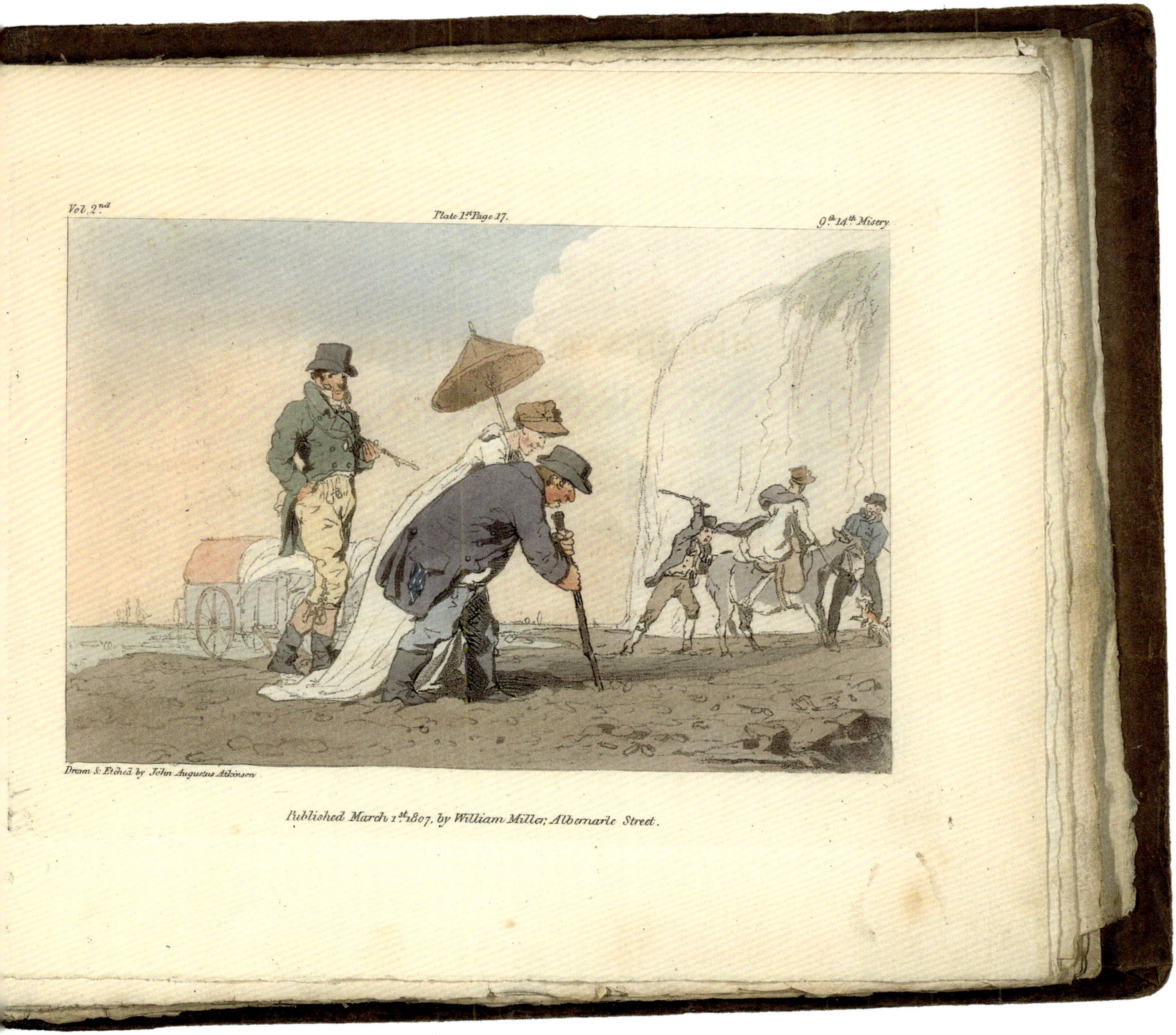
Vol. 2nd
Plate 1st Page 17.
9th 14th Misery
Drawn & Etched by John Augustus Atkinson
Published March 1st 1807, by William Miller, Albemarle Street.

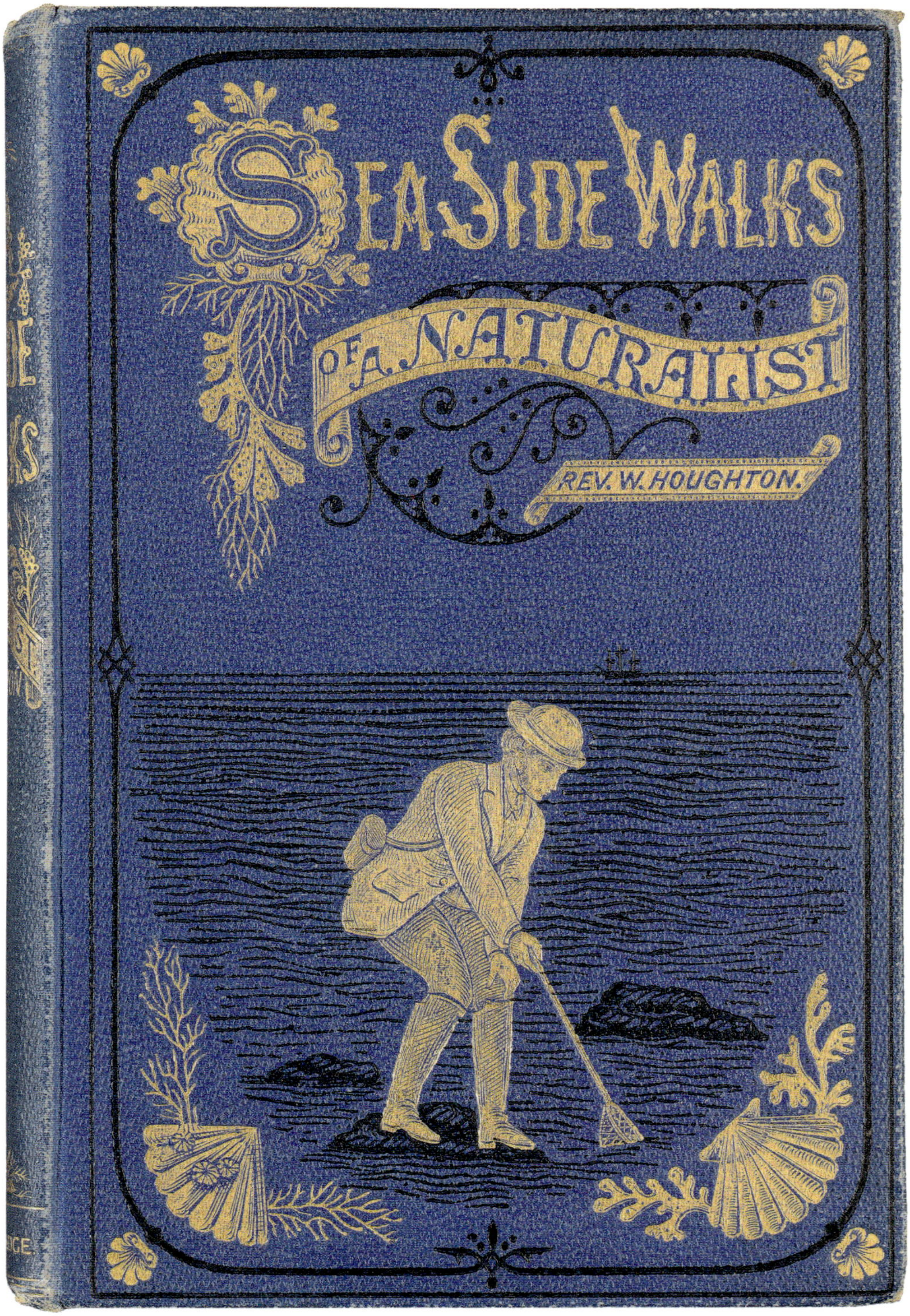

ABOVE AND OPPOSITE: William Houghton, cover, title page, and frontispiece, *Sea-Side Walks of a Naturalist with his Children*, London: Groombridge and Sons, 1870, chromolithograph. Yale Center for British Art, Paul Mellon Collection

RIGHT: Lady Mary Leighton, Dancing Crabs in *Sketchbook* (detail), 1824–1860, pen and brown ink. Yale Center for British Art, Paul Mellon Fund

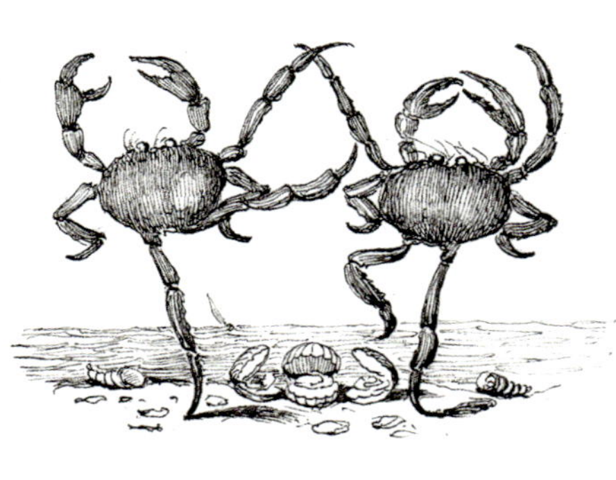

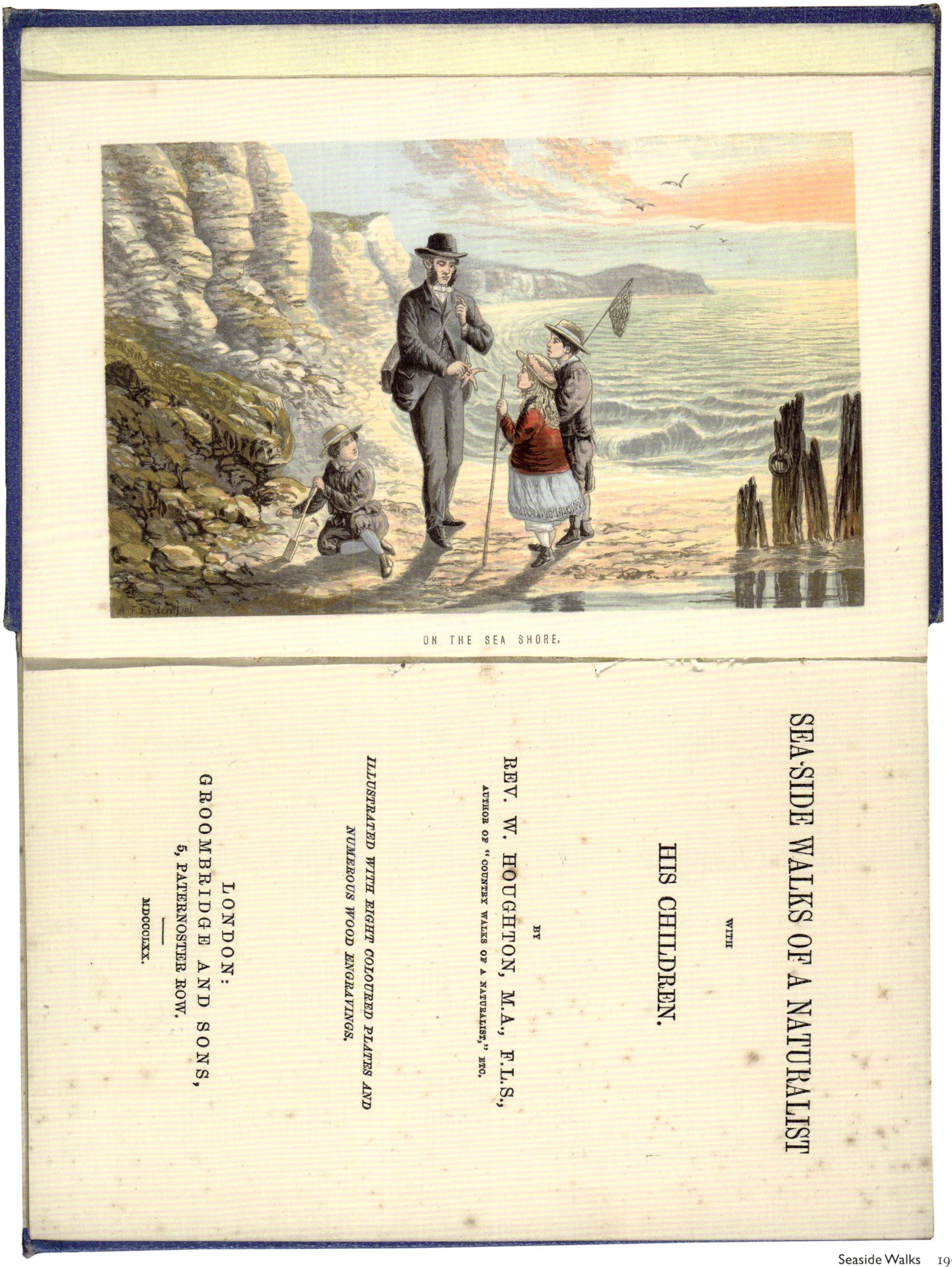
A. F. Lydon Del.
ON THE SEA SHORE.
SEA-SIDE WALKS OF A NATURALIST
WITH
HIS CHILDREN.
BY
REV. W. HOUGHTON, M.A., F.L.S.,
AUTHOR OF "COUNTRY WALKS OF A NATURALIST," ETC.
ILLUSTRATED WITH EIGHT COLOURED PLATES AND
NUMEROUS WOOD ENGRAVINGS.
LONDON:
GROOMBRIDGE AND SONS,
5, PATERNOSTER ROW.
MDCCCLXX.

Specimens of Sea Weed.

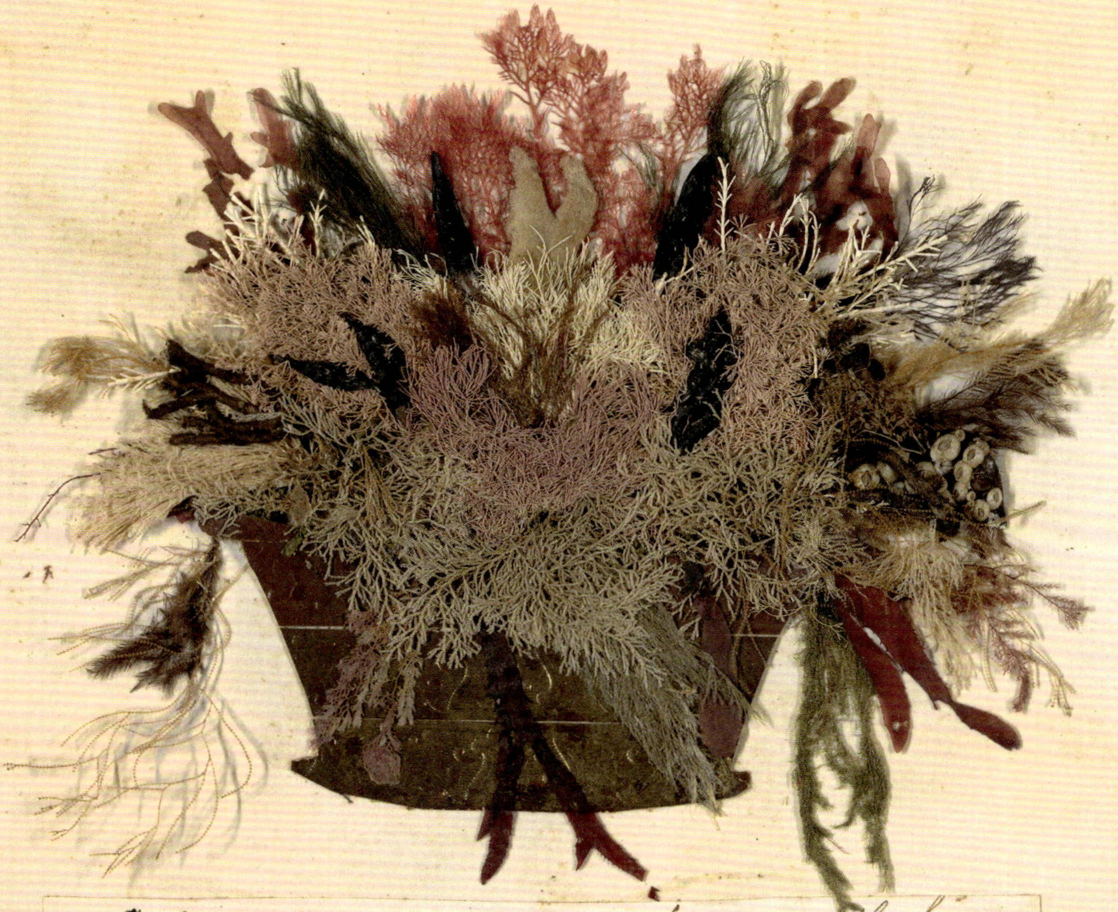

Call us not weeds we are flowers of the Sea.
For lovely & bright & gay tinted are we
And quite independent of sunshine or showers.
Then call us not weeds we are oceans gay flowers.

To be suitably dressed is one of the most important matters in setting forth to gather seaweeds. . . . Above all things it is desirable to protect the feet with stout boots, the so-called seaside boots of canvas or white leather being quite unsuitable for clambering over boulders to find rock-pools, or for wading in the marshy parts of a sandy shore where fresh water streams come down. Serviceable woollen garments that fit somewhat close are to be preferred to fashionable 'fly away' things which the wind will sport with unkindly, and which are sure to get well wetted when you begin to stoop and 'potter' about. A stout stick is a good friend, or, if that not be genteel enough, a strong umbrella on which you can lean as a stick, and the hooked handle of which may be serviceable to catch something the hand cannot possibly reach. . . . I have secured many a specimen by a smart thrust that I must have laboured hard for on my knees with hammer and chisel, at the risk, perhaps, of a too intimate acquaintance with the rock-pool in which it grew.

Shirley Hibberd

The Seaweed Collector: A Handy Guide to the Marine Botanist, London: Groombridge & Sons, 1872, pp. 6–7

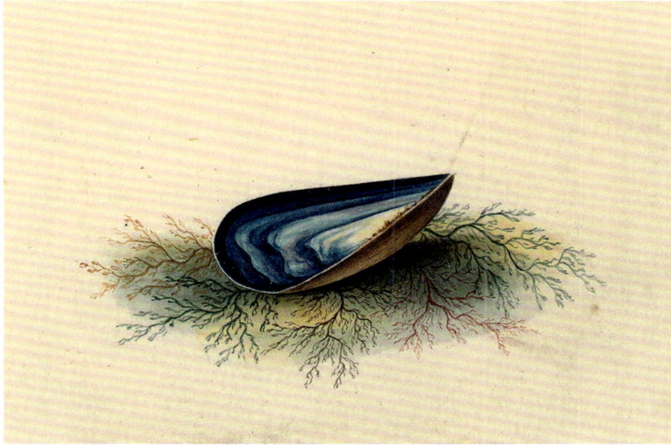

Mussel shell from *Commonplace Book, with Drawings of Insects*, ca. 1830, pen and ink with watercolor. Yale Center for British Art, Paul Mellon Fund

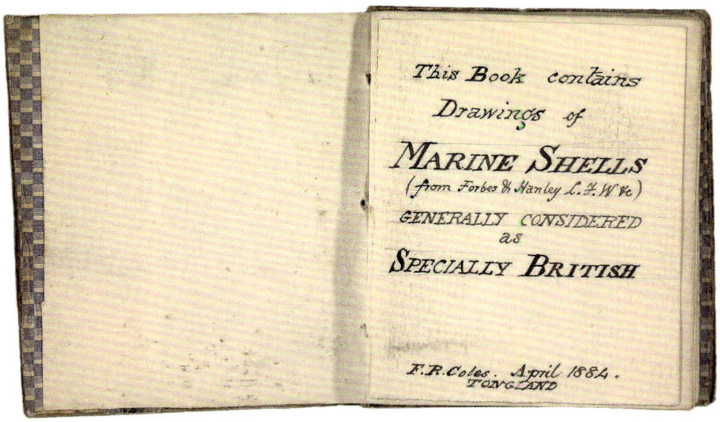

Frederick Rhenius Coles, *This Book Contains Drawings of Marine Shells (from Forbes & Hanley) Generally Considered as Specifically British*, Tongay, Scotland, 1884, graphite, pen and ink, with natural specimens. Yale Center for British Art, Paul Mellon Fund

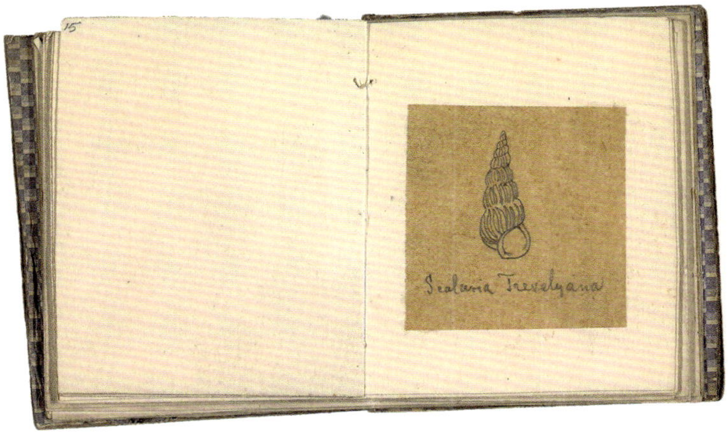

OPPOSITE: *Specimens of Sea Weed*, ca. 1840, natural specimens and hand-cut paper, with pen and ink labels. Yale Center for British Art, Paul Mellon Fund

OVERLEAF: Mandy Bonnell, *The Second Life of Shells*, with poem by Gabriel Gbadamosi, London: New North Press, 2001, etching. Yale Center for British Art, Friends of British Art Fund

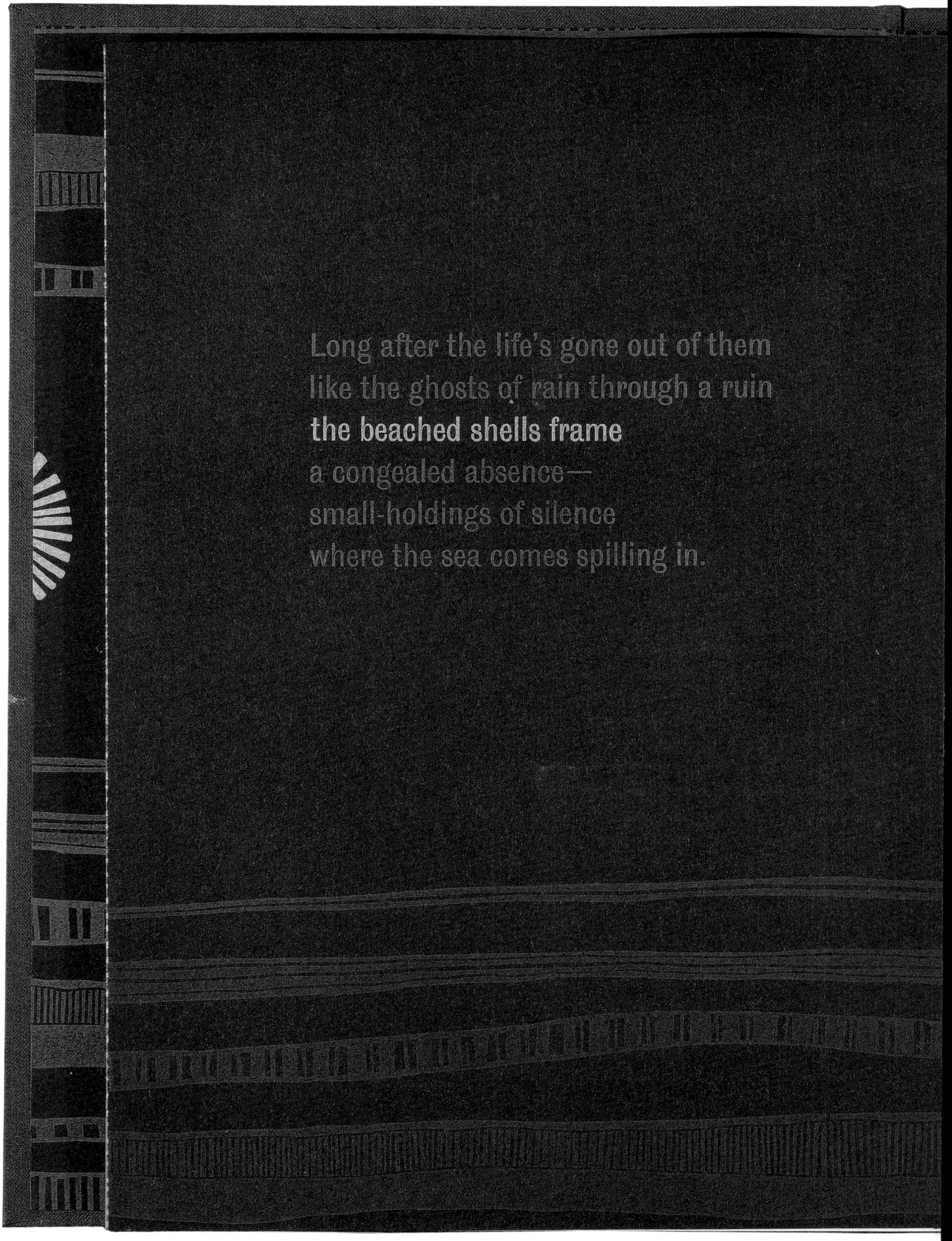
Long after the life's gone out of them
like the ghosts of rain through a ruin
the beached shells frame
a congealed absence—
small-holdings of silence
where the sea comes spilling in.

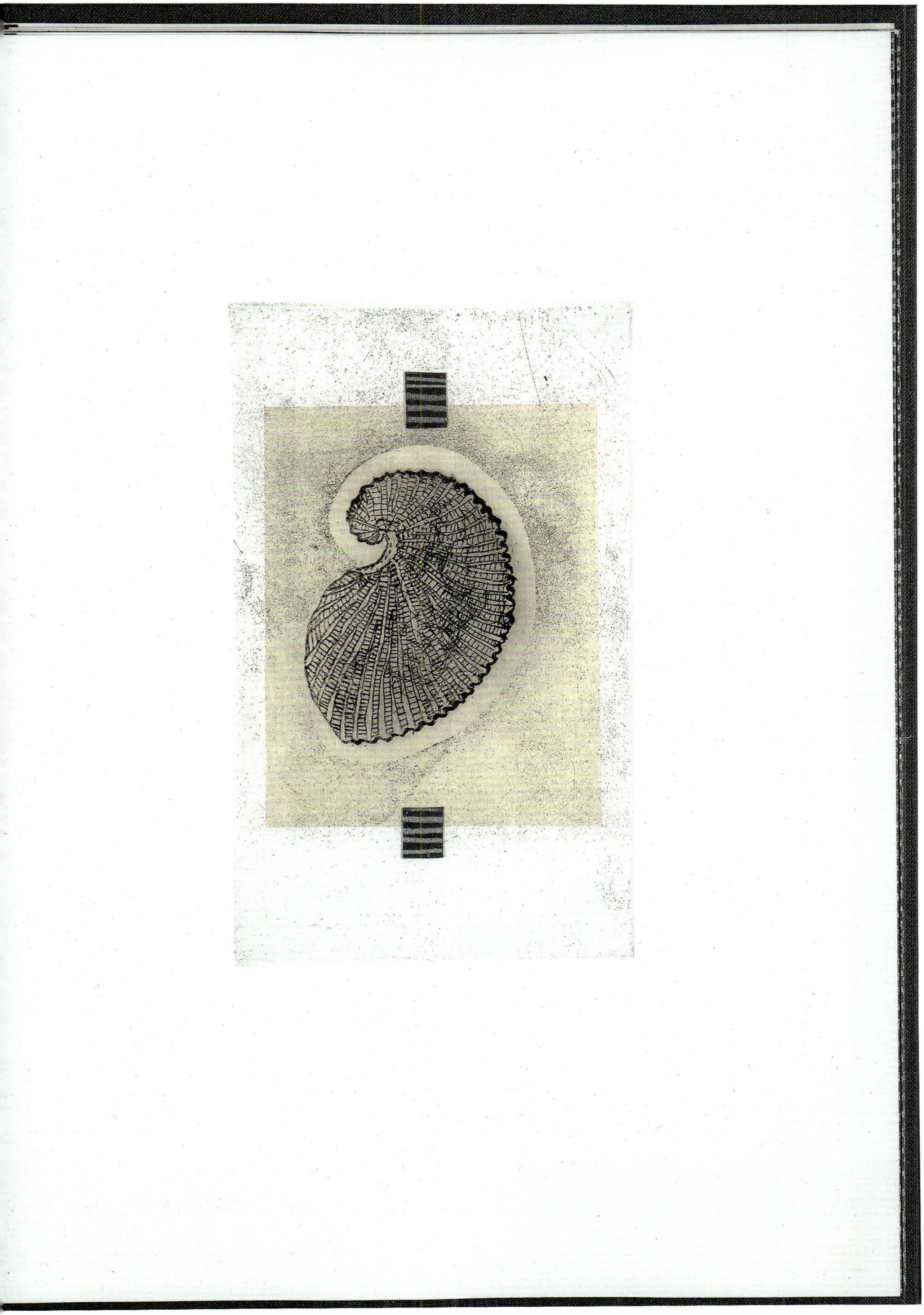

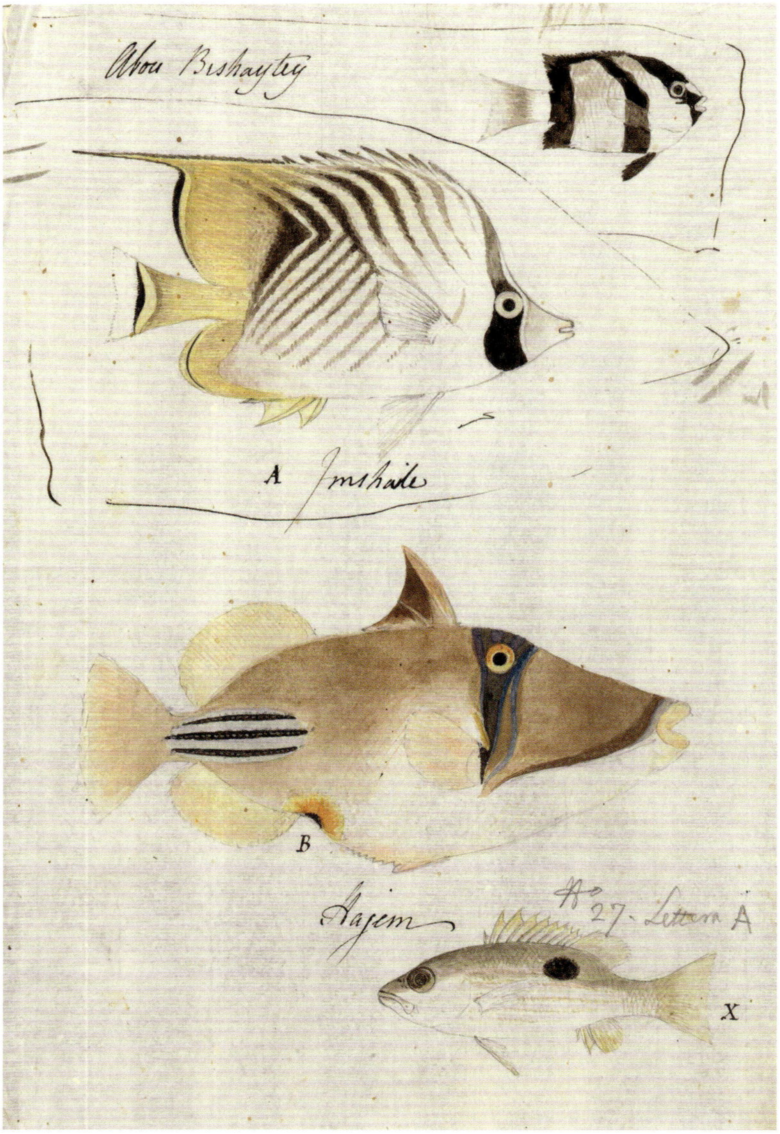

Luigi Balugani, *Four Fish*, ca. 1767–1773, watercolor and graphite with pen and ink. Yale Center for British Art, Paul Mellon Collection

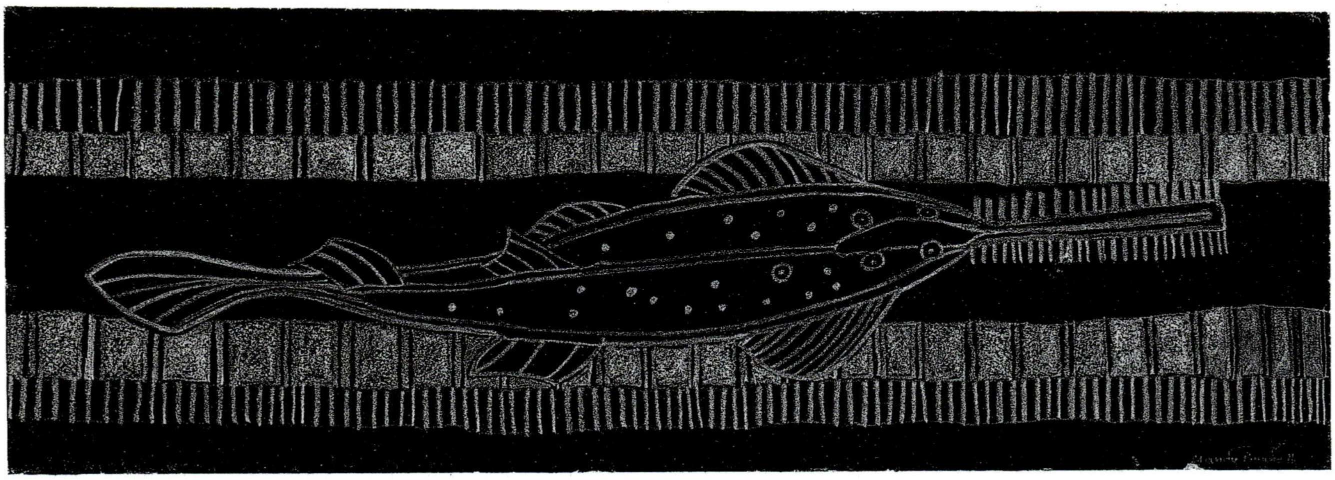

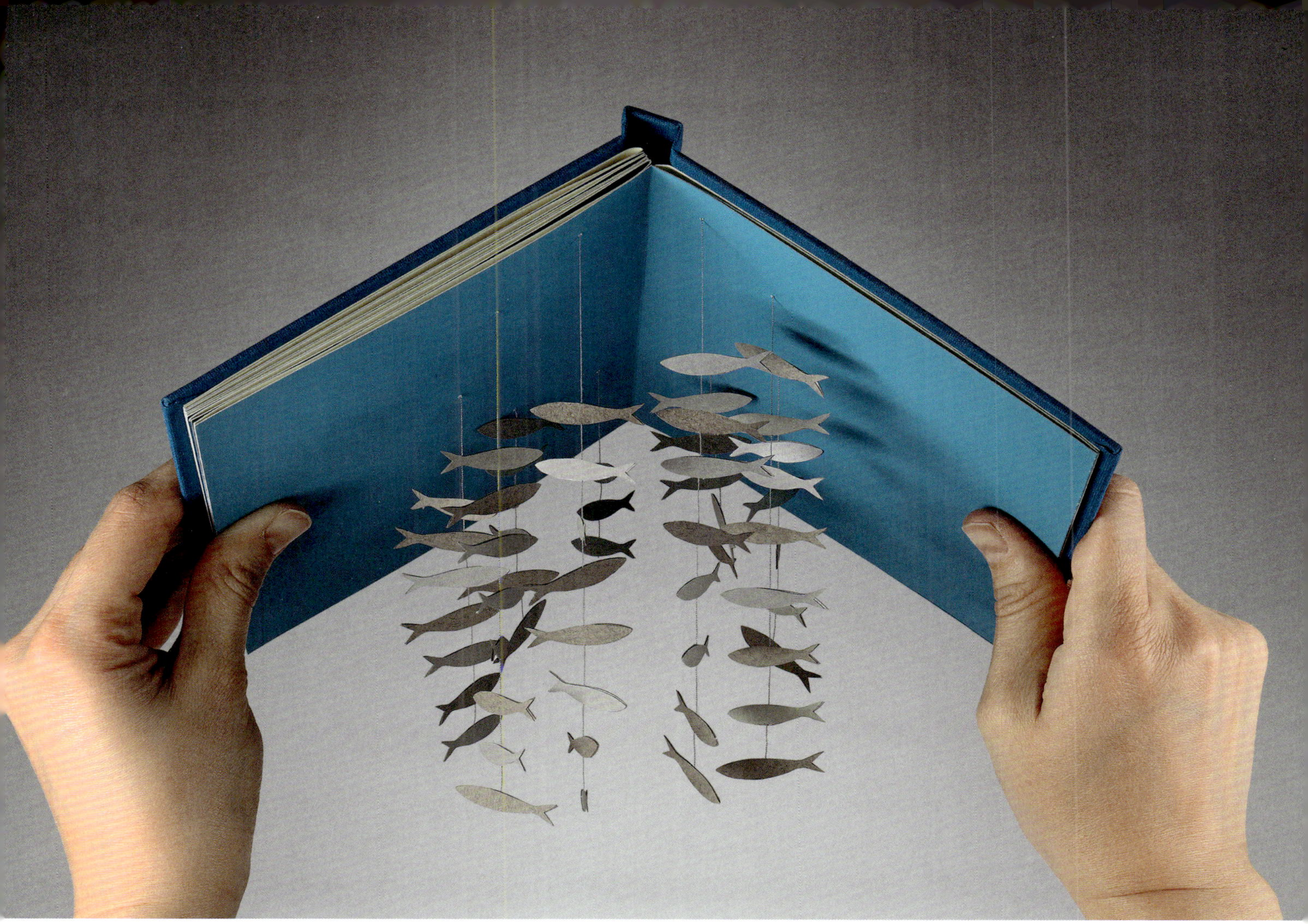

Chisato Tamabayashi, *Plunge*, London, 2010, screen print, hand-cut paper, and thread. Yale Center for British Art, Friends of British Art Fund

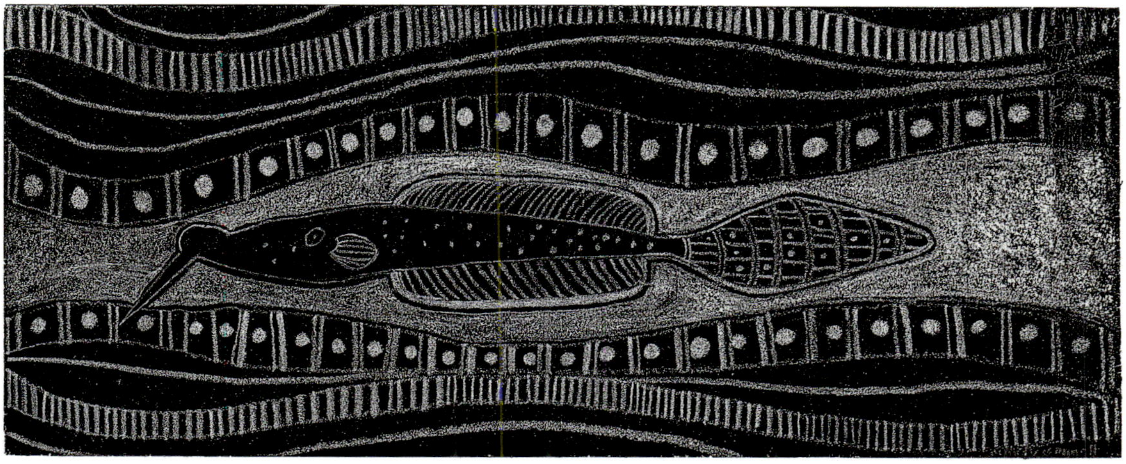

Mandy Bonnell, Original drawings for the woodblocks engraved by Bonnell for her *Coral Aquarium,* London, 1997, gray-colored pencil on black watercolored paper. Yale Center for British Art, Gift of the artist

PLATE VIII.

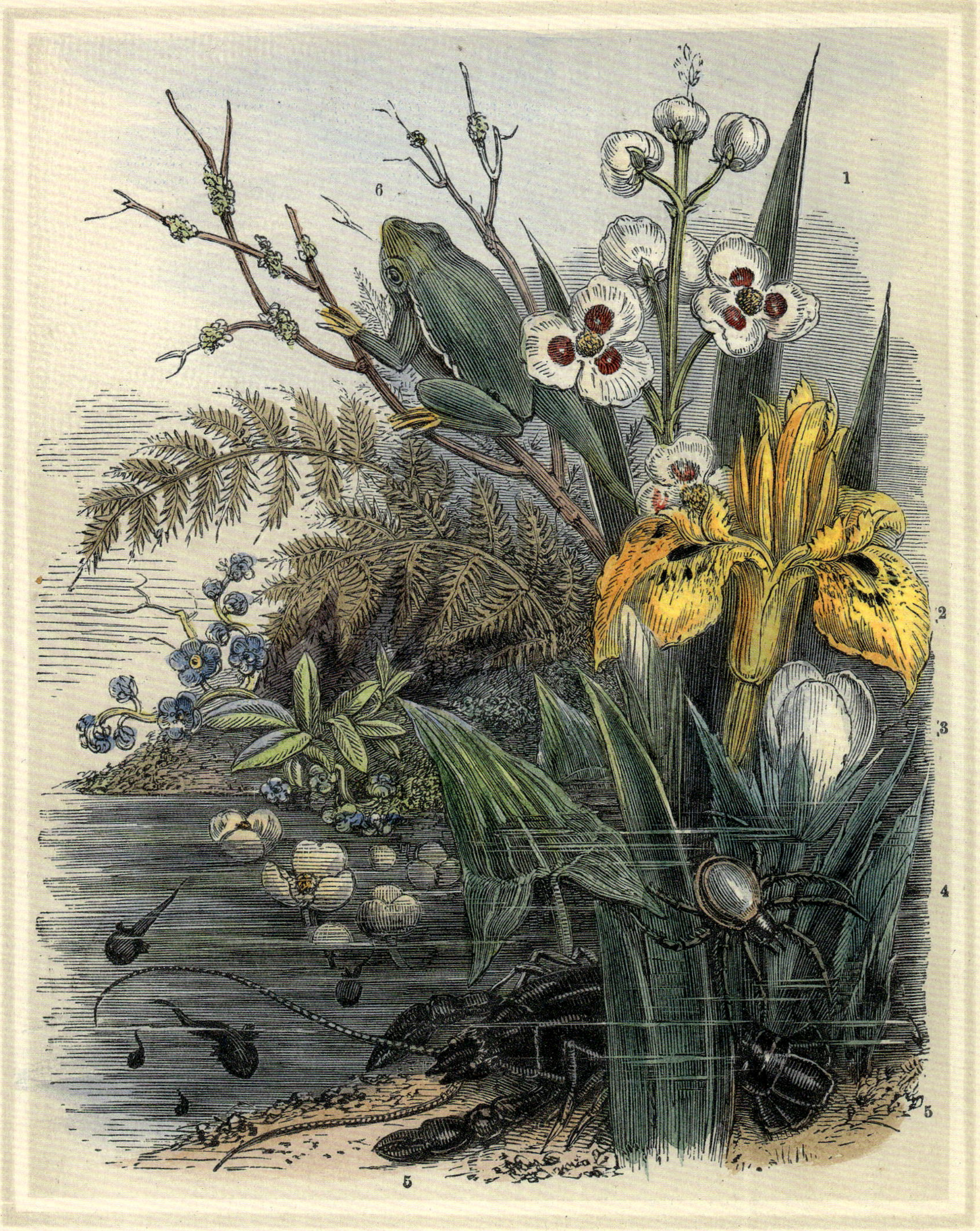

1. The Arrowhead (*Sagittaria sagittifolia*)
2. The Yellow Water Iris (*Iris pseudacorus*).
3. The Water Aloe, or Water Soldier (*Stratiotes aloides*).
4. The Water Spider (*Argyroneta aquatica*).
5. The Crayfish (*Astacus fluviatilis*).
6. The Tree Frog (*Rani arborea*).

Ponds & Streams

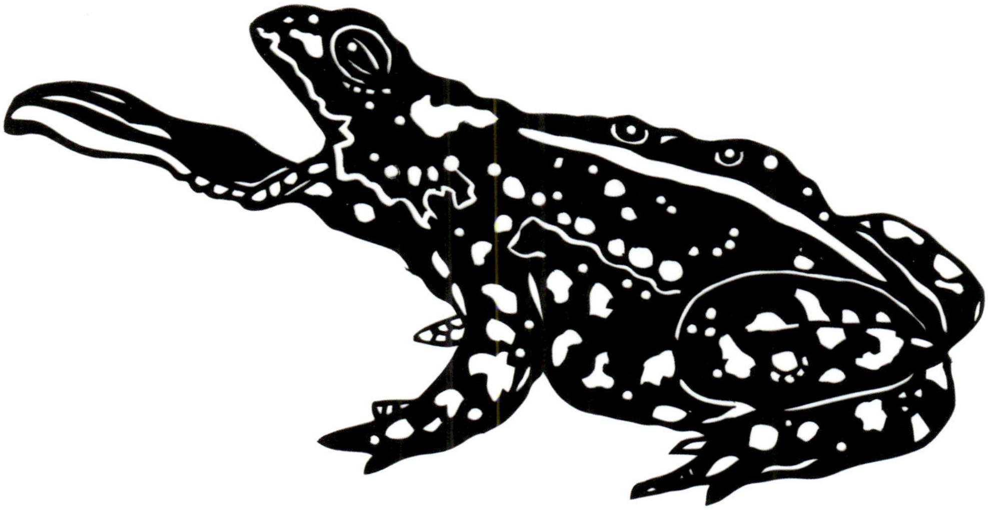

Ian Whitmore, hand-cut paper toad for *Othello: A Bestiary (with Floral Additions)*, Brighton: Borbonesa, 2011, screen print. Yale Center for British Art, Friends of British Art Fund

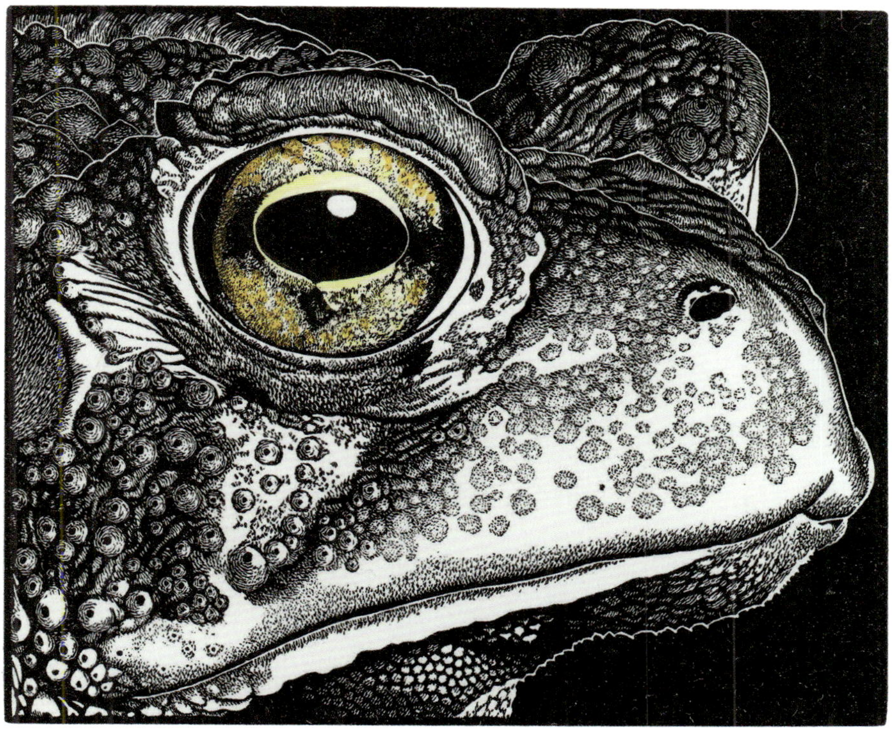

Abigail Rorer, *Bufo the First* from *Of Woodland Pools, Spring-holes & Ditches*, by Henry David Thoreau, Petersham, Massachusetts: Lone Oak Press, 2005, relief engraving. Collection of David Burnett

OPPOSITE: Henry Noel Humphreys, *Ocean and River Gardens*, London, 1857, hand-colored wood engraving. Yale Center for British Art, Paul Mellon Collection

OVERLEAF: Ken Campbell, *Skute Awabo*, London: Ken Campbell, 1992, letterpress, printed with colored and metallic inks. Yale Center for British Art, Friends of British Art Fund

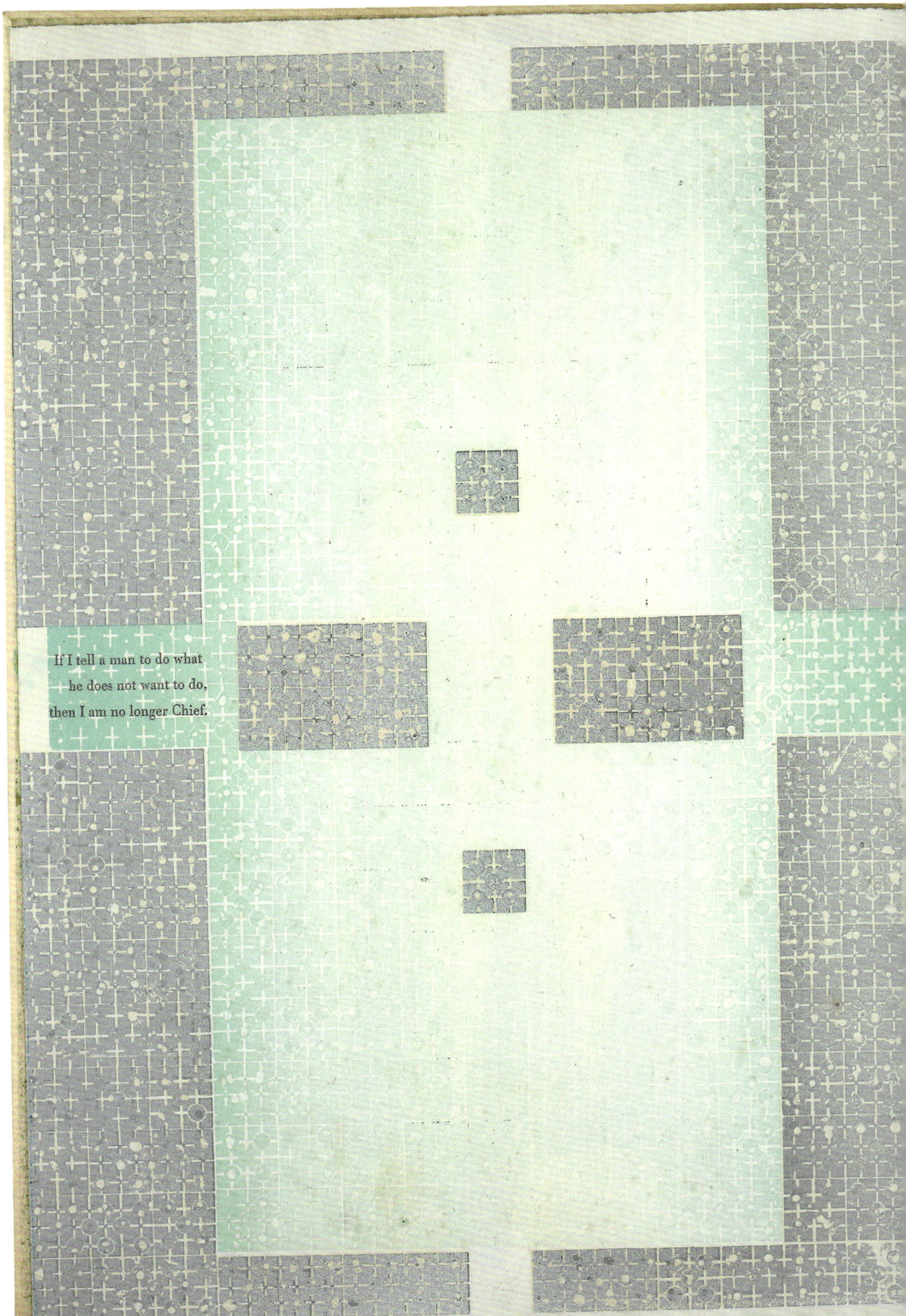
If I tell a man to do what
he does not want to do,
then I am no longer Chief.

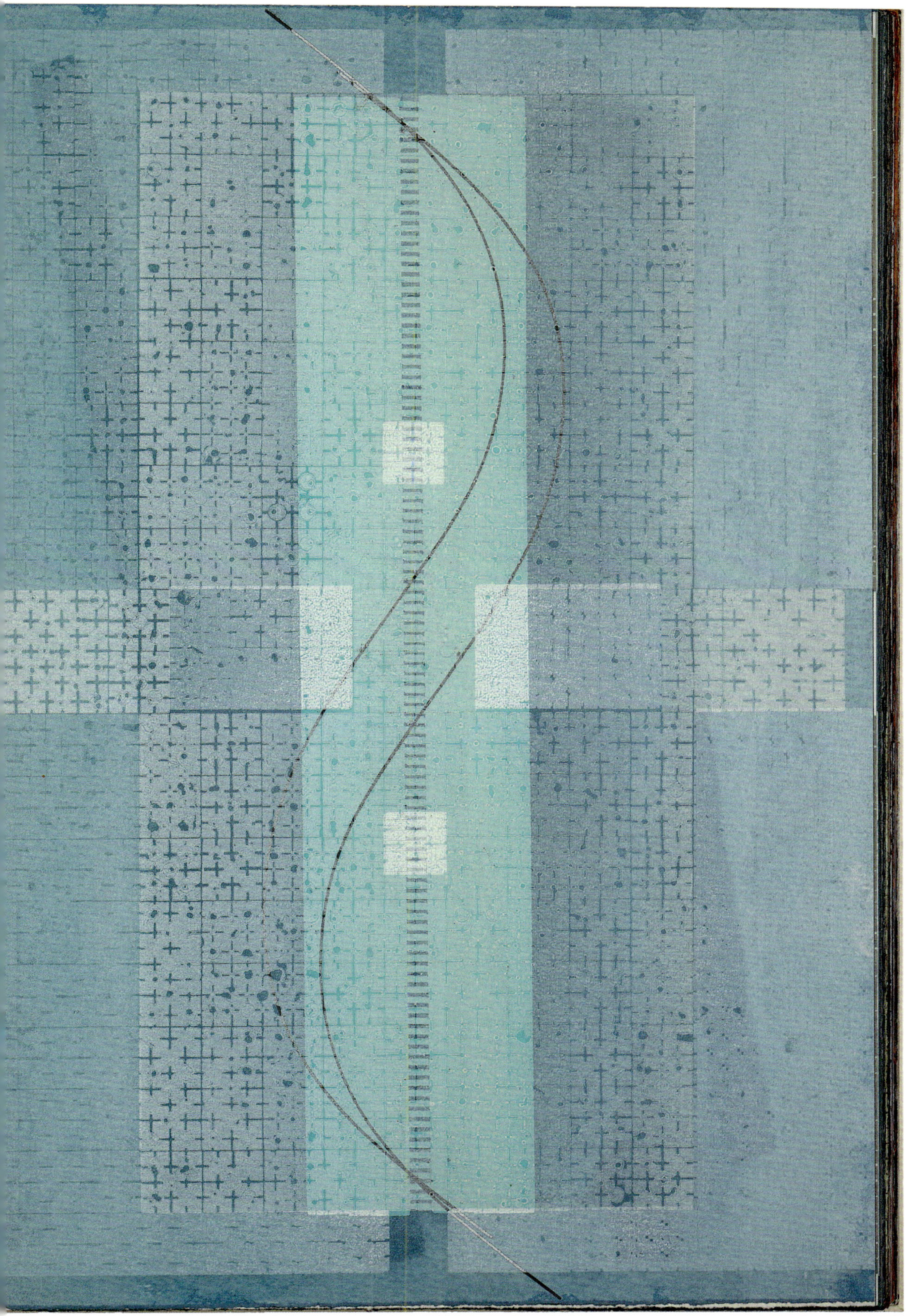

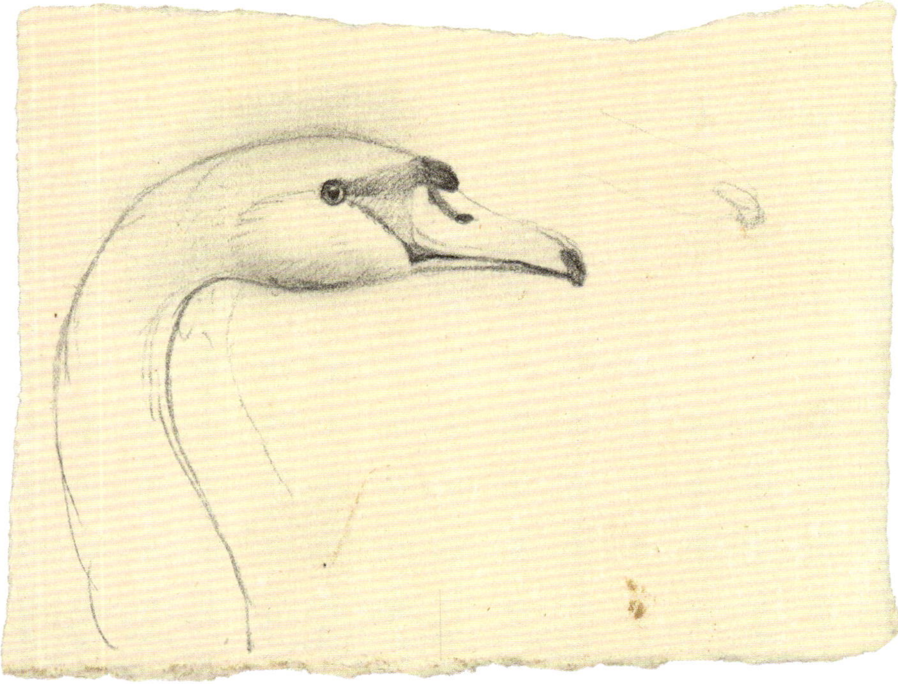

Possibly by James Sowerby, Sketch of *Head of a Swan*, ca. 1790, graphite.
Yale Center for British Art, Paul Mellon Collection

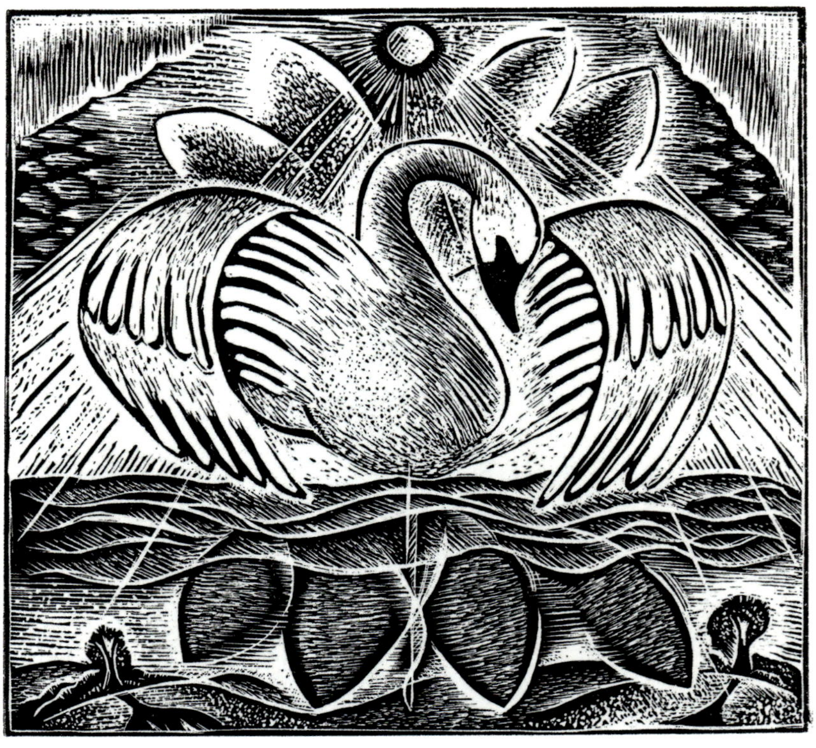

Angela Lemaire, *Swan*, 2011, wood engraving. Collection of David Burnett

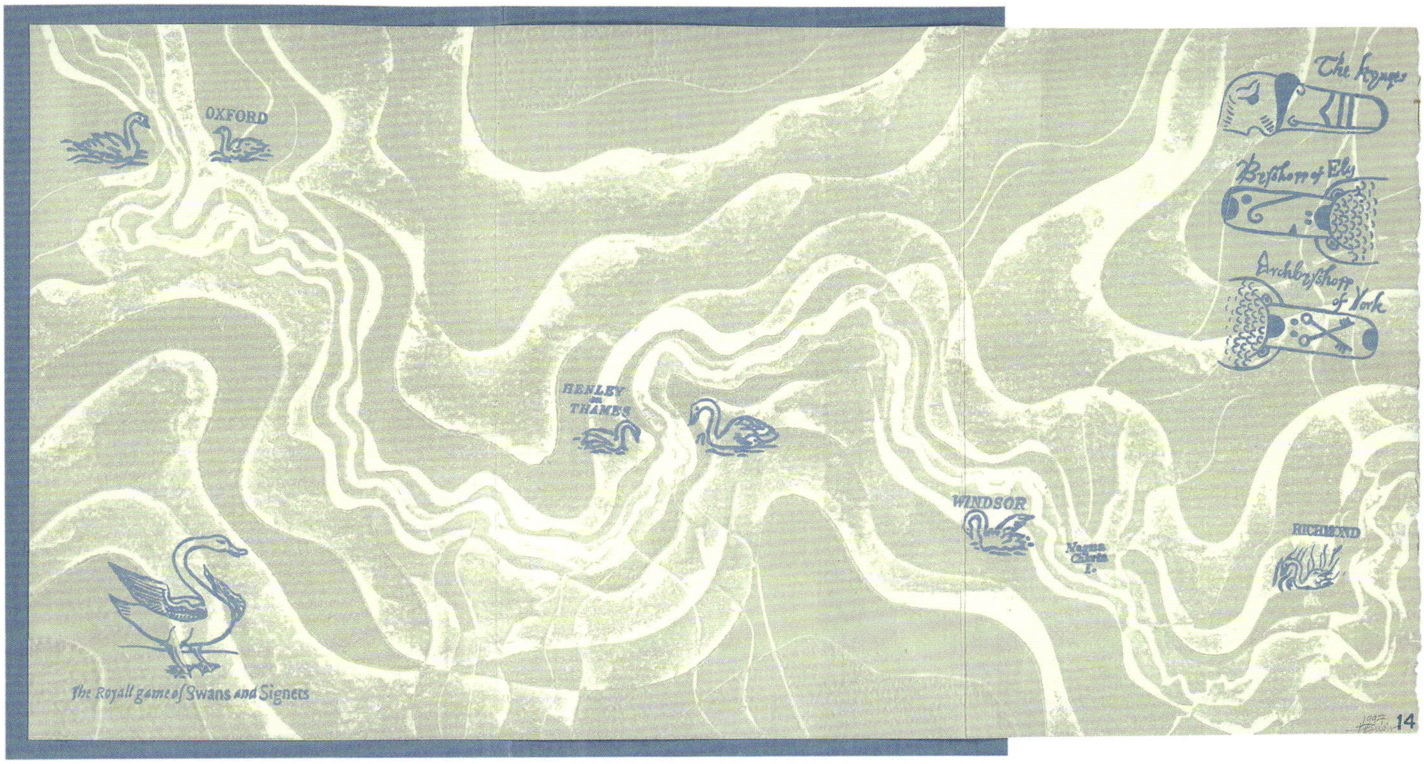

Tracey Bush, *Cygnea cantio: Swan Song*, London, 1997, blind relief print and rubber stamping. Yale Center for British Art, Friends of British Art Fund

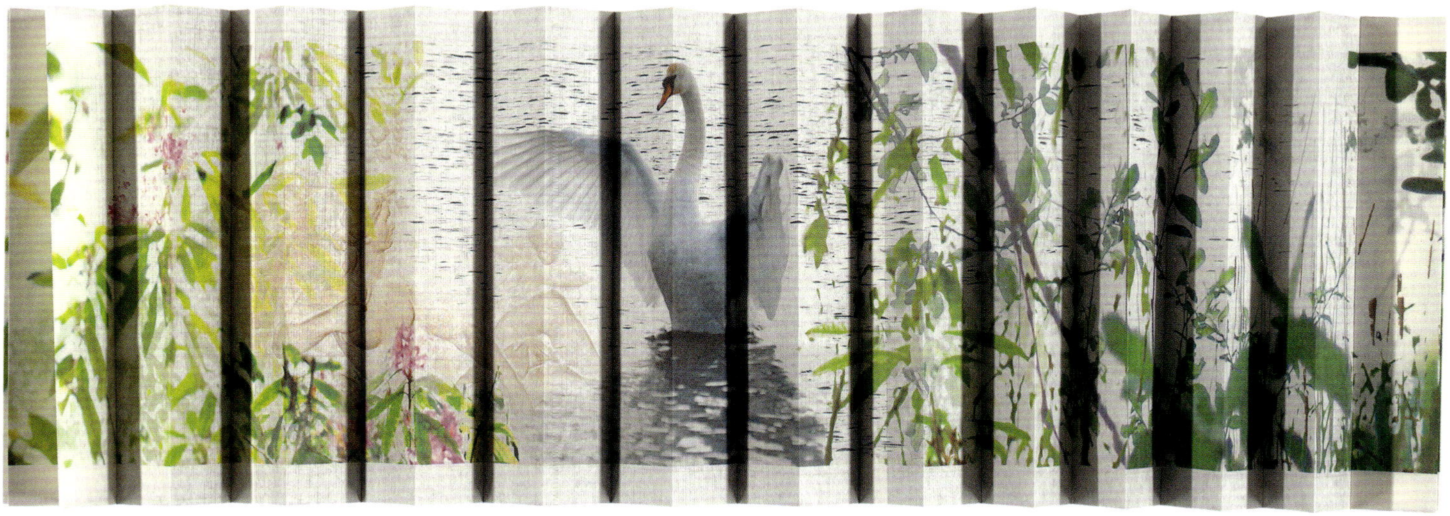

Helen Douglas, *Leda and the Swan*, Yarrow, Scotland: Weproductions, 2011, digital print on folded paper. Yale Center for British Art, Friends of British Art Fund

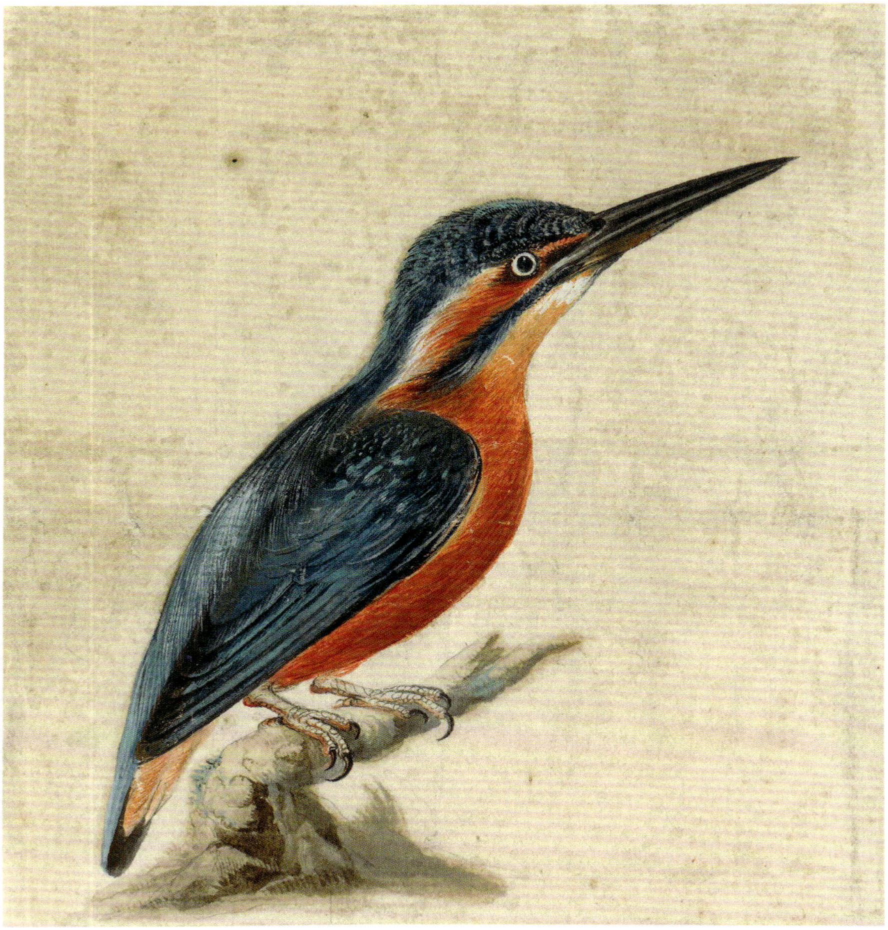

Thomas Atwood, *Kingfisher*, ca. 1761–1766, watercolor, gouache, and pen and brown ink. Yale Center for British Art, Paul Mellon Collection

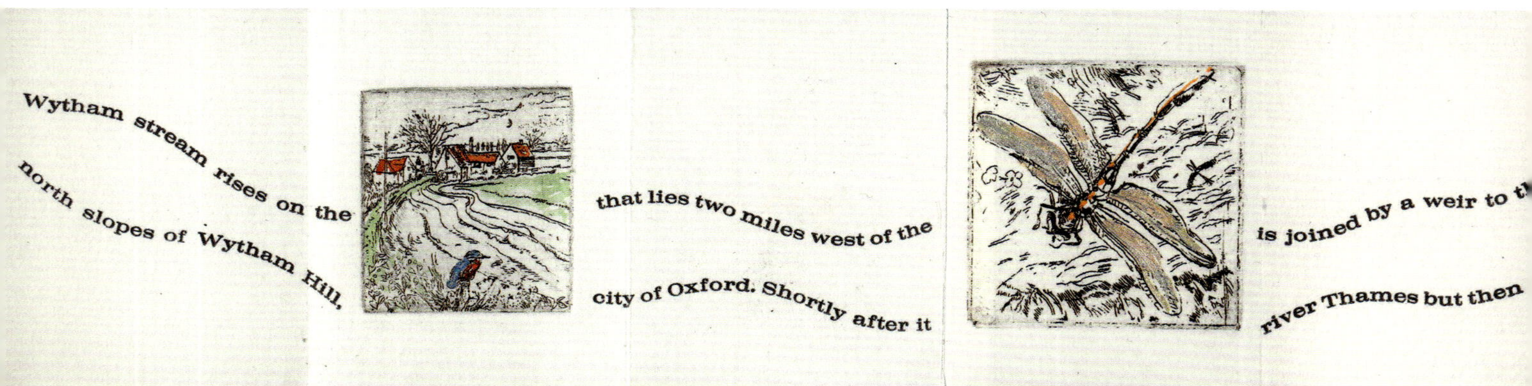

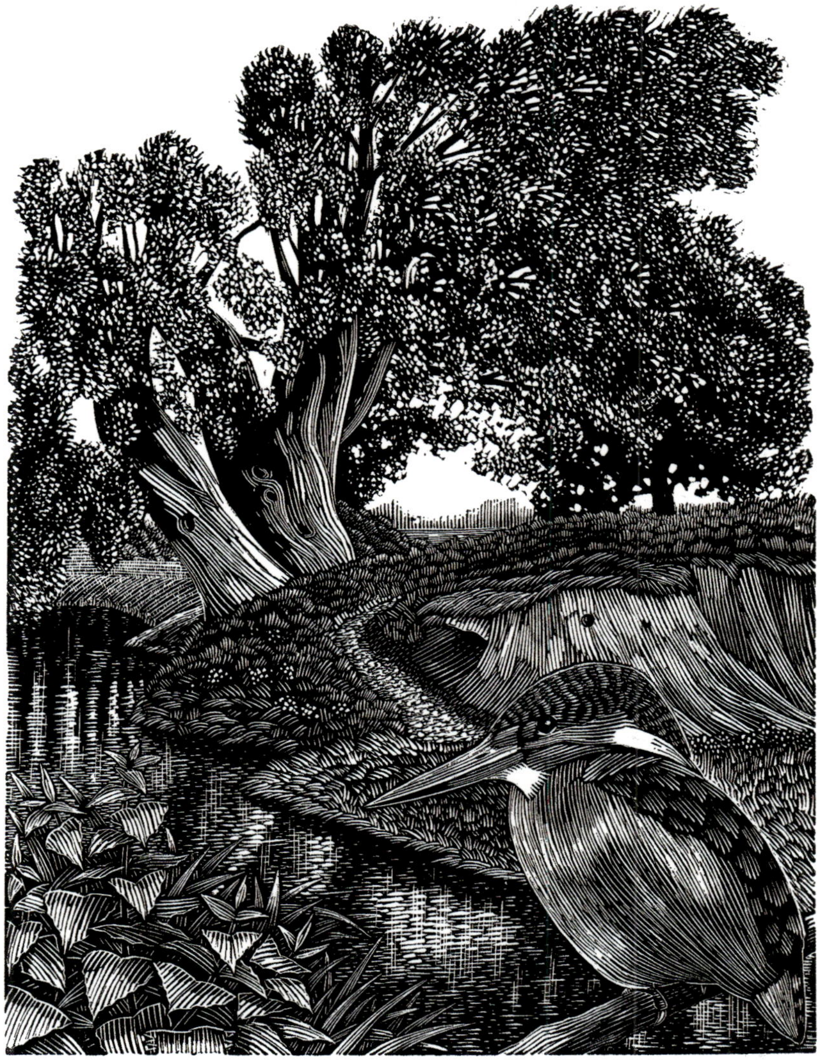

Andy English, *Kingfisher at Bourn Brook*, 2000, wood engraving.
Collection of David Burnett

Claire Lawson-Hall, *Wytham Stream*, Oxford: Alembic Press, 1993,
hand-colored etchings. Yale Center for British Art, Friends of British Art Fund

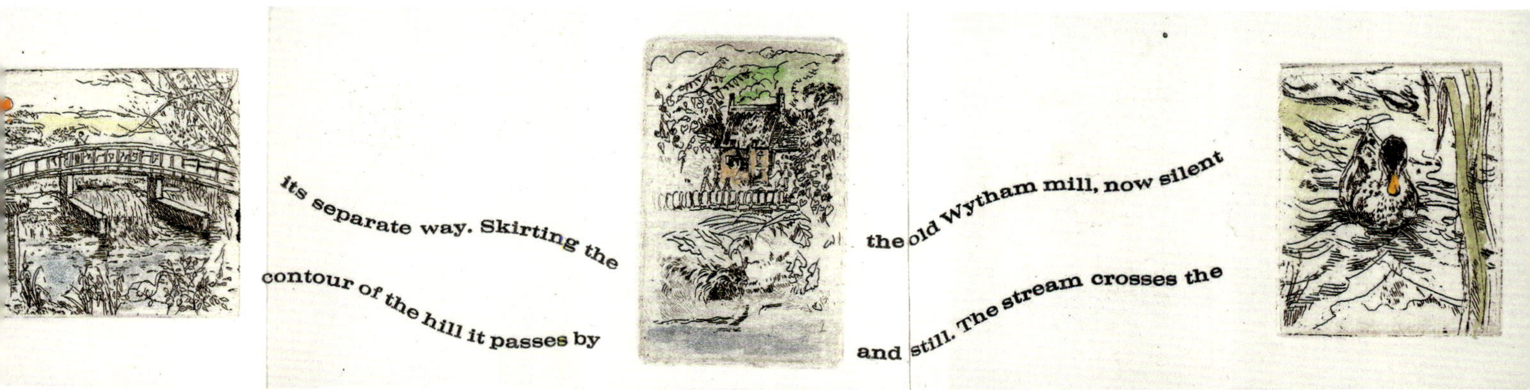

Little Sparta
Dunsyre
Scotland

Hans Waanders, *Perches*, Edinburgh: Morning Star, 2001.
Yale Center for British Art, Friends of British Art Fund

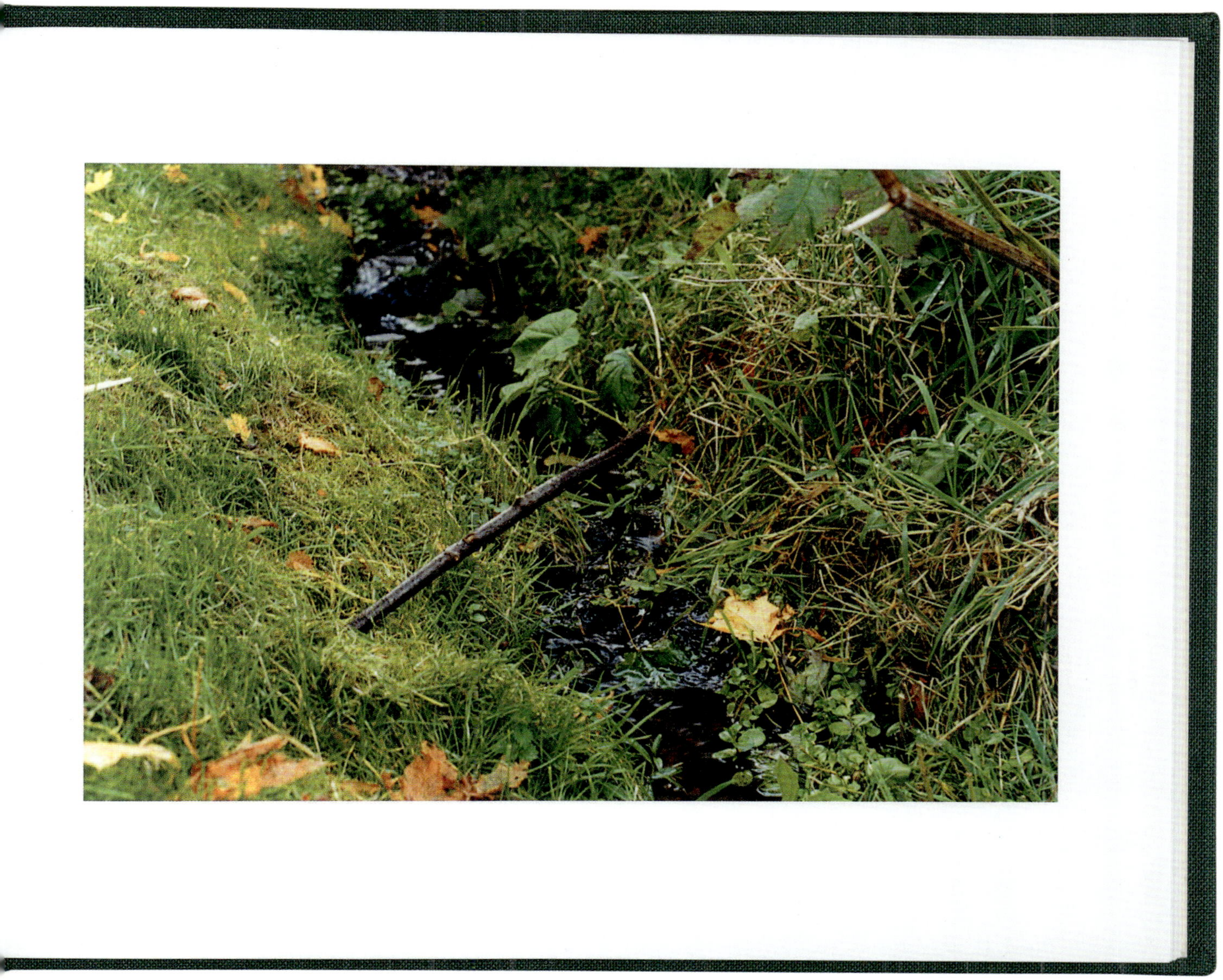

The Pond at Deuchar

Helen Douglas has been engaged with nature and with narratives for much of her life. Her formative years were spent in the Scottish Borders in a farming community, and she returned to live there as a working artist. Her chosen medium for embodying her expressive visual narratives has for many years been the photographic image used in conjunction with the book. In 2006, Douglas began to publish works in smaller editions, many of which questioned the familiar structure of a book or codex. Indeed, she recently has seized upon the historic hand scroll form to embody her ideas.

In 2010, Douglas made the hand scroll *poempondscroll*, in which images of the pond near which she lives were linked seamlessly, phrase by phrase, into a long narrative. This work led directly to *The Pond at Deuchar* in 2011, which was digitally printed on a Chinese paper and measured over fourteen meters long. The continuous digital photographic files used for printing this hand scroll have been directly transposed into the digital e-scroll. This process thereby has made something formerly published, of necessity, in limited numbers into an amazingly accessible and affordable work of art, remaining true to the "democratic" origins of artists' books.

Upon embarking on your spatial and temporal journey across *The Pond at Deuchar*, you first will have to negotiate your way past a pop-eyed speckled sentinel, the first of several frogs and toads that swim, glide, or lurk over the dark bed of the pond. This bed is layered with sunken, brown fallen leaves, while above it the pond's surface is interrupted by the green leaves of living water plants and sparks of color from flowers. A new development in the narrative is the appearance of calligraphic skeins of toad-spawn, the consequences of earlier frolics. As the season progresses, the spawn gives rise to a wriggling punctuation of tadpoles, dancing with their shadows, and joining other denizens of the pond, including water spiders, pond skaters, a newt, larvae, and maybe water crickets, in a ferment of fertility. Throughout the scrolled narrative the pond is embroidered by intricately engineered images of vegetation, sometimes above the pond, sometimes on the surface, sometimes reflected, sometimes underwater, often disturbed by exquisite ripples of light. In one passage the dark depths of the pond are intermingled with reflected blue sky, while the intense blue of a damselfly heralds the appearance of tiny specks of reflected blue speedwell, all contributing to a dance of light and color, surface and depth.

Clive Phillpot
WRITER AND CURATOR

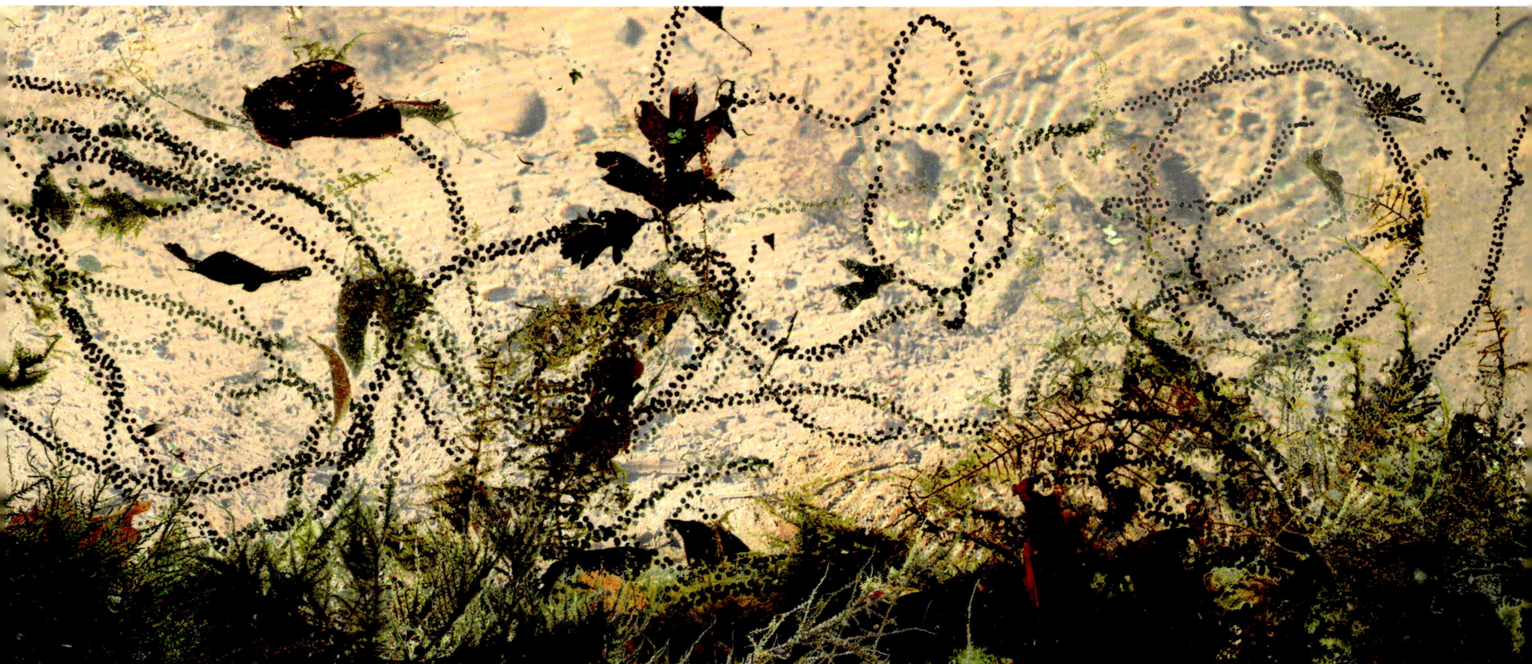

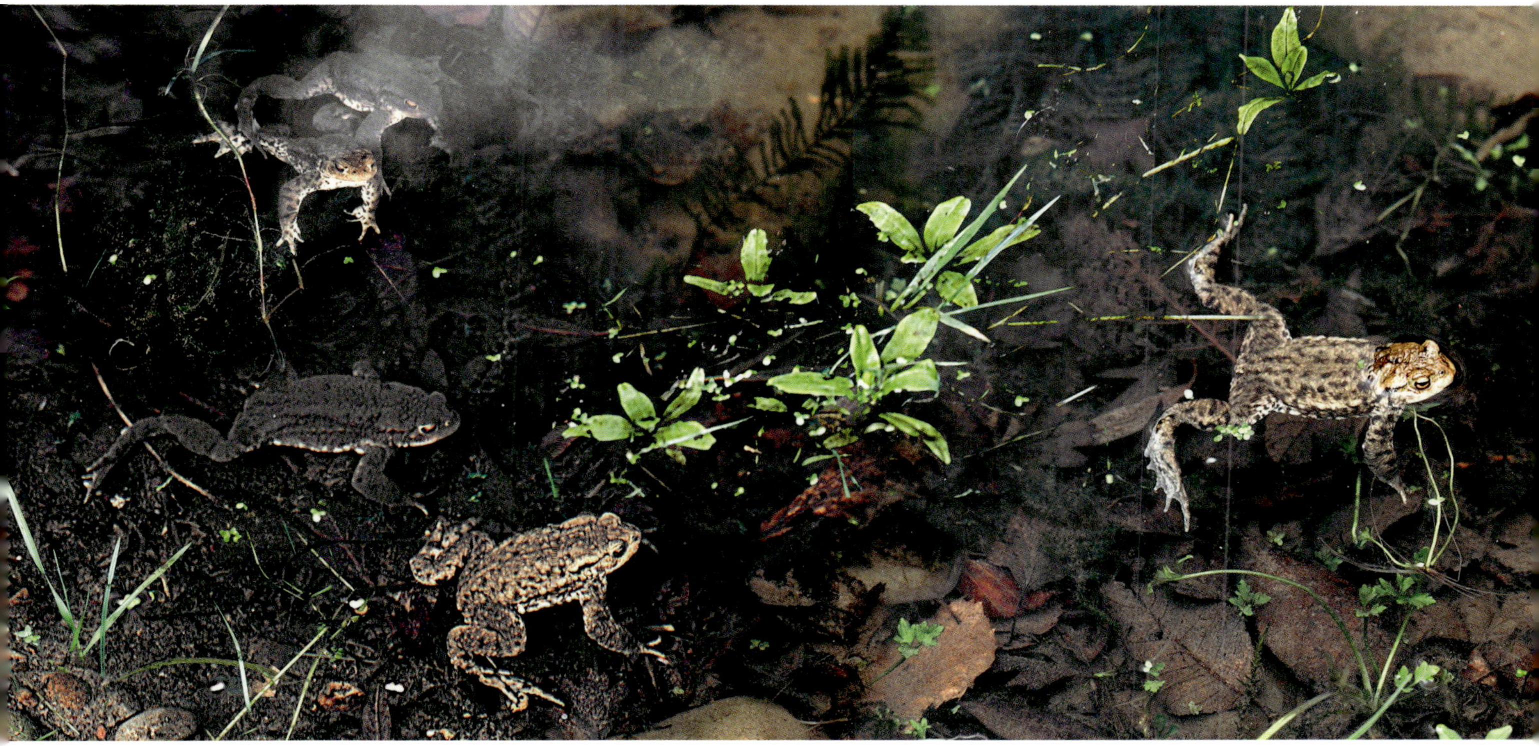

Helen Douglas, *The Pond at Deuchar E-Scroll*, London: Tate, 2013, *http://helendouglas.onlineculture.co.uk/ttp/ttp.html*. Collection of the artist

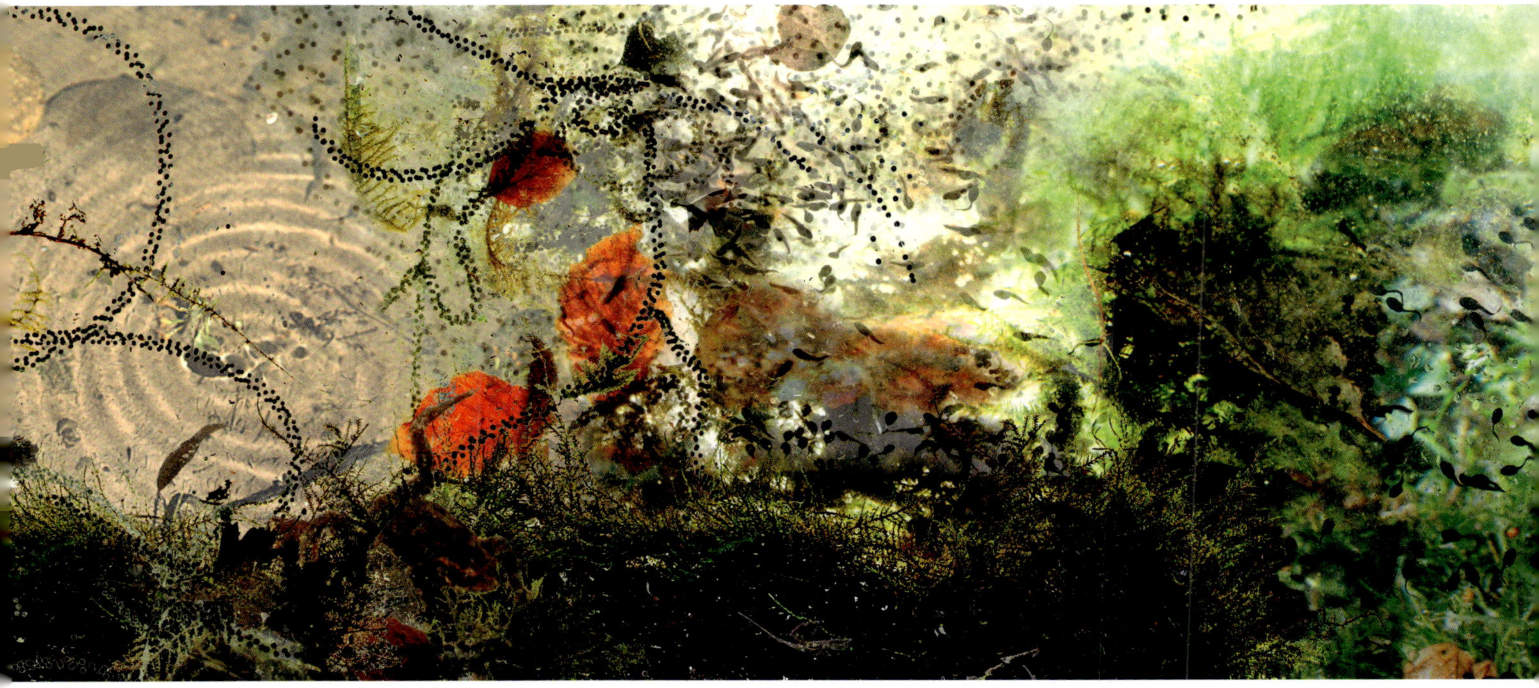

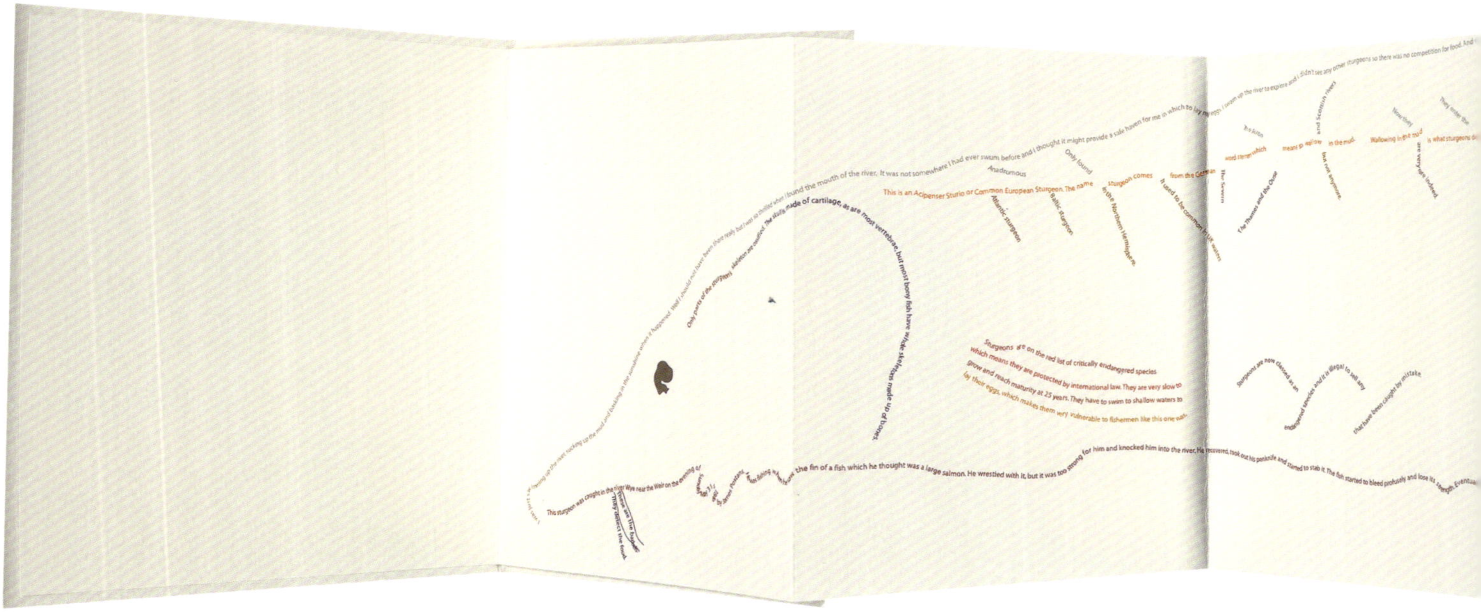

Pauline Lamont-Fisher, *Sturgeon*, Wormelow, England, 2011, inkjet print. Yale Center for British Art, Friends of British Art Fund

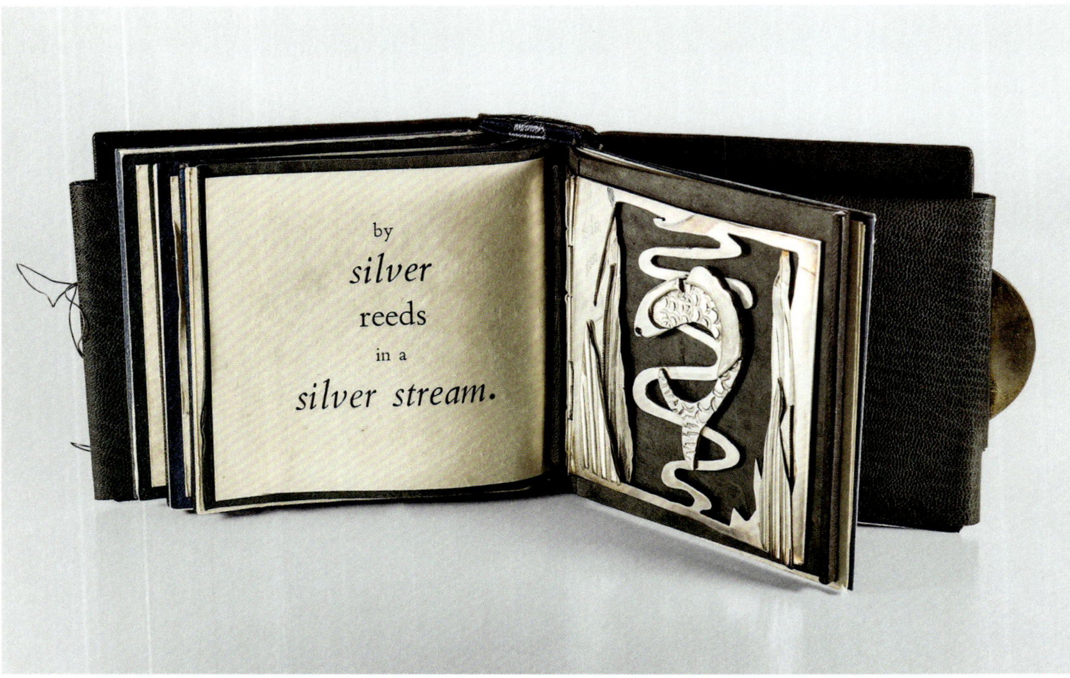

Susan Allix, *Silver*, London, 2005, miniature book, with sterling silver and wire. Collection of Neale and Margaret Albert

OPPOSITE: Sarah Sophia Banks, *Fishery Book of the River Witham in Lincolnshire*, watercolors by William Brand, 1784–1800, pen and ink with gouache and watercolor. Yale Center for British Art, Paul Mellon Collection

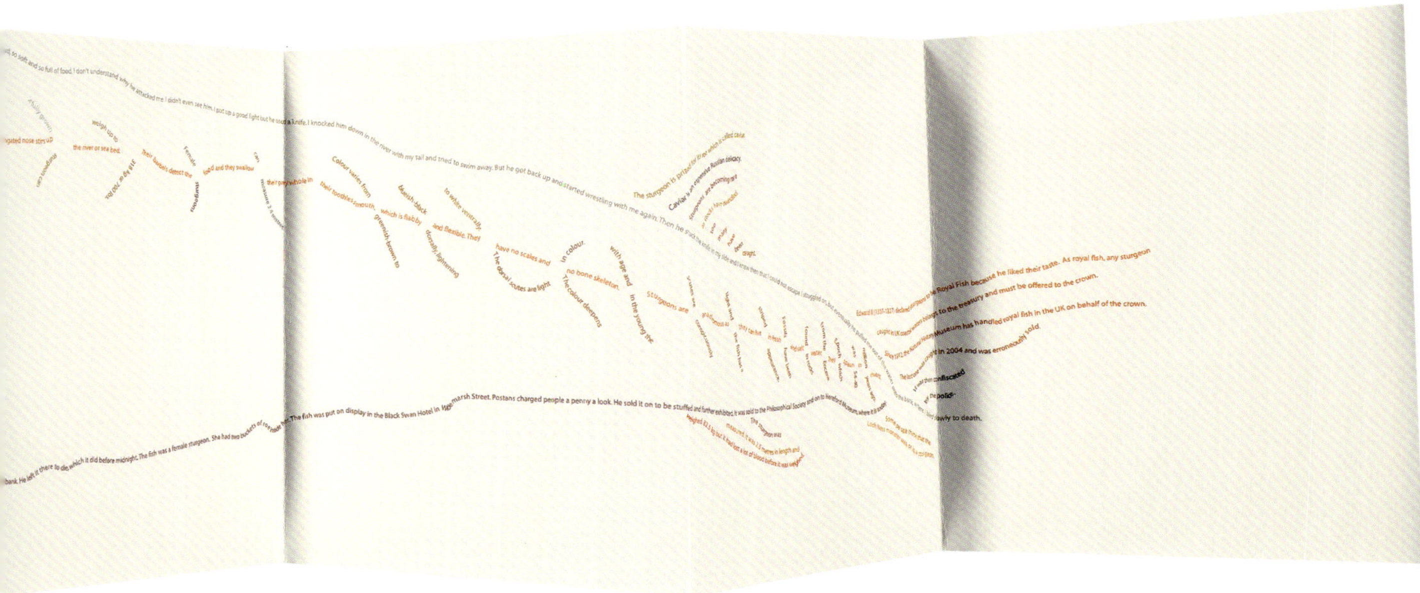

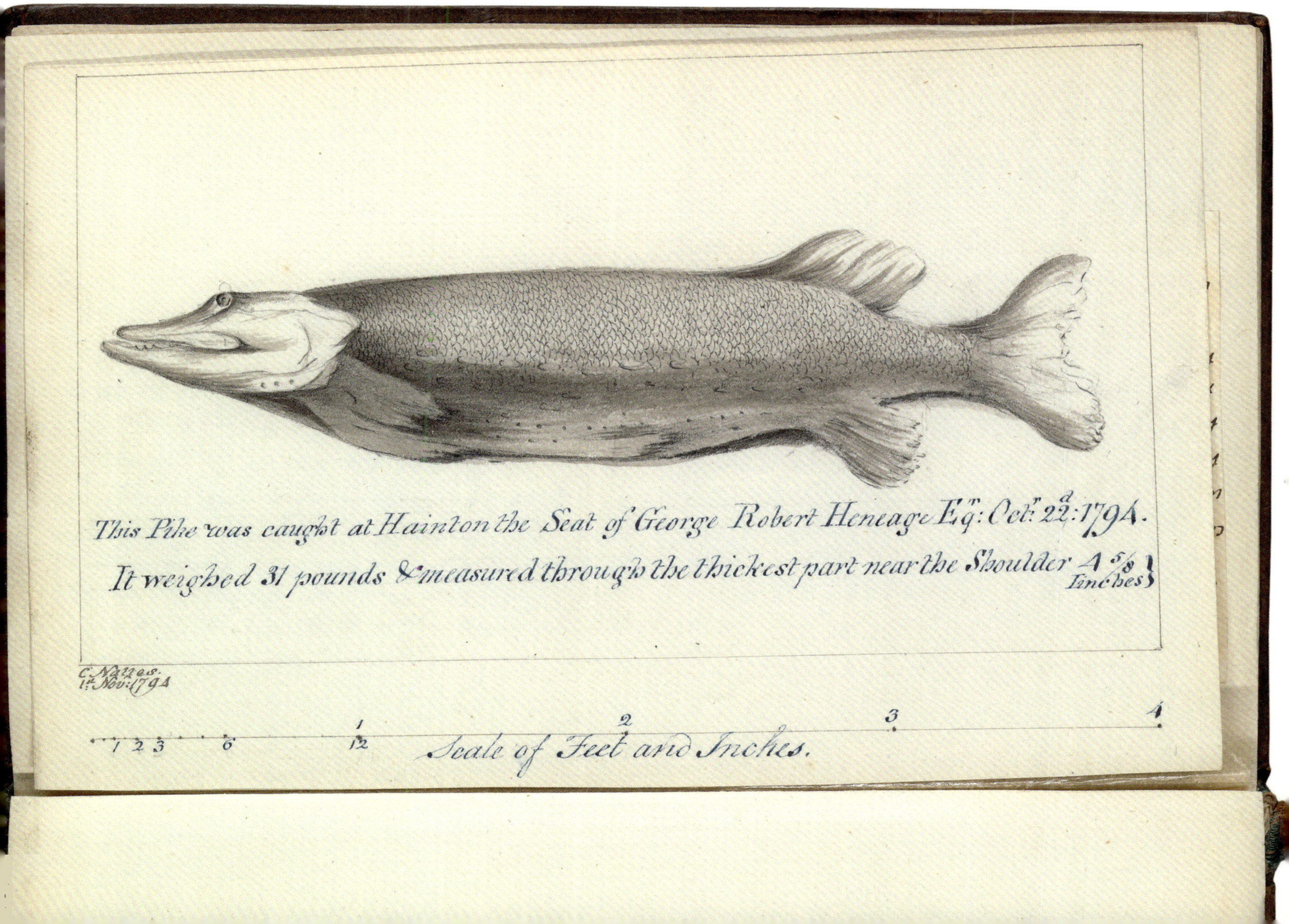
This Pike was caught at Hainton the Seat of George Robert Heneage Esq.r Oct.r 22.d 1794.
It weighed 31 pounds & measured through the thickest part near the Shoulder 4 5/8 Inches}
C. Nattes.
1st. Nov: 1794
1 2 3 6 1 12 2 3 4
Scale of Feet and Inches.

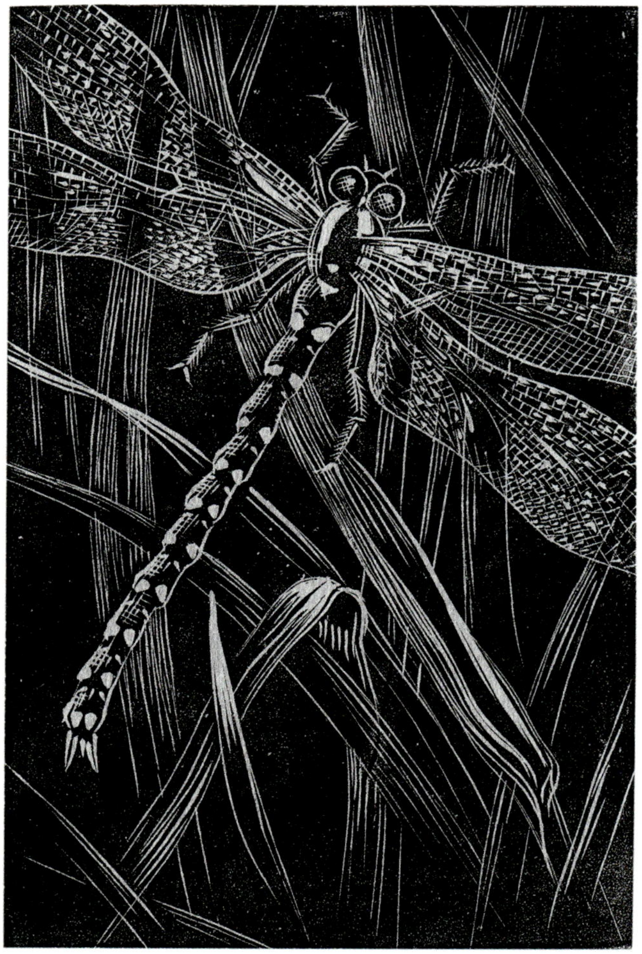

Sister Margaret Tournour, *Dragonfly*, 1991, wood engraving. Collection of David Burnett

Capturing Nature

For the past four summers, my family and I have spent a week at Yale's Outdoor Education Center (OEC) in East Lyme. A significant portion of our time at the OEC has been spent familiarizing ourselves with the plants, animals, and insects it has to offer. More than any of us, my seven-year-old daughter's enthusiasm for the natural world is boundless. She can spend an entire day hunting down frogs, turtles, and dragonflies. She catches them; I document her prizes with my camera phone and post the best examples on Instagram; and finally we negotiate the creature's release.

I have a personal affinity for collecting plants and flowers in a small pocket-sized field notebook, counting spiders, and watching the cormorant couple fly just above the water and dive for fish. The natural world is infinitely fascinating and entertaining. The only thing that separates people like my daughter and me from a nineteenth-century amateur naturalist is the way in which we document our discoveries. Beautifully hand-bound field notebooks filled with graphite and watercolor sketches and detailed notations about the where and what have been replaced with geo-tagged digital photographs and social media hash tags. We are no less curious than those amateur naturalists were two hundred or more years ago, and I hope that continues for centuries to come.

Sarah Welcome
AMATEUR NATURALIST

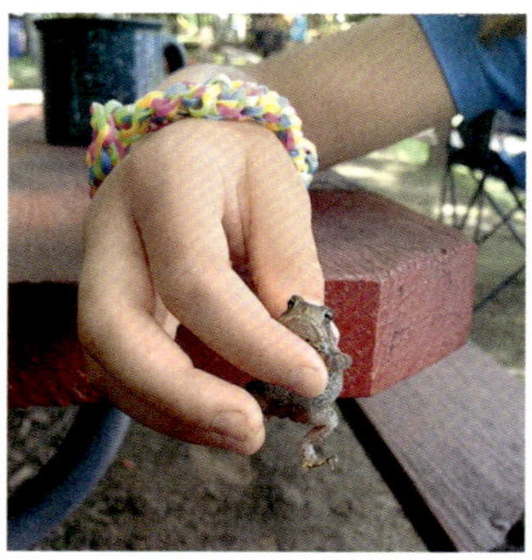

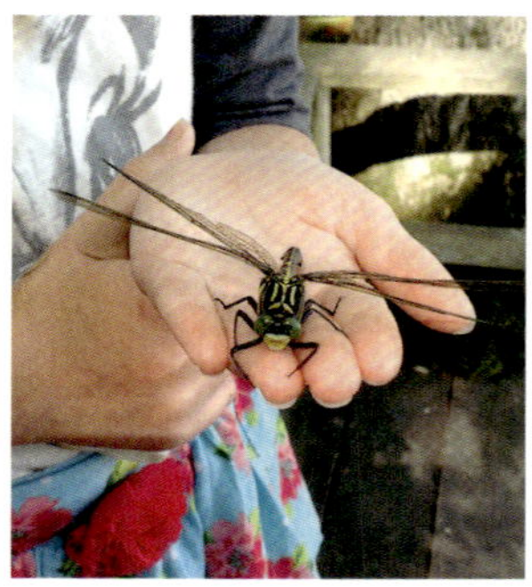

Young naturalists in East Lyme, Connecticut, summer 2013

OPPOSITE: George Edwards, *Pied Wagtail and Dragonfly*, ca. 1750, watercolor with pen and ink, heightened with gouache, over graphite. Yale Center for British Art, Paul Mellon Collection

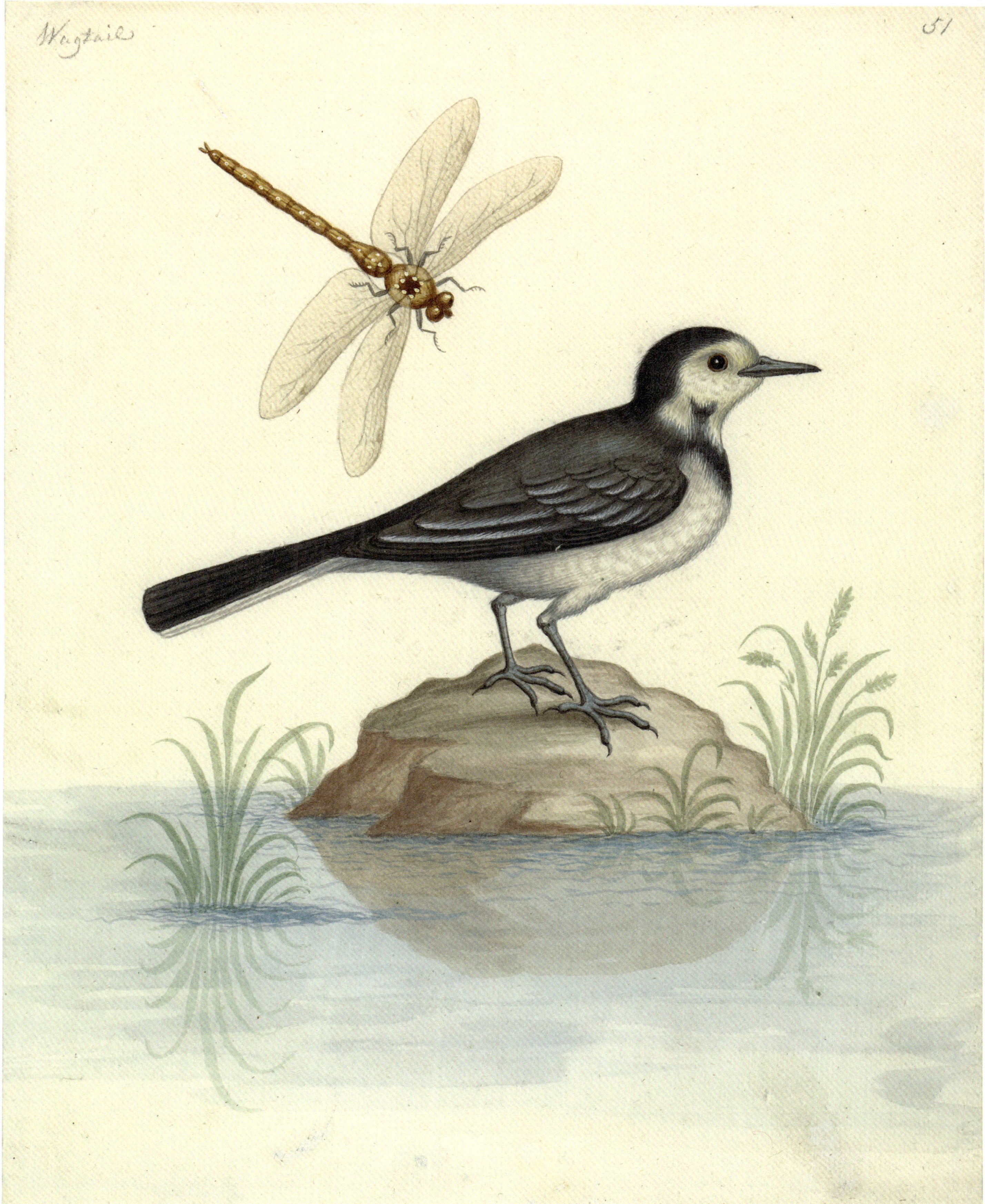
Wagtail
51

Close Observation

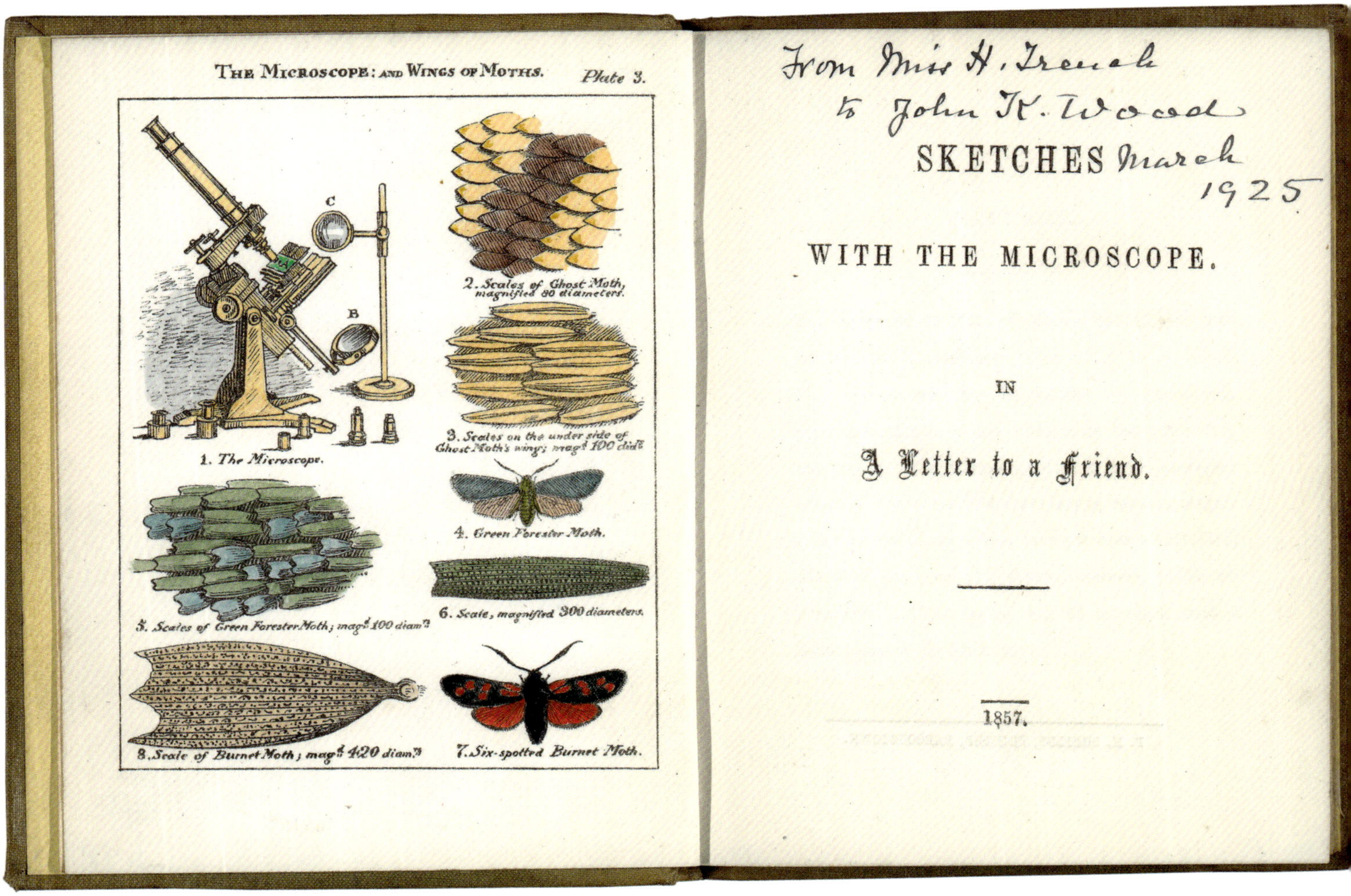

Mary Ward, *Sketches with the Microscope*, Parsontown, Ireland: F. H. Shields, 1857, hand-colored lithograph. Yale Center for British Art, Paul Mellon Collection

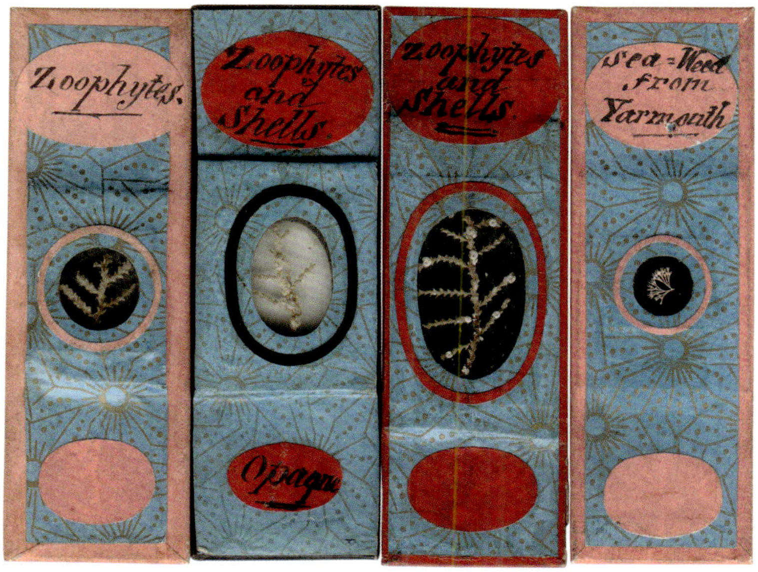

Microscope slides (zoophytes, shells, and seaweed from Yarmouth), ca. 1860, decorated paper, glass, pen and ink, with natural specimens. Division of Historical Scientific Instruments, Peabody Museum of Natural History, Yale University

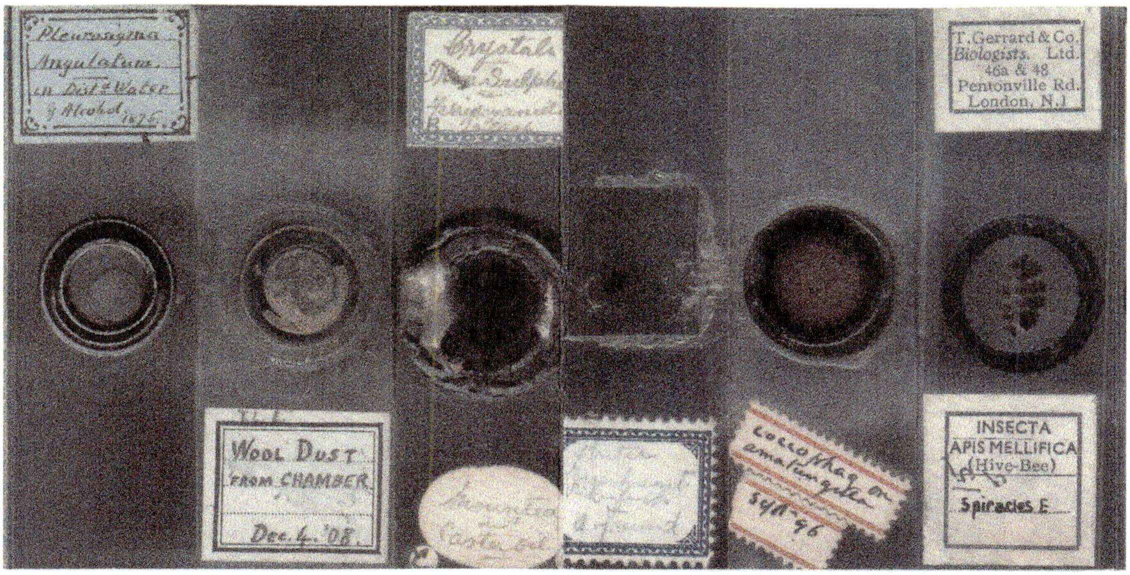

Sarah Bodman, *The Secret Garden*, Bristol, 2001, digital print. Yale Center for British Art, Friends of British Art Fund

OVERLEAF: Nicholas Phillips, *You're Never Alone in the Ocean: Life of Plankton*, London, 1981, pen and ink, glass, color etching, and blind relief etching. Yale Center for British Art, Friends of British Art Fund

SECOND OVERLEAF: William Spooner, *Spooner's Transformations, No. 2: The Microscope*, London: W. Spooner, 1838–1839, mounted hand-colored lithograph (second view showing transformation). Yale Center for British Art, Paul Mellon Collection

You're Never Alone In The Ocean   Life Of Plankton   9

Sample		Locality	Depth m.	Time	Temp. °C	Salinity ‰	Sex	Mating cycle	Size μ	Reliefograph® mag.	Hardy The Open Sea I.
1	*Luminosity : Noctiluca scintillans*	5.06 N 99.56 E	3	11.00 pm	20	34		33 hrs	44		Plate IIIa
2	*Transparency : Incognita felicitans*	24.43 S 43.53 W	444	11.00 am	21	32			77		
I	*Fluency : Lingua elegans*	33.33 N 180. E	4	7.00 pm	18	33	M	10-15 sec.		x 12	Plate IXc
3	*Thermal motions in the oceans*	37.39 N 160.21 E			7-23						p. 13 ff.
4	*Frantic in the Atlantic*	38.38 N 39.38 W	1	10.00 pm	12	31		3 sec.	350		
5	*Passive in the Pacific*	19.56 N 162.12 W	432	3.00 am	22	33		3 days.	25		
II	*Cheek to cheek off Martinique*	14.29 N 60.55 W	123	8.30 pm	22	32		2. hrs.		x 25	Plate XIII
6	*Amnesia in Indonesia*	8.32 S 104.31 E	57	5.00 am	23	34			115		
III	*Diatoms : floaters in the waters*	26.21 N 16.21 W	1-12	2.00 am	24	31		1-72 hrs.		x 75	Plate I a+b.
7	*Gregarious off the Galapagos*	0.07 S 90.30 W	32	8.00 pm	21	31		2 min.	320		
IV	*C. seen in the Caribbean*	16.56 N 79.19 W	9	10.00 am	23	33	F	10-15 sec.	38	x 15	Plate IV a+d.
8	*Ageing in the Aegean*	39.42 N 24.21 E	70	11.00 pm	20	31		12 yrs.	3		
9	*Capricious in Mauritius*	21.03 S 57.59 E	4	2.32 am	21	34		2.3 sec.	420		
10	*Ablutions in the Aleutians*	53.53 N 173.10 W	200	7.45 pm	7	33		24 hrs.	440		
V	*Reproducin' in the Aleutians*	51.03 N 175.11 W	201	3.00 pm	11	32		0.5 sec.		x 650	fig. 13.
11	*off Goa*	15.23 N 22.27 E	2	3.01 pm	24	35		0.25 sec.	38		nicholas phillips
12	*Dodecanese, do what you please*	36.37 N 26.27 E	7	12.00 pm	20	33		0.025 sec.	15-60		

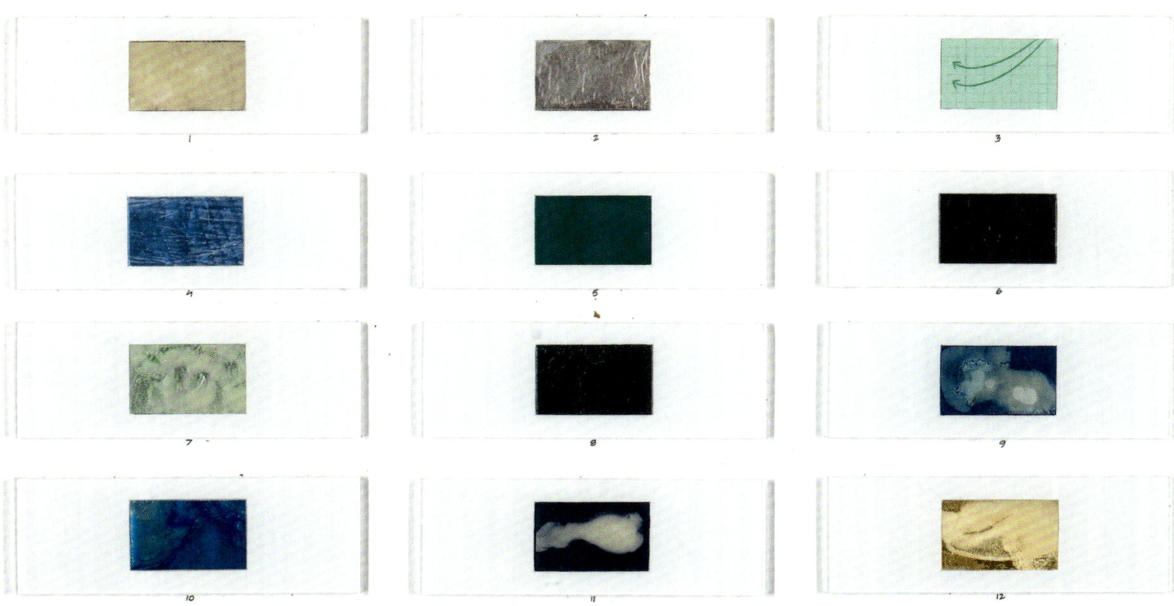

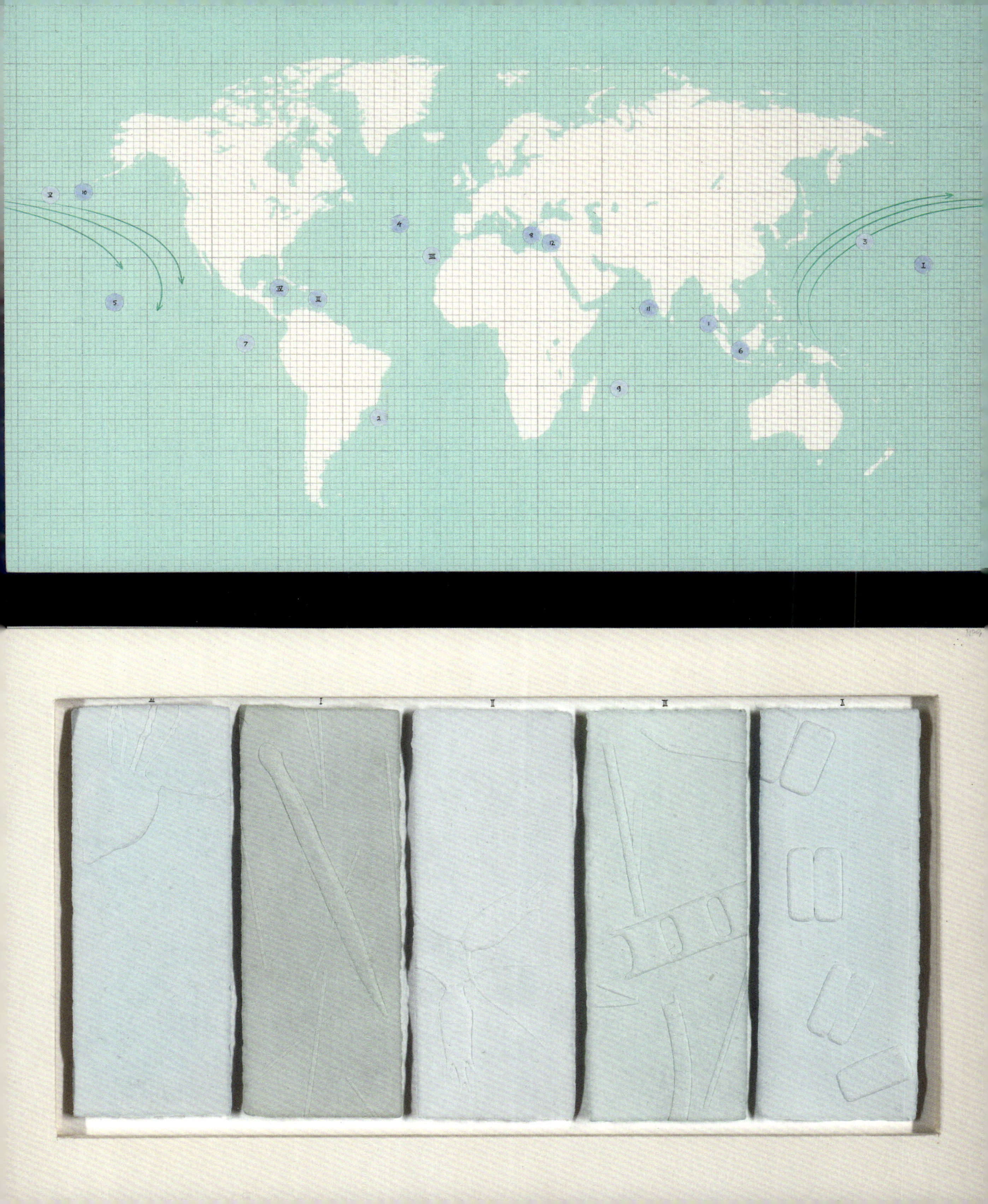

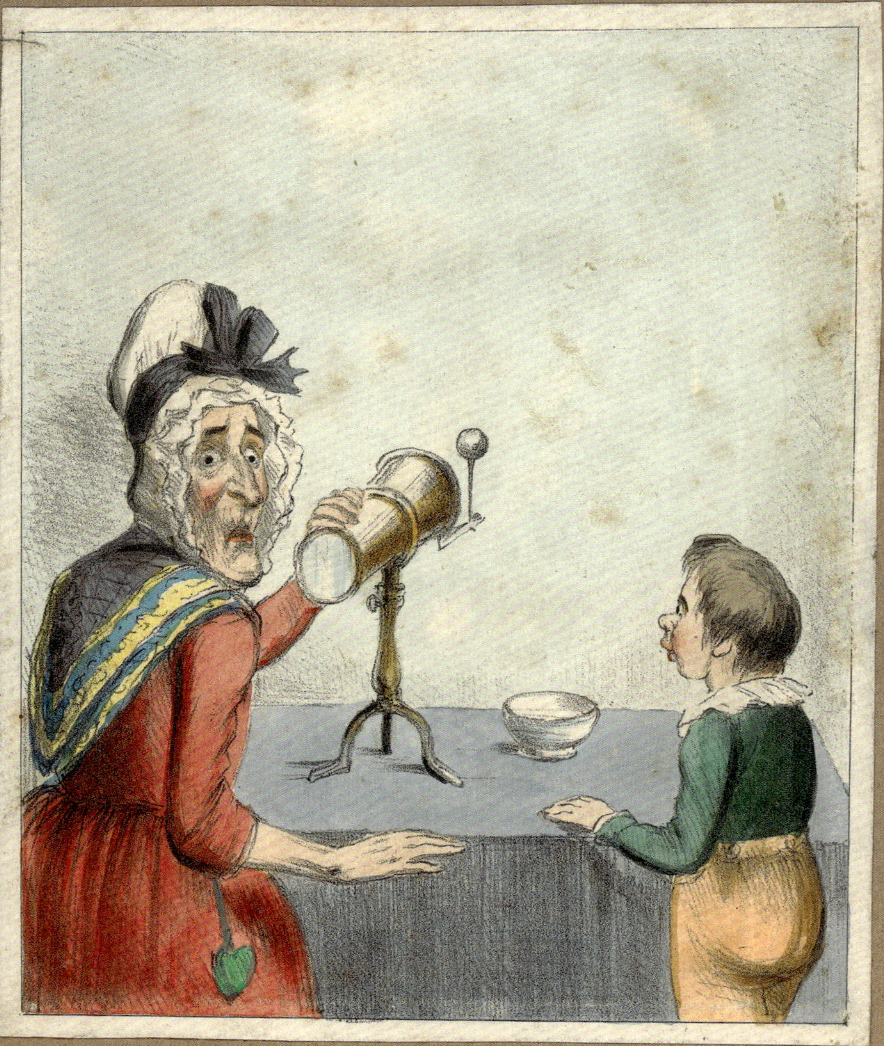

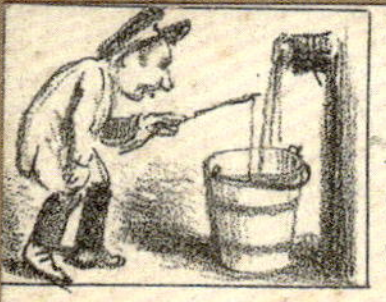

SPOONER'S TRANSFORMATIONS, No. 2

THE MICROSCOPE.

Displaying to the horror stricken old woman the wonderful inhabitants of a Drop of Water.

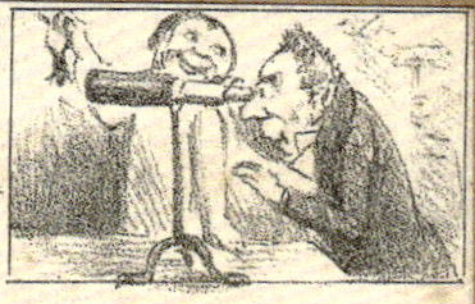

LONDON: WILLIAM SPOONER 377 STRAND

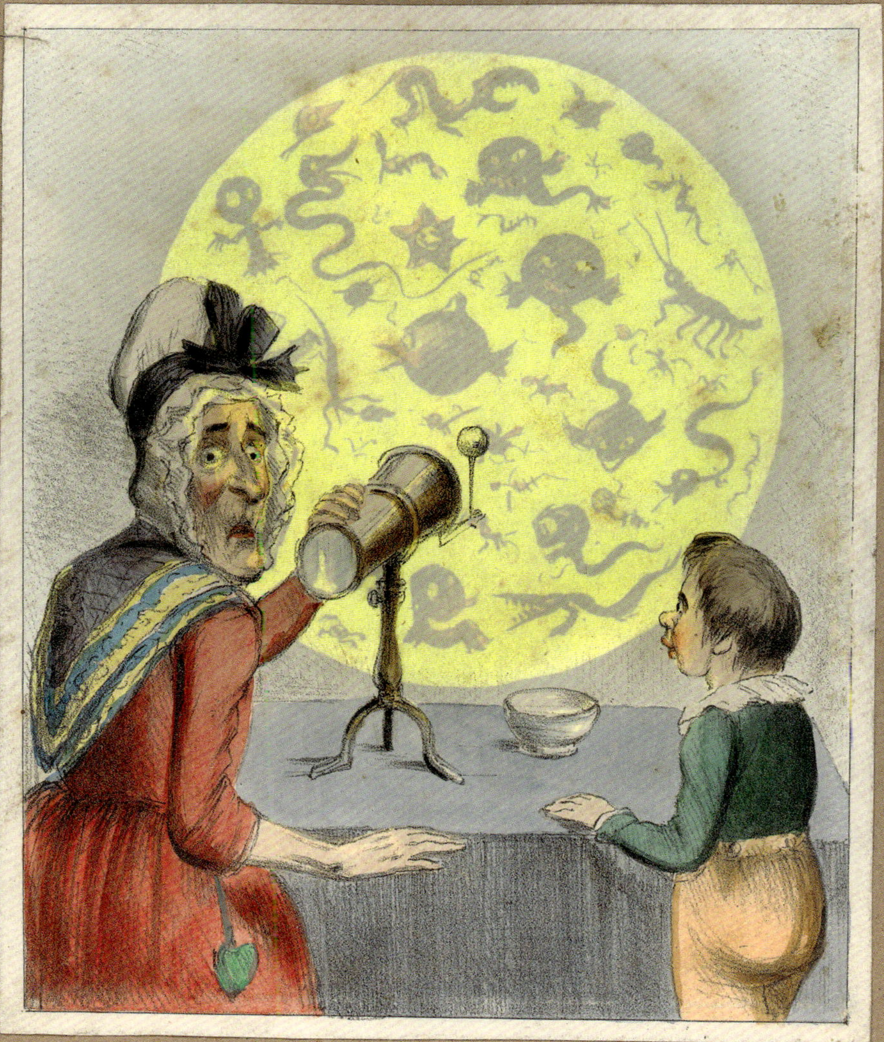

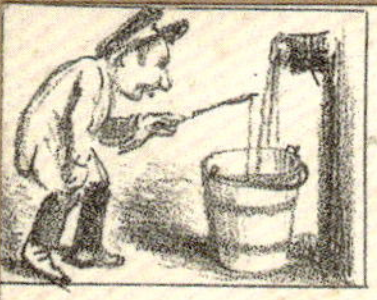

SPOONER'S TRANSFORMATIONS, No. 2

THE MICROSCOPE.

Displaying to the horror stricken old woman the wonderful inhabitants of a Drop of Water.

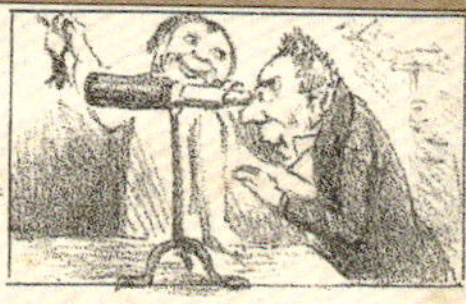

LONDON: WILLIAM SPOONER 377 STRAND

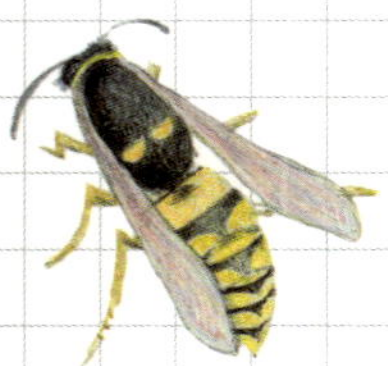

Chronological List of Works in the Exhibition

1500

Helmingham Herbal and Bestiary, ca. 1500, watercolor and gouache on parchment. Yale Center for British Art, Paul Mellon Collection

1600

Simon Basil, *Survey of Hertford Castle*, 1608, pen and ink, watercolor, and gouache on parchment. Yale Center for British Art, Paul Mellon Collection

Edward Topsell, *The History of Four-Footed Beasts and Serpents*, London: E. Cotes, for G. Sawbridge, 1658, wood cut. Yale Center for British Art, Paul Mellon Collection

John Dunstall, *The Therd Booke of Flowers Fruits Beastes Birds and Flies Exactly Drawne*, London: Peter Stent, 1661, etching. Yale Center for British Art, Paul Mellon Collection

Joshua Spencer, *Commonplace Book*, 1692–1729, pen and ink with watercolor. Archives of the Academy of Natural Sciences of Philadelphia, Drexel University (ANSP Coll. 426)

1700

Martin Engelbrecht, *Garden Pleasures*, Augsburg, ca. 1730, hand-cut paper with hand-colored engraving. Yale Center for British Art, Paul Mellon Collection

George Edwards, *The Long-Tailed Porcupine*, 1738, pen, gray ink, brown ink, brown wash, watercolor, and graphite. Yale Center for British Art, Paul Mellon Collection

George Edwards, *Pied Wagtail and Dragonfly*, ca. 1750, watercolor, with pen and ink, heightened with gouache, over graphite. Yale Center for British Art, Paul Mellon Collection

George Edwards, *Atlas Moth* and *Hercules Beetle* from *Album of Unpublished Watercolors for Gleanings of Natural History*, ca. 1758, watercolor with gouache. Yale Center for British Art, Paul Mellon Collection

Augustin Heckel, *The Florist, or, An Extensive and Curious Collection of Flowers, for the Imitation of Young Ladies, Either in Drawing, or in Needle-work*, London: Bowles and Son, 1759. Yale Center for British Art, Paul Mellon Collection

Thomas Atwood, *Kingfisher*, ca. 1761–1766, watercolor, gouache, and pen and brown ink. Yale Center for British Art, Paul Mellon Collection

Richard Brookes, *The Humble Bee and Insects* and *Fireflies*, 1760, pen and black ink, graphite, and gray wash. Yale Center for British Art, Paul Mellon Collection

Richard Brookes, *A New and Accurate System of Natural History*, London: J. Newbery, 1763. Beinecke Rare Book and Manuscript Library, Yale University

Peter Brown, *European Goldfinches (Carduelis carduelis) in a Snare*, 1766?, gouache on parchment. Yale Center for British Art, Paul Mellon Collection

Luigi Balugani, *Four Fish*, ca. 1767–1773, watercolor and graphite with pen and ink. Yale Center for British Art, Paul Mellon Collection

Luigi Balugani, *Fish, Zeezan Jidda*, ca. 1767–1773, watercolor and graphite with pen and ink. Yale Center for British Art, Paul Mellon Collection

James Bolton, Collection of drawings depicting specimens from the natural history cabinet of Anna Blackburne, ca. 1768, watercolor and gouache over graphite on parchment. Yale Center for British Art, Paul Mellon Fund, in honor of Jane and Richard C. Levin, President of Yale University (1993–2013)

Unknown artist, *Yellow-eyed Babbler (Chrysomma sinense) perched on Chinese hat plant (Holmskioldia sanguinea)*, India, ca. 1770, watercolor, heightened with gouache, with pen and ink over graphite. Yale Center for British Art, Paul Mellon Collection

Ann Lee, *Jatropha multifida*, 1771, watercolor and gouache with pen and brown ink over graphite on parchment. Yale Center for British Art, Paul Mellon Collection

George Humphrey, *Collecting and Preserving all Kinds of Natural Curiosities*, 1776, pen and ink with watercolor. Archives of the Academy of Natural Sciences of Philadelphia, Drexel University (ANSP Coll. 371)

John Bewick, *Magpie*, ca. 1777–1782, watercolor, graphite, and pen and black ink. Yale Center for British Art, Paul Mellon Collection

OPPOSITE: Mary Leslie, Wasp in *Sketchbook* (detail), ca. 1890, watercolor, graphite, and pen and ink. Yale Center for British Art, Paul Mellon Collection

Withering Botanical Microscope, London: William Withering, ca. 1785, brass and glass, in leather case. Division of Historical Scientific Instruments (YPM 040878), Peabody Museum of Natural History, Yale University

John Raphael Smith, *William Curtis (1746–1799), Entomologist and Botanist*, ca. 1787, pastel and gouache. Yale Center for British Art, Given by Lowell Libson in honor of Mrs. Rachel Lambert Mellon and the Oak Spring Garden Library

William Withering, *A Botanical Arrangement of British Plants*, Birmingham: Printed by M. Swinney for G. G. J. & J. Robinson, 1787–1792. Beinecke Rare Book and Manuscript Library, Yale University

William Lewin, *Common Hoopoe*, ca. 1789, watercolor and gouache over graphite. Yale Center for British Art, Paul Mellon Collection

William Booth Grey, *Ranunculus platanifolius ("Plane leav'd crowfoot")*, ca. 1790, hand-cut paper collage on laid paper prepared with watercolor. Yale Center for British Art, Paul Mellon Fund

William Booth Grey, *Lysimachia fenella / Viola neapolitan)*, ca. 1790, hand-cut paper collage on laid paper prepared with watercolor. Yale Center for British Art, Paul Mellon Fund

James Sowerby, *Bird egg, Song Thrush (Turdus philomelos)*, ca. 1790, watercolor over graphite. Yale Center for British Art, Paul Mellon Collection

Possibly by James Sowerby, *Head of a Swan* and *Three Swans*, ca. 1790, graphite. Yale Center for British Art, Paul Mellon Collection

Erasmus Darwin, *The Botanic Garden; A Poem, in Two Parts. Part I. Containing The Economy of Vegetation. Part II. The Loves of the Plants*, London: Printed for J. Johnson, 1791. Yale Center for British Art, Paul Mellon Fund

Naturalist's Pocket Microscope, ca. 1795, brass and ivory, in leather case. Division of Historical Scientific Instruments (YPM 040876), Peabody Museum of Natural History, Yale University

Thomas Bewick, *History of British Birds*, Newcastle upon Tyne: Printed by S. Hodgson, 1797 [i.e. 1798]–1804. Yale Center for British Art, Paul Mellon Collection

1800

William Fordyce Mavor, *The Lady's and Gentleman's Botanical Pocket Book*, London: Vernor & Hood, 1800. Beinecke Rare Book and Manuscript Library, Yale University

Album of British Grasses, England, ca. 1800, natural specimens with pen and ink labels. Yale Center for British Art, Paul Mellon Fund

Henry Kingsbury after Martin Hoffman, *Linnaeus in Laplander Costume*, 1805, mezzotint. Archives of the Academy of Natural Sciences of Philadelphia, Drexel University (ANSP Coll. 2009–005)

Album of Drawings of English Moths, Butterflies, Flowers, and Mollusks, 1805–1822, pen and ink and watercolor. Yale Center for British Art, Paul Mellon Fund

John Galpine, *A Synoptical Compend of British Botany*, Salisbury: Printed by B.C. Collins, for S. Bagster and J. White, 1806. Yale Center for British Art, Paul Mellon Collection

Thomas Rowlandson, *Butterfly Hunting*, ca. 1806, watercolor with pen and gray ink over graphite. Yale Center for British Art, Paul Mellon Collection

John Atkinson, *Sixteen Scenes taken from the Miseries of Human Life*, London: Wm. Miller, 1807, hand-colored etching. Yale Center for British Art, Paul Mellon Collection

James Sowerby, *A Botanical Drawing-Book, or, An Easy Introduction to Drawing Flowers According to Nature*, London: Richard Taylor & Co., 1807. Yale Center for British Art, Paul Mellon Collection

Robert John Thornton, *A New Family Herbal, or, Popular Account of the Natures and Properties of the Various Plants Used In Medicine, Diet, and the Arts,* London: Richard Phillips, 1810. Yale Center for British Art, Paul Mellon Collection

Mary Ann Burges, *Lepidoptera of the British Isles*, ca. 1810, watercolor and gouache, with pen and ink. Yale Center for British Art, Paul Mellon Fund

Thomas Bewick, *A Cabinet of Natural History: Natural History of Fishes; Natural History of Reptiles, Serpents, and Insects; Natural History of British Birds*, Alnwick: Printed and Sold by W. Davison, 1814. Yale Center for British Art, Paul Mellon Collection

Character of Forest Trees, ca. 1817, pen and ink. Yale Center for British Art, Paul Mellon Fund

George Graves, *The Naturalist's Pocket-Book, or, Tourist's Companion, Being a Brief Introduction to the Different Branches of Natural History, with Approved Methods for Collecting and Preserving the Various Productions of Nature*, London: Longman, 1818. Yale Center for British Art, Paul Mellon Fund

George Cruikshank, *Landing the Treasure, or, The Results of the Polar Expedition!!!*, London, 1819, etching. Private collection

Botanical Field Microscope, 1820, brass and glass, in wood box. Division of Historical Scientific Instruments (YPM 040874), Peabody Museum of Natural History, Yale University

Cut-Paper Basket, England, ca. 1820, hand-cut paper with watercolor, stitched with silk thread. Yale Center for British Art, Paul Mellon Fund

Flower Vase, England, ca. 1820, quillwork. Yale Center for British Art, Paul Mellon Fund

Album of Watercolors of British Birds, ca. 1820–1830, watercolor with pen and ink labels. Yale Center for British Art, Paul Mellon Fund

Musée des dames et des demoiselles [Les oiseaux, Les insectes, Les papillons, Les fleurs, Les fruits, Les minéraux et les coquillages], Paris: Chez Marcilly Ainé, ca. 1825, set of 6 books in decorated box. Yale Center for British Art, Paul Mellon Collection

Sarah Mary Fitton, *Conversations on Botany*, London: Longman, Rees, Orme, Brown, and Green, 1828. Yale Center for British Art, Paul Mellon Collection

John Leonard Knapp, *The Journal of a Naturalist*, London: John Murray, 1829. Yale Center for British Art, Paul Mellon Collection

Ranunculus acris (Tall Buttercup), collected by Gottlieb Wilhelm Bischoff (1797–1854) in the area of Heidelberg, Germany, 1829, natural specimen mounted on paper. Division of Botany (YU.066744), Peabody Museum of Natural History, Yale University

Nathaniel Whittock, *The Art of Drawing and Colouring from Nature: Flowers, Fruit, and Shells*, London: Isaac Taylor Hinton, 1829. Yale Center for British Art, Paul Mellon Collection

Commonplace Book, with Drawings of Insects, ca. 1830, pen and ink with watercolor. Yale Center for British Art, Paul Mellon Fund

Drum Microscope, England, ca. 1830, brass, glass, and ivory, in mahogany box. Division of Historical Scientific Instruments (YPM 040844), Peabody Museum of Natural History, Yale University

Fossils and Shells, collected by Etheldred Benett at Bradford, Wiltshire, before 1831, natural specimens mounted on boards. Academy of Natural Sciences of Philadelphia, Drexel University

Ellen W., *Album of Cut-Paper Flowers*, ca. 1835, hand-cut paper. Yale Center for British Art, Paul Mellon Fund

James Duncan, *The Natural History of British Moths, Sphinxes, &c.*, *Naturalist's Library*, edited by Sir William Jardine, vol. 4, Edinburgh: W.H. Lizars, 1836. Beinecke Rare Book and Manuscript Library, Yale University

William Spooner, *Spooner's Transformations, No. 2: The Microscope*, London: William Spooner, 1838–1839, hand-colored lithograph. Yale Center for British Art, Paul Mellon Collection

Abel Ingpen, *Instructions for Collecting, Rearing, and Preserving British and Foreign Insects*, London: W. Smith, 1839. Yale Center for British Art, Paul Mellon Collection

James Francis Stephens, *Haustellata Lepidoptera Diurna*, ca. 1839, pen and ink with watercolors. Yale Center for British Art, Paul Mellon Fund

Children's Magical Flowers, ca. 1840, cut and folded paper with watercolor and pen and ink. Yale Center for British Art, Paul Mellon Fund

Specimens of Sea Weed, ca. 1840, natural specimens and hand-cut paper, with pen and ink labels. Yale Center for British Art, Paul Mellon Fund

John Lindley, *Ladies' Botany, or, A Familiar Introduction to the Study of the Natural System of Botany*, London: Ridgway, ca. 1842. Yale Center for British Art, Paul Mellon Collection

Silene maritima (Sea campion), collected by John Stuart Mill, at Southampton Water, England, August, 1844, natural specimens mounted on paper. Academy of Natural Sciences of Philadelphia, Drexel University

Design for Embroidered Waistcoat, Britain or France, ca. 1845, watercolor and gouache on black coated paper. Yale Center for British Art, Paul Mellon Collection

Madame Le Lacheur, *Novel and Interesting Exhibition of Fancy Shell-Work!*, London, ca. 1845. Yale Center for British Art, Paul Mellon Fund

William Gardiner, *Twenty Lessons on British Mosses, or, First Steps to a Knowledge of That Beautiful Tribe of Plants*, Dundee, Scotland: Printed by J. Duff, 1846, with natural specimens. Yale Center for British Art, Paul Mellon Collection

Peter De Wint, *Ash Tree in Ashstead Park*, ca. 1847, black chalk, heightened with white. Yale Center for British Art, Paul Mellon Collection

William Henry Hunt, *Apples, Grapes, and a Cob-Nut*, ca. 1850, watercolor and gouache over graphite. Yale Center for British Art, Paul Mellon Collection

Papaver rhoeas (Corn Poppy), collected by Gottlieb Wilhelm Bischoff (1797–1854) or colleague, somewhere in Europe before 1850, natural specimen mounted on paper. Division of Botany (YU.066743), Peabody Museum of Natural History, Yale University

Anna Elizabeth, *Vase of Flowers*, Boston: G. W. Cottrell, 1851, miniature book, with needlepoint cover by Sue Bakker, Peebles, Scotland, 1993, silk thread on sixty-count gauze fabric. Collection of Neale and Margaret Albert

Mary Leslie, Letter to Charles Robert Leslie, 9 June 1852, pen and ink, with watercolor. Yale Center for British Art, Paul Mellon Collection

Anne Pratt, *Wild Flowers*, London: Society for Promoting Christian Knowledge, 1852–1853. Yale Center for British Art, Paul Mellon Collection

Henry Gardiner Adams, *Nests and Eggs of Familiar British Birds: Described and Illustrated, with an Account of the Haunts and Habits of the Feathered Architects and Their Times and Modes of Building*, London: Groombridge and Sons, 1855. Yale Center for British Art, Paul Mellon Collection

Henry Bradbury, *Allosorus crispus* from Thomas Moore's *The Ferns of Great Britain and Ireland*, London: Bradbury and Evans, 1855, colored nature print. Private collection

Samuel Palmer, *Brambles*, 1856, graphite. Yale Center for British Art, Paul Mellon Collection

Album of Dried Flower and Seaweed Specimens, 1856–1863, natural specimens with pen and ink. Yale Center for British Art, Paul Mellon Fund

Mary Ward, *Sketches with the Microscope*, Parsontown, Ireland: F. H. Shields, 1857, hand-colored lithographs. Yale Center for British Art, Paul Mellon Collection

Flora: An Album of Pressed Floral Arrangements, ca. 1860, natural specimens. Yale Center for British Art, Paul Mellon Fund

The Language of Flowers, England, ca. 1860, chromolithograph, cut and folded paper, and envelope. Yale Center for British Art, Paul Mellon Fund

Microscope slides (zoophytes, shells, and seaweed from Yarmouth), ca. 1860, decorated paper, glass, pen and ink, with natural specimens. Division of Historical Scientific Instruments, Peabody Museum of Natural History, Yale University

Henry Seebohm, *Ornithological Notebook and Diary*, Sheffield?, 1860–1866, pen and ink, with natural specimens. Yale Center for British Art, Paul Mellon Fund

Phebe Lankester, *Wild Flowers Worth Notice*, London: Robert Hardwicke, 1861. Yale Center for British Art, Paul Mellon Collection

Miss Rowe, *Collection of Botanical Specimens*, Liverpool, 1861, natural specimens in envelopes with watercolors, housed in mahogany box. Yale Center for British Art, Paul Mellon Fund

John Gould and Henry Constantine Richter, *Rook (Corvus frugilegus)*, ca. 1862, for John Gould's *Birds of Great Britain*, London: Taylor, 1873, watercolor and gouache. Private collection

Joseph Merrin, *Butterflying with the Poets: A Picture of the Political Aspect of Butterfly Life*, London: Longman, Green, Longman, Roberts, and Green, 1864, nature prints. Yale Center for British Art, Paul Mellon Collection

Cut-Paper Featherwork Illustrations of Gould's Birds, ca. 1865, hand-cut paper, with watercolor and pen and black ink, in handmade moiré-patterned and decorated paper box. Yale Center for British Art, Paul Mellon Fund

John George Wood, *Common Objects of the Microscope*, London: George Routledge and Sons, 1865. Yale Center for British Art, Paul Mellon Collection

If Truest Love Can Gain Respect, London: De La Rue, ca. 1866, lace-paper valentine, with hand coloring, pen and ink, and natural specimen. Yale Center for British Art, Paul Mellon Collection

William Houghton, *Country Walks of a Naturalist, with his Children*, London: Groombridge and Sons, 1869. Yale Center for British Art, Paul Mellon Collection

Album of Comic Sketches, England, 1869–1873, pen and ink. Yale Center for British Art, Paul Mellon Fund

The Art of Modelling Flowers in Paper to Imitate Nature, London: Hutton & Co., ca. 1870. Yale Center for British Art, Paul Mellon Fund

William Houghton, *Sea-Side Walks of a Naturalist with his Children*, London: Groombridge and Sons, 1870. Yale Center for British Art, Paul Mellon Collection

Shirley Hibberd, *The Seaweed Collector, a Handy Guide to the Marine Botanist, Suggesting What to Look for, and Where to Go, in the Study of the British Algæ, and the British Sponges*, London: Groombridge and Sons, 1872. Yale Center for British Art, Paul Mellon Collection

Mary Gibbs Shapter, *Notes on Various Trees*, England, 1879–1880, pen and ink, graphite, and natural specimens. Yale Center for British Art, Paul Mellon Fund

Album of Bird Illustrations, Made from Feathers, Great Britain, ca. 1880, natural specimens, with watercolor. Yale Center for British Art, Paul Mellon Fund

Robert Brendel, Model of *Brassica napus*, Breslau, ca. 1880, papier-mâché, wood, and plaster. Yale Center for British Art, Paul Mellon Fund

Drum Microscope, France, ca. 1880, brass and glass, in wood box. Division of Historical Scientific Instruments (YPM 041015), Peabody Museum of Natural History, Yale University

Album of Pressed Wild Flowers, 1882, natural specimens with pen and ink labels, Yale Center for British Art, Paul Mellon Fund

Frederick Rhenius Coles, *This Book Contains Drawings of Marine Shells (from Forbes & Hanley) Generally Considered as Specifically British*, Tongay, Scotland, 1884, graphite, pen and ink, with natural specimens. Yale Center for British Art, Paul Mellon Fund

*Taraxacum officinale (*Dandelion*)*, collected by Francis Wilson Starmer in Glastonbury, Connecticut, 1884, natural specimen mounted on paper. Division of Botany (CBS.35636), Peabody Museum of Natural History, Yale University

Mary Leslie, *Sketchbook*, ca. 1890, watercolor, graphite, pen and ink. Yale Center for British Art, Paul Mellon Collection

1900

Rex Vicat Cole, *The Artistic Anatomy of Trees*, London: Seeley, 1916, with original drawings and photographs. Private collection

Robert Sargent Austin, *Autumn*, 1922, etching and drypoint. Yale Center for British Art, Yale Art Gallery Collection, The G. Allen Smith Collection

Robert Sargent Austin, *Early Spring, Gloucestershire*, 1922, etching and drypoint. Yale Center for British Art, Yale Art Gallery Collection, The G. Allen Smith Collection

Gilbert White, *The Natural History of Selborne,* with wood engravings by Eric Fitch Daglish, New York: E. P. Dutton & Co., 1930. Yale Center for British Art, Paul Mellon Collection

Agnes Miller Parker, *Foxglove, Honeysuckle, and Fronds*, 1936, wood engraving. Collection of David Burnett

Herbert Ernest Bates, *Through the Woods: The English Woodland, April to April*, with wood engravings by Agnes Miller Parker, New York: Macmillan, 1936. Yale Center for British Art, Gift of Janet and James Sale, Yale BA 1960

Tom Eckersley, *Watch or Catch*, London: London Transport, 1954, lithograph. Yale Center for British Art, Gift of Henry S. Hacker, Yale BA, 1965

Robert Tavener, *Country Walks: Summer*, London: London Transport, 1958, lithograph. Yale Center for British Art, Gift of Henry S. Hacker, Yale BA, 1965

Ian Hamilton Finlay, *From an Inland Garden,* Dunsyre, Scotland: Wild Hawthorn Press, 1971. Yale Center for British Art, Friends of British Art Fund

Thomas A. Clark and Laurie Clark, *Gatherings*, Nailsworth, Gloucestershire: Moschatel Press, 1978, line block engraving. Yale Center for British Art, Friends of British Art Fund

Thomas A. Clark and Laurie Clark, *Proverbs of the Meadow*, Nailsworth, Gloucestershire: Moschatel Press, 1979, line block engraving. Yale Center for British Art, Friends of British Art Fund

Thomas A. Clark and Laurie Clark, *Short Tour of the Highlands,* Nailsworth, Gloucestershire: Moschatel Press, 1979, line block engraving. Yale Center for British Art, Friends of British Art Fund

Richard Long, *A Walk Past Standing Stones*, London: John Roberts Press for Anthony d'Offay, 1979. Collection of Bryan Fuermann

George Tute, *Bather in Landscape*, ca. 1979, wood engraving. Collection of David Burnett

Thomas A. Clark, *Proverbs of the Mountain,* Nailsworth, Gloucestershire: Moschatel Press, 1980, line block engraving. Yale Center for British Art, Friends of British Art Fund

John Heath-Stubbs, *Buzz Buzz: Ten Insect Poems,* with wood engravings by Richard Shirley Smith, Sidcot, Somerset, England: Gruffyground Press, 1981. Yale Center for British Art, Gift of Roger W. Moss

Nicholas Phillips, *You're Never Alone in the Ocean: Life of Plankton*, London, 1981, pen and ink, glass, and color etching. Yale Center for British Art, Friends of British Art Fund

Hamish Fulton, *Camp Fire*, Eindhoven: Stedelijk Van Abbemuseum, ca. 1985. Yale Center for British Art, Friends of British Art Fund

Hamish Fulton, *Fourteen Coast to Coast Walks, British Isles, 1971–87* and *Seven Winds* from *Fourteen Works*, offset lithographs, London: Paragon Press, 1982–1989. Yale Center for British Art, Friends of British Art Fund

Colin Hall, *Garsington Manor*, London: New Pyramid Press, 1988, hand-colored linocuts. Yale Center for British Art, Friends of British Art Fund

Robert Maude, *English Oak Tree and Wheat Field*, London, 1988. Yale Center for British Art, Friends of British Art Fund

Donald Myall, *Blackbird*, 1988, wood engraving. Collection of David Burnett

Donald Myall, *Hedgehog*, 1988, wood engraving. Collection of David Burnett

Ken Campbell, *Father's Garden*, Oxford, 1989, letterpress, printed in multi-color and metallic inks. Yale Center for British Art, Friends of British Art Fund

Colin Sackett, *Black Bob*, London: Coracle Press, 1989, line block engraving, with original line block. Yale Center for British Art, Friends of British Art Fund

Rebecca Salter, *Lake District Sketchbook*, 1990, pastels. Yale Center for British Art, Gift of the artist in memory of her father Samuel Salter (1922–2005)

Sister Margaret Tournour, *Dragonfly*, 1991, wood engraving. Collection of David Burnett

Ian Hamilton Finlay, Collection of pamphlets, invitation cards, and other pieces of ephemera, 1991–2000, Collection of Eileen Hogan and Cathy Courtney

Ken Campbell, *Skute Awabo*, London, 1992, letterpress, printed with colored and metallic inks. Yale Center for British Art, Friends of British Art Fund

Claire Lawson-Hall, *Wytham Stream*, Oxford: Alembic Press, 1993, hand-colored etchings. Yale Center for British Art, Friends of British Art Fund

David Burnett, *Twelve Poems,* with wood engravings by Sister Margaret Tournour, Bath: Old School Press, 1994. Collection of David Burnett

John Bevis, *An A–Z of Bird Song*, Docking, Norfolk: Coracle Press and St. Paulinus, 1995. Yale Center for British Art, Friends of British Art Fund

John Dilnot, *Good and Bad Apples,* Brighton, 1995. Yale Center for British Art, Friends of British Art Fund

Ron King, *Hick, Hack, Hock*, London: Circle Press, 1995, stone and blind intaglio relief. Yale Center for British Art, Friends of British Art Fund

Sister Margaret Tournour, *Dandelions*, 1995, wood engraving. Collection of David Burnett

Sister Margaret Tournour, *Prickles,* ca. 1995, wood engraving. Collection of David Burnett

Sister Margaret Tournour, *Teasel,* ca. 1995, wood engraving. Collection of David Burnett

Sister Margaret Tournour, *Caterpillars (Vapourer moth),* ca. 1995, wood engraving. Collection of David Burnett

Victoria Bean, *One Fox*, London, 1996, letterpress, screen print, and wood. Yale Center for British Art, Friends of British Art Fund

Sarah Bodman, *Flowers for Friends and Enemies (How to Say It with Flowers)*, Bristol, 1996, screen print, with natural specimens. Yale Center for British Art, Friends of British Art Fund

Childe Roland, *Elm the Elm = Fel llwyf Awelog*, Llangollen, Wales, 1996, inkjet print, with enameled bird lapel pin, in corrugated cardboard covers. Yale Center for British Art, Friends of British Art Fund

Thomas Joshua Cooper, *A Handful of Stones*, Norfolk: Coracle Press, 1996. Yale Center for British Art, Friends of British Art Fund

A Printmaker's Flora: An Anthology of the Names of British Wild Flowers, Illustrated by Twenty-one Printmakers, Dartington Hall, Devon: Dartington Printmakers, 1996. Yale Center for British Art, Friends of British Art Fund

Sister Margaret Tournour, *Paradise Tree*, 1996, wood engraving. Collection of David Burnett

Karen Bleitz, *Dolly*, London: Circle Press, 1997, fret-sawed painted wood and paper. Yale Center for British Art, Friends of British Art Fund

Mandy Bonnell, *Coral Aquarium*, poem by Gabriel Gbadamosi, London: New North Press, 1997, wood engravings, with original drawings and wood blocks. Yale Center for British Art, Friends of British Art Fund

David Burnett, *Six Poems*, with wood engravings by Sister Margaret Tournour, Durham: Black Cygnet Press, 1995. Collection of David Burnett

Tracey Bush, *The Pocket Book of Sea Creatures*, London, 1997, rubber stamping and postage stamps. Yale Center for British Art, Friends of British Art Fund

Tracey Bush, *Cygnea cantio: Swan Song*, London, 1997, blind relief print and rubber stamping. Yale Center for British Art, Friends of British Art Fund

Eileen Hogan, *Four Sketchbooks, Recording Little Sparta, Scotland*, 1997–2013, graphite, pen and ink, and watercolor, with natural specimens. Collection of the artist

Sarah Bodman, *Nature trail*, Bristol, 1998, screen prints, with natural specimens and found objects. Yale Center for British Art, Friends of British Art Fund

Ian Hamilton Finlay, *Little Sparta: der Garten = The Garden*, Dunsyre, Scotland: Wild Hawthorn Press for the Landesmuseum, 1998. Yale Center for British Art, Gift of Ian Hamilton Finlay and Pia Maria Simig

Ron King, *1/4 Hollow Log Book*, London: Circle Press, 1998, cut log, bound with linen. Yale Center for British Art, Friends of British Art Fund

Ron King, *Termite Grid*, London: Circle Press, 1998, cut and drilled log, bound with linen. Yale Center for British Art, Friends of British Art Fund

Ron King, *Tic Tac Toe*, London: Circle Press, 1998, cut and drilled log, bound with linen. Yale Center for British Art, Friends of British Art Fund

Rosalind Atkins, *Shelter*, 1999, wood engraving. Collection of David Burnett

Mark Dion and J. Morgan Puett, *The Ladies Field Club of York*, York: Impressions Gallery, 1999. Yale Center for British Art, Friends of British Art Fund

Anne Lydiat, *Without*, Sheffield, England, 1999, cut paper. Yale Center for British Art, Gift of the artist

2000

David Burnett, *Boxwood and Graver*, with wood engravings by Sister Margaret Tournour, Durham: Black Cygnet Press, 2000. Collection of David Burnett

Andy English, *Kingfisher at Bourn Brook*, 2000, wood engraving. Collection of David Burnett

Hamish Fulton, *Wild Life: Walks in the Cairngorms*, Edinburgh: Pocketbooks, Morning Star, 2000. Yale Center for British Art, Friends of British Art Fund

Marc Quinn, *Garden2*, London: Paragon Press, 2000, pigment print. Yale Center for British Art, Friends of British Art Fund

Sarah Bodman, *The Secret Garden*, Bristol, 2001, digital print. Yale Center for British Art, Friends of British Art Fund

Mandy Bonnell, *The Second Life of Shells*, with poem by Gabriel Gbadamosi, London: New North Press, 2001, etching. Yale Center for British Art, Friends of British Art Fund

Hans Waanders, *Perches*, Edinburgh: Morning Star, 2001. Yale Center for British Art, Friends of British Art Fund

Julie Cockburn, *Beetles Book*, England, 2002, altered book. Yale Center for British Art, Friends of British Art Fund

David Hackman, *Winter Skeletons*, 2002, wood engraving. Collection of David Burnett

Jane Hyslop, *Wild Plants Collected in Midlothian*, Midlothian, Scotland, 2002, etching. Yale Center for British Art, Friends of British Art Fund

Michael Landy, *Creeping Buttercup,* one of twelve prints from *Nourishment*, London: Paragon Press, 2002, etching. Yale Center for British Art, Friends of British Art Fund

Stuart Mugridge, *Seven Short Walks*, Ambleside, England: Grizedale Books, 2002, screen prints with magnifying glass. Yale Center for British Art, Friends of British Art Fund

Anne Rook, *The Book of 4021 & 4020: (Golden Delicious)*, London: MM Visual Catering, 2002, digital prints. Yale Center for British Art, Friends of British Art Fund

John Norris Wood, *An Alphabet in Praise of Frogs and Toads*, Church Hanborough, Oxford: Inky Parrot Press, 2002. Collection of David Burnett

Tracey Bush, *British Butterflies*, London, 2003, hand-cut paper, in album. Yale Center for British Art, Friends of British Art Fund

Tracey Bush, *Butterflies of Britain*, London, 2003, hand-cut paper, in wooden box. Yale Center for British Art, Friends of British Art Fund

John Dilnot, *A Little Book of Garden Birds*, Brighton, 2003. Yale Center for British Art, Friends of British Art Fund

Judith Rothchild, *Study of Two Pears*, Octon, France: Verdigris Press, 2003, mezzotints. Collection of David Burnett

Clare Bryan, *Beech*, England, 2004, hand-cut paper. Yale Center for British Art, Friends of British Art Fund

Tracey Bush, *British Butterflies: Museum Box, no. 1*, London, 2004, hand-cut paper specimens, mounted in glass-topped box. Yale Center for British Art, Friends of British Art Fund

Eileen Hogan, *Four Studies of Ian Hamilton Finlay*, 2004, oil on panel. Collection of the artist

Eileen Hogan, *Two Studies of Ian Hamilton Finlay*, 2004, oil on paper. Collection of the artist

Norman Ackroyd, *The Furthest Lands: Ten Etchings from the Edge of the British Isles*, London, 2005, aquatints. Yale Center for British Art, Gift of Jason Bacon Yale BA 1956

Susan Allix, *Silver*, London, 2005, miniature book, with sterling silver and wire. Collection of Neale and Margaret Albert

Tracey Bush, *Florilegium Sketchbook*, 2005, hand-cut paper packaging collage, with gold leaf, and pen and ink, with natural specimen. Collection of the artist

Tracey Bush, *Herbarium Notebook*, 2005, pen and ink, with natural specimens and hand-cut paper packaging collages. Collection of the artist

Julie Cockburn, *Flower Book: White Tea-Rose*, 2005, altered book with watercolor. Yale Center for British Art, Friends of British Art Fund

John Dilnot, *Bad Apples*, Brighton, 2005. Yale Center for British Art, Friends of British Art Fund

Jane Hyslop, *Cobweb I*, Midlothian, Scotland, 2005, digital print, hand-cut paper, and thread. Yale Center for British Art, Friends of British Art Fund

Paul Johnson, *Ursula's Garden*, Cheadle Hulme, England: Book Art Project, 2005, digital print, cut and folded paper. Yale Center for British Art, Friends of British Art Fund

Richard Long, *Untitled*, 2005, River Avon mud on black card. Collection of Bryan Fuermann

Andrew Norris, *POETree*, Zagreb, Croatia, 2005, natural specimens and rubber stamping. Yale Center for British Art, Friends of British Art Fund

Abigail Rorer, *Bufo the First* from *Of Woodland Pools, Spring-holes & Ditches* by Henry David Thoreau, Petersham, Massachusetts: The Lone Oak Press, 2005, relief engraving. Collection of David Burnett

Abigail Rorer, *Love Song* from *Of Woodland Pools, Spring-holes & Ditches* by Henry David Thoreau, Petersham, Massachusett: The Lone Oak Press, 2005, relief engraving. Collection of David Burnett

Abigail Rorer, *Skunk Cabbage* from *Of Woodland Pools, Spring-holes & Ditches* by Henry David Thoreau, Petersham, Massachusetts: The Lone Oak Press, 2005, relief engraving. Collection of David Burnett

Mandy Bonnell, *Antmothbeetlemillipedespider*, poems by Gabriel Gbadamosi, wood engravings, with original drawing, London: EMH Arts, 2006. Yale Center for British Art, Friends of British Art Fund

Clare Bryan, *Canopy (Sycamore, Beech, Horse Chestnut)*, 2006, hand-cut paper. Yale Center for British Art, Friends of British Art Fund

Tracey Bush, *Notebook*, 2006, pen and ink, collaged newspaper articles, sweet wrappers, and other pieces of ephemera. Collection of the artist

Tracey Bush, Herbarium sheets for *Nine Wild Plants: Dandelion, Common Poppy, Buttercup*, 2006, hand-cut paper packaging collages with pen and ink. Yale Center for British Art, Friends of British Art Fund

Tracey Bush, *Nine Wild Plants*, London, 2006, digital print, unique proof copy. Collection of the artist

Helen Douglas, *Field Notes*, Yarrow, Scotland: Weproductions, 2006, digital print, Yale Center for British Art, Friends of British Art Fund

Pauline Lamont-Fisher, *Daylight*, Wormelow, England, 2006, digital print. Yale Center for British Art, Friends of British Art Fund

Andrew Norris, *Walking Songs, 7 Walks, 7 Views*, Zagreb, Croatia, 2006, ink jet print. Yale Center for British Art, Gift of the artist

Lucy May Schofield, *Roadkill*, Manchester, 2006, ink jet print, with hand coloring. Yale Center for British Art, Friends of British Art Fund

Jem Southam, *Upton Pyne: January 1997*, diptych, printed 2006, chromogenic dye coupler prints. Yale Center for British Art, Purchased with contributions from family and friends in memory of Evelyn Silver (1925–2006), Yale Center for British Art docent

Clare Bryan, *Forest Islands*, London: CB Works, 2007, digital prints. Yale Center for British Art, Friends of British Art Fund

Chisato Tamabayashi, *Branches*, London, 2007, screen print, hand-cut and folded paper. Yale Center for British Art, Friends of British Art Fund

Lin Charlston, *Seeking Trees: Reproduction of an Altered Book*, England, 2008, facsimile of altered book. Yale Center for British Art, Friends of British Art Fund

John Dilnot, *64 of My Favourite Urban Trees*, Brighton, 2008. Yale Center for British Art, Friends of British Art Fund

John Dilnot, *Winter Buds*, Brighton, 2008, screen print. Yale Center for British Art, Friends of British Art Fund

Sarah Morpeth, *Crow Landscape*, Elsdon, Northumberland, 2008, hand-cut paper and wire. Yale Center for British Art, Friends of British Art Fund

Sarah Morpeth, *Torquil's Book*, Elsdon, Northumberland, 2008, hand-cut paper and wire. Yale Center for British Art, Friends of British Art Fund

Steve Perfect, *Wind Flowers*, London, 2008. Yale Center for British Art, Friends of British Art Fund

Diana Porteus, *Where the Bee Sucks*, Oxford, 2008, hand-cut and folded paper. Gift of David Burnett

Mercedes Baliarda, *They are Better Left Unspoken*, London, 2009, digital print on translucent paper. Collection of Otto Saumarez Smith

Mandy Bonnell, *Sketchbook of Moth Drawings*, 2009–2014, graphite. Collection of the artist

John Dilnot, *Weeds and Pests*, Brighton, 2009, screen print. Yale Center for British Art, Friends of British Art Fund

Jane Hyslop, *Herbarium series: Dandelion* and *Buttercup*, Midlothian, Scotland, 2009, hand-colored etchings and hand-cut and folded paper. Yale Center for British Art, Friends of British Art Fund

Ian Stephens, *Crossbill*, 2009, wood engraving, Collection of David Burnett

Eileen Hogan, *UL 238, FH 172, and FR 59 (Beehives)*, 2008, oil and charcoal on panel. Private collection

Paula Macgregor, *Hollyhock*, Colchester, Essex, 2009, digital prints, with natural specimens. Yale Center for British Art, Friends of British Art Fund

Gavin Pretor-Pinney, *The Cloud Collector's Handbook*, London: Sceptre, 2009. Yale Center for British Art, Friends of British Art Fund

Workshop of Tulsi Das Nimbark, *Hoopoe on a Branch*, Jaipur, India, 2009, gouache on printed leaf of text. Yale Center for British Art, Gift of Holly Shaffer

Mandy Bonnell, *Bethany Seed*, 2010, graphite. Collection of the artist

Mandy Bonnell, *Palm Seed (Shela, Lamu Island, Kenya)*, 2010, graphite. Collection of the artist

Mandy Bonnell, *Sketchbook Recording Shela, Lamu Island, Kenya*, 2010–2014, graphite. Collection of the artist

Laurie Clark, *100 Buttercups*, Fife, Scotland: WAX366, 2010, line block engravings, with colored pencil drawing. Yale Center for British Art, Friends of British Art Fund

Laurie Clark, *Warblers*, Pittenweem, Fife, Scotland: Moschatel Press, 2010, line block engraving. Yale Center for British Art, Friends of British Art Fund

Helen Douglas, *Poempondscroll*, Yarrow, Scotland: Weproductions, 2010, digital print. Yale Center for British Art, Friends of British Art Fund

Felicity Ford, *I Hear: Warborough, Overy, Dorchester, Shillingford: A Circular Walk Exploring the Landscape and the Ambient Sounds Surrounding a Section of the A4074 Road*, 2010, screen prints. Yale Center for British Art, Friends of British Art Fund

Jane Hyslop, *Collection III*, Midlothian, Scotland, 2010, digital print. Yale Center for British Art, Friends of British Art Fund

Liz Machin, *Working Class Botanists*, Manchester, 2010, inkjet print, with original photograph. Yale Center for British Art, Friends of British Art Fund

Made Collective, *Surplus to Requirement: the Herbarium Special Edition Box*, Manchester, 2010, with cut and folded paper, postage stamps, photographic negative, ribbon, lace, rubber band, natural specimens, and DVD, in glass-topped specimen box. Yale Center for British Art, Friends of British Art Fund

Colin See-Paynton, *Twitter of Linnets*, Powys, Wales, 2010, wood engraving. Collection of David Burnett

Chisato Tamabayashi, *Plunge*, London, 2010, screen print, hand-cut paper, and thread. Yale Center for British Art, Friends of British Art Fund

Charlotte Verity, *Emerging Snowdrops*, 2010, oil on canvas, with etching, manuscript notebooks, sketchbooks, and drawings. Collection of the artist

Jane Wildgoose, *Shellflower*, London, 2010, shellwork flower made with white clam shells, in glass box. Yale Center for British Art, Purchased with a gift from Mark and Adam Eaker, Yale BA 2007, in memory of Charles A. Ryskamp, Yale PhD 1956

Mandy Bonnell, *Moth (Shela, Lamu Island, Kenya)*, 2011, graphite. Collection of the artist

Helen Douglas. *Leda and the Swan*, Yarrow, Scotland: Weproductions, 2011, digital print on folded paper. Yale Center for British Art, Friends of British Art Fund

Sergio Galaviz and Katriona Chapman, *The Garden*, London: Tomatito Press, 2011. Yale Center for British Art, Friends of British Art Fund

Pauline Lamont-Fisher, *Sturgeon*, Wormelow, England, 2011, inkjet print. Yale Center for British Art, Friends of British Art Fund

Angela Lemaire, *Swan*, 2011, wood engraving. Collection of David Burnett

Sarah Morpeth, *Circle Concertina Book*, Elsdon, Northumberland, 2011, hand-cut paper and thread. Yale Center for British Art, Friends of British Art Fund

Sarah Morpeth, *Crow*, Elsdon, Northumberland, 2011, cut and folded paper, thread, and digital prints. Yale Center for British Art, Friends of British Art Fund

Sarah Morpeth, *Owl*, Elsdon, Northumberland, 2011, cut and folded paper and digital prints. Yale Center for British Art, Friends of British Art Fund

Jonathan Raisin and Elizabeth Willow. *Making Time: Thoughts from a Garden, a Series of 50*, Lancaster, England, 2011. Yale Center for British Art, Friends of British Art Fund

Colin Sackett, *Observer's Marginalia*, Axminster, Devon, 2011. Yale Center for British Art, Friends of British Art Fund

Ian Whitmore, hand-cut paper toad for *Othello: A Bestiary (with Floral Additions)*, Brighton: Borbonesa, 2011, screenprint. Yale Center for British Art, Friends of British Art Fund

Mandy Bonnell, *Bethany Seeds*, with poem by Gabriel Gbadamosi, London: EMH Arts, 2012, wood engraving. Yale Center for British Art, Friends of British Art Fund

Mandy Bonnell, *Lichen*, poem by Gabriel Gbadamosi, London, 2012, graphite and hand-cut paper in twenty folded packets, in cloth-covered box with hand-cut paper interior. Yale Center for British Art, Friends of British Art Fund

Mandy Bonnell, *Rose Mallow and Seed*, 2012, drypoint, with collage cut-out and hand coloring. Collection of the artist

Mandy Bonnell, *Wild Flowers Worth Notice*, London, 2012, graphite drawings in cloth-covered box. Yale Center for British Art, Gift of the artist

Caroline Coode, *Rana temporaria*, Newcastle upon Tyne, 2012, wood engraving and leaf print on handmade Nepalese paper. Yale Center for British Art, Gift of David Burnett

Eileen Hogan, *Bountiful UL 238*; *Sweet Promise FH 172; Golden Gain FR 59*, 2012, oil on paper. Collection of the artist

Eileen Hogan, *From Shadow n. The Hour Hand*, 2012, oil on paper. Collection of the artist

Eileen Hogan, *Goose Hut*, 2012, oil on paper, mounted on card. Collection of the artist

Eileen Hogan, *Ian Hamilton Finlay*, 2012, oil and charcoal on paper, mounted on card, Collection of the artist

Julie Johnstone, *Landscape*, Edinburgh, Essence Press, 2012, torn paper. Yale Center for British Art, Gift of David Burnett

John Dunbar Kilburn, *The Golden Plaice*, Falmouth, 2012, inkjet print, hand-cut and folded paper. Yale Center for British Art, Friends of British Art Fund

Yoojin Kim, *Penguin* from *Flightless Avian Osteology*, Baltimore, Maryland, 2012, digital print and cut paper. Collection of the artist

Dizzy Pragnell, *A is for Apple*, Canterbury, Kent, 2012, collagraph and fruit papyrus, with pen and ink. Yale Center for British Art, Friends of British Art Fund

Mandy Bonnell, *Eloise Butler's Wildflower Garden, Est. 1907*, London, 2013, graphite, letterpress, hand-cut paper, and natural specimens. Collection of the artist

Jacqueline Butler, *Pseudo Abstract*, Manchester, 2013, digital print. Yale Center for British Art, Friends of British Art Fund

Jennifer Collier, *Bird House*, Stafford, England, 2013, altered book, paper, wood, and thread. Yale Center for British Art, Friends of British Art Fund

John Dilnot, *Bishop Wood*, Brighton, England, 2013, sticks, printed map, and cut paper, in glass-topped wooden box. Yale Center for British Art, Friends of British Art Fund

John Dilnot, *Moth Collection*, Brighton, England, 2013, screen print and cut paper in glass-topped wooden box. Yale Center for British Art, Friends of British Art Fund

John Dilnot, *Night Flight*, Brighton, England, 2013, screen print. Yale Center for British Art, Friends of British Art Fund

Helen Douglas, *The Pond at Deuchar E-Scroll*, London: Tate, 2013, *http://helendouglas.onlineculture.co.uk/ttp/ttp.html*. Collection of the artist

Eileen Hogan, *UL 238, FH 172, and FR 59*, 2013, charcoal and oil on paper. Collection of the artist

Eileen Hogan, *UL 238, FH 172, and FR 59 (Beehives)*, 2013, oil, charcoal, and pastel on paper, mounted on card. Collection of the artist

Eileen Hogan, *Kettle Drum*, 2013, oil on paper. Collection of the artist

Elizabeth Murray Jones and Steve Carlton, *What a Hoot!*, Manchester: Young Explorer, 2013, inkjet print, string, and metal badges. Yale Center for British Art, Friends of British Art Fund

Ron King, *Going, Going . . . Gone*, with poems by Richard Price, West Sussex, England: Circle Press, 2013, blind embossed images printed relief from sawn and engraved masonite and wire. Yale Center for British Art, Friends of British Art Fund

Chris McCabe and Maria Viotides, *Pharmapoetica: A Dispensary of Poetry*, London: Pedestrian Publishing, 2013. Yale Center for British Art, Friends of British Art Friend

Dizzy Pragnell, *Beetroot*, Canterbury, Kent, 2013, vegetable papyrus with leaf of printed text, Yale Center for British Art, Friends of British Art Fund

Dizzy Pragnell, *Provenance: Vegetable Box (Cabbage, Asparagus, Mushroom, Aubergine, and Beetroot)*, Canterbury, Kent, 2013, vegetable papyrus with leaves of printed text, housed in folded paper wrapper tied with string. Yale Center for British Art, Friends of British Art Fund

Anne Rook, *Book of Sixteen Pears*, London: M.M. Visual Catering, 2013, inkjet print. Yale Center for British Art, Friends of British Art Fund

Astrid Weigel, *Flower Bed*, Newmilns, Ayrshire, Scotland, 2013, hand-colored screen print. Yale Center for British Art, Friends of British Art Fund

Eiko Yoshida, *Apple*, London, 2013, hand-cut paper and twig. Yale Center for British Art, Friends of British Art Fund

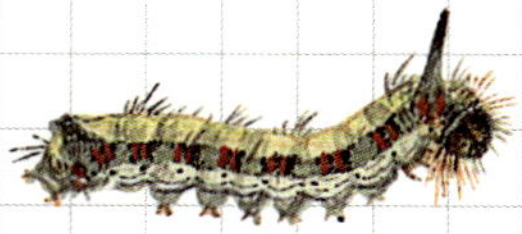

Acknowledgments

This project would not have come to fruition without the warm and whole-hearted support of Amy Meyers, Director of the Yale Center for British Art. It could not have been completed without the invaluable assistance of my colleagues in the Center's Department of Rare Books and Manuscripts, particularly Sarah Welcome, Senior Curatorial Assistant, who has remained cheerful and focused throughout, keeping us from losing the plot at crucial moments. Francis Lapka, Catalogue Librarian, and Barbara Aniballi, Senior Cataloguing Assistant, did "emergency cataloging" of new acquisitions of artists' books at a moment's notice. Sarah Torgesen (Yale BA 2014) and Darcy Tuttle (Yale BA 2016) assisted in making innumerable lists of titles, taking measurements, and helping in countless ways. Katherine Chabla, Senior Curatorial Assistant, managed the task of keeping track of the identification of the more mysterious fossils, insects, plants, shells, and birds with the invaluable help of colleagues at the Yale Peabody Museum of Natural History (Susan Butts, Senior Collections Manager,Division of Invertebrate Paleontology; Raymond Pupedis, Senior Collections Manager, Division of Entomology; Patrick Sweeney, Collections Manager, Division of Botany; Eric Lazo-Wasem, Senior Collections Manager, Division of Invertebrate Zoology; and Kristof Zyskowski, Collections Manager, Ornithology and Mammalogy, Division of Vertebrate Zoology). Thomas Lentz, Professor Emeritus of Cell Biology at the Yale School of Medicine, recently donated his microscopy collection to the Yale Peabody Museum, and Dr. Lentz himself provided crucial guidance in our selection process.

I would like to thank E. C. Shroeder, Director of the Beinecke Rare Book and Manuscript Library, for loans of a number of key historic works to the exhibition. I am grateful to Derek Briggs, Director of the Yale Peabody Museum of Natural History, for the loans of objects from the Peabody, including herbarium sheets, natural specimens, and a number of precious microscopes from the Lentz Collection in the Division of Historical Scientific Instruments. I would like to thank the Academy of Natural Sciences of Philadelphia, Drexel University for their loans; they include the fascinating commonplace books of George Humphrey and Joshua Spencer, fossils collected by Etheldred Benett, and one of the herbarium sheets assembled by John Stuart Mill. I also would like to thank Neale and Margaret Albert, Tim Barringer, Mandy Bonnell, David Burnett, Tracey Bush, Cathy Courtney, Bryan Fuermann, Eileen Hogan, Yoojin Kim, Randy Lerner, Robert Peck, and Charlotte Verity, all of whom have lent works from their personal collections.

Tim Goodhue, Registrar, and Corey Myers, Associate Registrar, handled the registrarial issues with their usual aplomb. They worked with Rebecca Hatcher, Preservation Coordination Librarian, Beinecke Rare Book and Manuscript Library, and with Annette Van Aken, Project Registrar of the Yale Peabody Museum of Natural History. Rick Johnson, Installation Manger, and his team, including Kevin Derken, Associate Installation Manager, and Rachel Hellerich, Rick Omonte, Greg Shea, and Dylan Vitale worked unstintingly and cheerfully to install the works in the exhibition (although the chocolate may have helped!). As always, Stephen Saitas provided critical advice and support, and was responsible for designing cases and installation mounts, as well as the overall look of the exhibition.

I would like to thank Beth Miller, Associate Director for Advancement and External Affairs, and her team for their support, but particularly Julienne Richardson who did a superb job with all aspects of the public relations. Lyn Bell Rose designed all signage, labels, and other printed material related to the exhibition with spectacular distinction.

I am grateful to Soyeon Choi, Head Paper Conservator at the Center, who performed exquisite and sensitive conservation work on a number of historic and contemporary objects, and to Mary Regan-Yttre, Conservation Assistant, who as always, took such an interest in the works themselves while carefully matting and framing them.

I would like to thank Linda Friedlaender, Curator of Education, for her enthusiastic support for this project,

OPPOSITE: James Francis Stephens, Caterpillar in *Haustellata Lepidoptera Diurna* (detail), ca. 1839, pen and ink and watercolor. Yale Center for British Art, Paul Mellon Collection

and her colleagues Cyra Levenson and Jaime Ursic for their interest. I am, as always, indebted to the extraordinary docents and student guides at the Center, who are responsible for bringing our curatorial projects to life for our visitors.

Eleanor Hughes, Associate Director for Exhibitions and Publications, adeptly supervised the details of the catalogue production and exhibition organization and implementation in her peerless fashion. Christopher Lotis, Editor, reviewed the essays and text of the catalogue as well as the exhibition materials with characteristic thoughtfulness, and Shaunee Cole, Publications Assistant, secured images and permissions for photographs in her usual competent manner. Robin Hoffman, Postdoctoral Research Associate, shepherded the proofing process with extraordinary patience and creativity.

The catalogue itself was beautifully and creatively designed by Miko McGinty. This was our second collaboration, bringing our mutual interests full circle: I supervised an exhibition she curated at the Center on contemporary British book arts for her Yale undergraduate senior project some twenty years ago. It has been an enormous pleasure to work with her and her team—Rita Jules, Tina Henderson, and Anjali Pala—at every stage of this project.

I would like to thank Sally Salvesen, the Center's publisher at Yale University Press London, who has never wavered in her support of this project from the first moment we spoke of it some years ago. She has worked closely with the printers at Conti Tipocolor in Florence ensuring that every last detail is perfect (including the collecting pocket that had to be assembled and attached by hand!).

Colleagues in the Center's Imaging and Rights department were brilliant. Richard Caspole photographed the works from the Center's collection that appear in this publication, dealing creatively and thoughtfully with the challenges of some of the more unusual three-dimensional bookworks. Maria Singer, Imaging and Rights Assistant, processed the administrative details of the project splendidly. Kurt Heumiller, Digital Imaging Technician, worked tirelessly to get the images just right. We also thank the excellent color separators Professional Graphics, who made it all work.

I am so grateful to Robert Peck, Senior Fellow, Academy of Natural Sciences of Philadelphia, Drexel University; Molly Duggins, Lecturer, Department of Art History and Theory, National Art School, Sydney, Australia; and poet David Burnett, of Edinburgh, Scotland, for their superb contributions; as well as to the seventeen different artists, critics, historians, gardeners, and amateur naturalists who were persuaded to contribute to this publication: Tim Barringer, Mandy Bonnell, Clare Bryan, Tracey Bush, Laurie Clark, John Dilnot, Lisa Ford, Eileen Hogan, Jane Hyslop, Ron King, Clive Phillpot, Martin Postle, Dizzy Pragnell, Colin Sackett, Rebecca Salter, Patrick Sweeney, and Sarah Welcome. Their words add immeasurable worth to the book.

I would like to thank Linda L. Brownrigg, who has championed the book arts so ardently in both the United States and the United Kingdom. We truly appreciate her encouragement and support of this exhibition. I am so grateful to the artists whom I have gotten to know over the past two decades and whose work is now part of the collections as well as represented in the exhibition and the catalogue. Ron King, artist and publisher of Circle Press (with his wife, the sculptor Willow Legge), and Ken Campbell have been especially generous and supportive beyond all measure in their dealings with me and the Center over many years. The archives of both Circle Press and Ken Campbell are in the Center's collections and play a crucial role in understanding artists' books of the twentieth century. In the last few years, as I have concentrated on building the collections of artists' books that focus on the natural world, I have had the pleasure of forming some wonderful relationships with incredibly talented artists, including Mandy Bonnell, who has spent so much time at the Center looking at the work of Miss Rowe and other Victorian self-taught naturalists. It has been such a pleasure to see the extraordinary new artistic creations that have resulted from her close engagement with our historic collections. I am thankful to Emma Hill, Director of the Eagle Gallery in London, for introducing Mandy's work (and that of Julie Cockburn) to me.

I owe Brian Allen, Tracey Bush, John Christie, Helen Douglas, Tony Fothergill, Rebecca Salter, Romilly

Saumarez-Smith, Kim Sloan, Charlotte Verity, and Jane Wildgoose so many thanks for their friendship and support. Eileen Hogan, artist, writer, and Professor of Fine Art, CCW (Camberwell, Chelsea, and Wimbledon Colleges of Art, University of the Arts London) has worked with me on many aspects of this exhibition and helped with the complications of organization and selection. Cathy Courtney, Project Director for National Life Stories at the British Library, and responsible for the Artists Lives series there, has interviewed a number of the artists featured in this exhibition, including Ken Campbell, Ron King, and Ian Hamilton Finlay. Most recently she has recorded conversations with Ralph Irving, gardener at Little Sparta and a key collaborator with Finlay. Her writings on the book arts from the early 1980s onward have been crucial to my understanding of the field.

I would like to thank botanist Henry J. Noltie, of the Royal Botanic Garden Edinburgh, who has always been so generous about sharing his knowledge of the natural world. I had a delightful time with Margot Calder, who allowed me to take a few photographs in her beautiful English garden (though situated in Chester, Connecticut). I am grateful for the enthusiastic interest and support of Cheryl Tipp, Curator, Natural Sounds, Sound and Vision, at the British Library, who has compiled a selection of bird songs and sounds of the British countryside for the exhibition. Thanks are due also to Michael Stocking, Managing Director, Armadillo Systems, who produced two innovative digital bookwork projects on display in this exhibition: one developed with Eileen Hogan and focused on her work related to Ian Hamilton Finlay's garden Little Sparta, and another with Helen Douglas and her e-scroll *The Pond at Deuchar*.

I also would like to acknowledge the very generous promised gift of twentieth-century illustrated books and wood engravings by David Burnett. He writes about the collection in his essay, and we are thrilled that many of the works he discusses will be in the exhibition and thereafter will remain part of the Center's collections. His understanding of the natural world has been an inspiration. A number of works by wood engraver and artist Richard Shirley Smith, part of the large collection of his work donated to the Center by Roger Moss, will be in the exhibition as well, and I acknowledge Dr. Moss once again for his generosity.

Two research fellowships, one in the Research Department at the Victoria and Albert Museum and another granted under the auspices of CCW (Camberwell, Chelsea, and Wimbledon Colleges of Art, University of the Arts London) Graduate School and Tate Research, allowed me the opportunity to look at literally hundreds of artists' books at the V&A, Tate Britain, and Chelsea College of Art, and to have valuable discussions with librarians and curators responsible for building and cataloguing those collections. I would like to thank Oriana Baddeley, Dean of Research for the University of the Arts London, for her support. I am also very grateful to colleagues at the V&A, particularly Rowan Watson, Senior Curator, Manuscripts; Elizabeth James, Senior Librarian, National Art Library; Julius Bryant, Keeper, Word and Image Department; and Liz Miller, Deputy Head, Research Department, who were all so generous with their time during my stay in London.

I would like to thank Maria White, former Chief Cataloguer for Tate Library, who was responsible for that library's artists' books collection. She introduced me to numerous examples by the book artists she was acquiring for Tate, and we had many useful conversations about the cataloguing of artists' books. I am also grateful to Gustavo Grandal Montero, Special Collections Librarian at Chelsea College of Art & Design, for his generosity in opening the artists' books collection at Chelsea to me—the important assemblage formed by Clive Phillpot and Stephen Bury and now being augmented so thoughtfully by Gustavo.

Finally, I would like to dedicate this publication to the memory of Honora and Warren Fairman, Thomas Miner, and Greer Allen.

Elisabeth Fairman
April 2014

Index of Names

Photography Credits

All photography by Richard Caspole, Yale Center for British Art, unless otherwise noted.

Courtesy of the Academy of Natural Sciences of Philadelphia, Drexel University: figures 22, 27, 30, 32, 34
Courtesy of Beinecke Rare Book and Manuscript Library, Yale University: figure 28
© Trustees of the British Museum: figure 17
Courtesy of Clare Bryan: page 133
Photographs by John Christie: page 131
Courtesy of Helen Douglas: pages 216–17
Photographs by Elisabeth Fairman: figure 7; pages 12 (above), 150, 189
Courtesy of the Estate of Ian Hamilton Finlay: page 190
Photographs by Tom Grill: pages 100, 218
Courtesy of Houghton Library, Harvard University: figure 35
Photograph by Martine La Roche: page 186
Courtesy of the Lewis Walpole Library, Yale Univesity: figure 18
Photographs by John McKenzie: figures 63–69, 71–75; pages 64, 126, 149, 154, 179 (below), 180, 207 (below), 210 (below), 213 (above), 220 (above)
Photographs by George Meyrick: page 82
Courtesy of the Division of Botany Peabody Museum of Natural History, Yale University: YU.066743, figure 8, page 93; CBS.35636, page 91; YU.066744, page 95
Courtesy of the Division of Historical Scientific Instruments (HSI. 041190), Peabody Museum of Natural History, Yale University: figure 58; page 223
Private collection: pages 124, 125
Photograph by Rosamond Purcell: figure 27
Photograph by Colin Sackett: page 183
Photographs by Sarah Welcome: page 220 (below)
Photographs by David Wood: pages 187 (below), 188, 190 (above), 191
Courtesy of the Yale University Art Gallery: figure 39

Clive Phillpot, "The Pond at Deuchar" reprinted from "Postscript: *The Pond at Deuchar* E-scroll," http://www.tate.org.uk/about/projects/transforming-artist-books/digital-books/helen-douglas-postscript. © Tate, London 2013 Private collection: figures 25, 37

OVERLEAF: Mandy Bonnell, "Poppy" (detail) from *Wild Flowers Worth Notice*, London, 2012, graphite. Yale Center for British Art, Gift of the artist

COLOPHON

Designed by Miko McGinty

Type composed in Adobe Caslon and Gill Sans by Tina Henderson
Color separations by Professional Graphics, Rockford, Illinois
Printed on Gardapat Kiara paper and bound by Conti Tipocolor, Florence, Italy

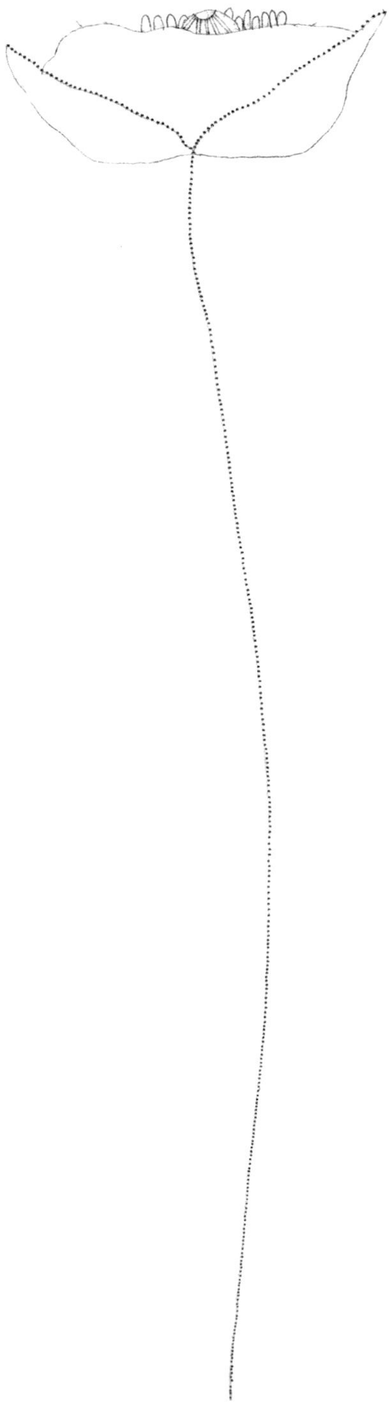

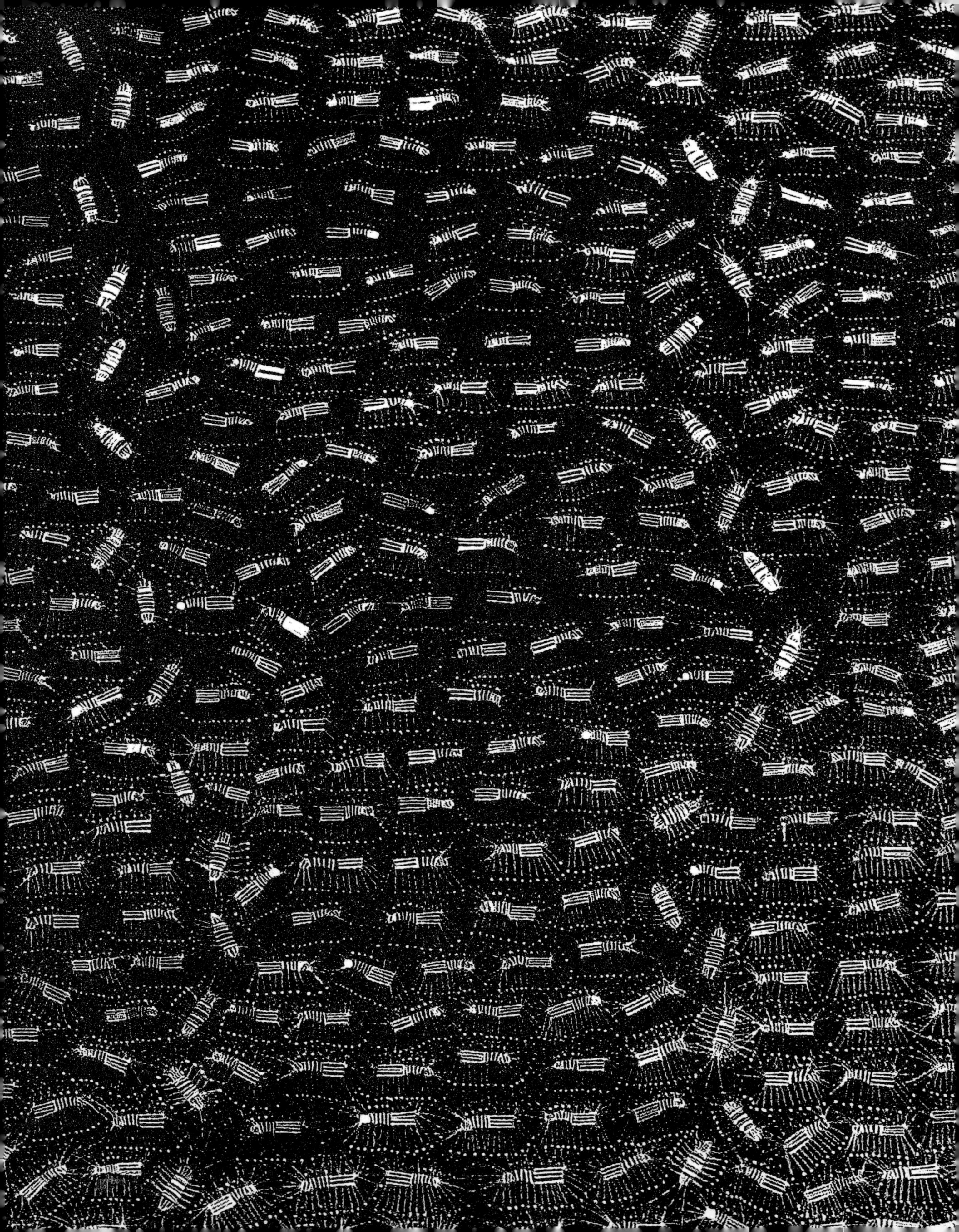